RSME Springer Series

Volume 11

As of 2015, RSME - Real Sociedad Matemática Española - and Springer cooperate in order to publish works by authors and volume editors under the auspices of a co-branded series of publications including advanced textbooks, Lecture Notes, collections of surveys resulting from international workshops and Summer Schools, SpringerBriefs, monographs as well as contributed volumes and conference proceedings. The works in the series are written in English only, aiming to offer high level research results in the fields of pure and applied mathematics to a global readership of students, researchers, professionals, and policymakers.

José Bonet • David Jornet • Pablo Sevilla-Peris

Function Spaces and Operators between them

Real Sociedad Matemática
Española

Springer

José Bonet
Instituto Universitario de Matemática Pura y Aplicada
Universitat Politècnica de València
Valencia, Spain

David Jornet
Instituto Universitario de Matemática Pura y Aplicada
Universitat Politècnica de València
Valencia, Spain

Pablo Sevilla-Peris
Instituto Universitario de Matemática Pura y Aplicada
Universitat Politècnica de València
Valencia, Spain

ISSN 2509-8888 ISSN 2509-8896 (electronic)
RSME Springer Series
ISBN 978-3-031-41601-9 ISBN 978-3-031-41602-6 (eBook)
https://doi.org/10.1007/978-3-031-41602-6

This Springer imprint is published by the registered company Springer Nature Switzerland AG
The registered company address is: Gewerbestrasse 11, 6330 Cham, Switzerland

Preface

The aim of this lecture notes is to present, in a unified and reasonably self-contained way, certain aspects of functional analysis which are needed to treat function spaces whose topology is not derived from a single norm, their topological duals and continuous linear operators between those spaces. We treat spaces of continuous, analytic, and smooth functions as well as sequence spaces. Operators of differentiation, integration, composition, and multiplication as well as partial differential operators between those spaces are studied. A brief introduction to Laurent Schwartz's theory of distributions and to Lars Hörmander's approach to linear partial differential operators is presented. These two topics constituted certainly two of the cornerstones of Mathematics in the twentieth century.

Our lecture notes had their origin in a course, entitled *Operators between spaces of holomorphic and differentiable functions*, which aimed at serving as an introduction to the topics mentioned above and related ones. Functional analysis is the proper language in which to formulate and understand these theories, and it is the guiding theme of the course. It has been taught several times over the last years as a part of the syllabus of the joint Masters Programme at Universitat Politècnica de València and Universitat de València. Having our notes as a starting point, we decided to polish, enrich, and expand them. Some topics have been enlarged; others that were not part of the original course have been added, and they could eventually become the content of different courses, with new points of view.

The text is addressed to students at a master level, or even undergraduate at the last semesters, since only knowledge on real and complex analysis is assumed. We have intended to be as self-contained as possible, and wherever an external citation is needed, we try to be as precise as we can. Our aim is to be an introduction to topics in, or connected with, different aspects of functional analysis. Many of them are in some sense classical, but we tried to show a unified direct approach; some others are new. This is why parts of these lectures might be of some interest even for researchers in related areas of functional analysis or operator theory.

The first three chapters of the text are devoted to establish the basic concepts and facts needed later. We begin with some elementary facts about pointwise and uniform convergence of sequences of functions. These are probably known to the reader but, since this is basic all along the text, we feel that it is good to dedicate some time to refresh these concepts. As motivating examples, we give a short look at uniform convergence for different situations: power series, Fourier series, and Dirichlet series. In Chaps. 2 and 3, we give an introduction to the basics of functional analysis that we are going to need later. We give some basic properties of Hausdorff locally convex spaces, looking also at properties like metrisability and normability. As first examples, we give the space of continuous functions endowed with the topology of uniform convergence on the compact subsets and Köthe echelon spaces. We then move to the study of duality theory and continuous linear operators. We define the dual of a locally convex space and give an introduction to the weak topology defined by a dual pair. We prove the fundamental theorems of the theory: Hahn-Banach, Alaouglu-Bourbaki, bipolar theorem, Mackey-Arens, and Banach-Steinhaus. We also make a short introduction to the general theory of continuous linear operators (Banach-Schauder theorem, transpose, ...). All the notions and results in this part are absolutely classical, and can also be found in other references. We believe, however, that our approach and point of view is direct and relatively novel. We have no aim to be comprehensive or to develop the whole theory. There are already a number of excellent monographs that serve this purpose for the interested reader. Our aim here is to give a short, concise introduction to what will be needed later in the text. In each case, we try to get to the point as fast as possible.

In the following two chapters, we introduce what are the objects of our main interest: spaces of functions and operators between them. We define and study the spaces of holomorphic complex functions and of infinitely differentiable real functions. We analyse their locally convex structure. For the space of holomorphic functions, we give a representation as a sequence space, and a description of the dual. Immediately after, we introduce the classical operators between spaces of functions which we are interested in: composition, multiplication, derivation, integration, and (mostly) partial differential operators. Once we have done this, we look at some dynamical properties of the operators. We give a brief introduction to the general theory of transitive and mean ergodic operators. Once again, we do not intend to be comprehensive, but to serve as a sort of gateway to this theory. Some readers might find a motivation in our exposition and get interested in the topic. Precise references are given to help with a more complete study of this theory. For some classical operators (multiples of the backward shift, composition, and partial differential operators), we classify when are they transitive/hipercyclic or power bounded/mean ergodic. This aspect is totally new and cannot be found in other monographs.

In the last chapter of the text, we deal with the theory of distributions of Schwartz and linear partial differential operators. First, we give an introduction to the theory of distributions. We consider the spaces of test functions and of distributions and define the basic operations there. Then we consider the spaces of rapidly decreasing functions and of tempered distributions, paying special attention to the Fourier transform defined on these spaces. We finish by looking at partial differential operators defined on the spaces that we have been considering all along the text.

We introduce the notion of fundamental solution of such an operator and prove the Malgrange–Ehrenpreis theorem on the existence of fundamental solutions.

All in all the manuscript deals with different topics. Many of them are classical and can be found sparse in different texts. The novelty of our approach lies mainly on two facts. First of all, we show all these topics together in an accessible way, stressing the connection between them. Second, we keep it always at a level that is accessible to beginners and young researchers. A list of exercises with different levels of difficulty is included at the end of each chapter. As we have said, several parts of this text have already been tested in different courses at a masters level, and different combinations of chapters would lead to other possible courses, equally interesting. Our aim is not to build and describe a whole, complete theory, but to serve as an introduction to some aspects that we believe are interesting. We wish to guide any reader that wishes to enter in some of these topics in their first steps. Our hope is that they learn interesting aspects of functional analysis and become interested to broaden their knowledge about function and sequence spaces and operators between them. To this end, the bibliography collects several monographs of different levels, both classical and more modern, where the interested reader will find deep, comprehensive exposés of the topics we deal with here. At the end of each chapter, we give a list of references where the interested reader can delve into the subject.

Acknowledgement The first and second authors were supported by the project PID2020-119457GB-100, and the third author by the project PID2021-122126NB-C33. Both projects are funded by MCIN/AEI/10.13039/501100011033 and by 'ERFD A way of making Europe'.

València, Spain
18 June 2023

José Bonet
David Jornet
Pablo Sevilla-Peris

Contents

List of Symbols

$\langle \cdot, \cdot \rangle$	Bilinear mapping defining a dual pair
(x, y)	Scalar product
$\|x\|$	Euclidean norm on $\mathbb{R}^d$
$\Subset$	Compact subset
χ_A	Characteristic function of A
$\mathbb{1}$	Constant sequence or function equal to 1
$f * g$	Convolution of functions
$u * \psi$	Convolution of a distribution and a function
$\overline{A}$	Closure of A
$\mathring{A}$	Interior of A
$A^{\perp}$	Orthogonal of A
A°	Polar of A
$\beta(E', E)$	Strong topology on the dual of E
$B(a, r)$	Open ball of centre a and radius r in $\mathbb{R}^N$
$\overline{B(a, r)}$	Closed ball of centre a and radius r in $\mathbb{R}^N$
$B_d(x, \varepsilon)$	Ball of centre x and radius ε defined by the distance d
$B_p(x, \varepsilon)$	Ball of centre x and radius ε defined by the seminrm p
B_X	Open unit ball of a normed space X
$C(z_0, r)$	Circle of centre z_0 and radius r in $\mathbb{C}$
$C[a, b]$	Space of continuous functions on the interval $[a, b]$
$C(K)$	Space of continuous functions on the compact set K
$C(\Omega)$	Space of continuous functions on the open set Ω
$C^k(\Omega)$	Space of C^k-functions on the open set Ω
$C^{\infty}(\Omega)$	Space of C^{∞}-functions on the open set Ω
$C(\mathbb{T})$	Space of 2π-periodic continuous functions $f : \mathbb{R} \to \mathbb{C}$
c_0	Space of null sequences
C_{φ}	Composition operator defined by φ
δ, δ_a	Dirac 'delta' functions
$\mathbb{D}(z_0, r)$	Open disc of centre z_0 and radius r in $\mathbb{C}$
$\mathbb{D}$	Open unit disc of $\mathbb{C}$
$\mathscr{D}(K)$	Space of C^{∞}-functions whose support is contained in K

$\mathscr{D}(\Omega)$	Space of test functions on Ω
$\mathscr{D}'(\Omega)$	Space of distributions on Ω
D_n	Dirichlet kernel
$d(x, A)$	Distance between a point and a set
e_n	Canonical vectors
E^*	Algebric dual of E
E'	Topological dual of E
E'_β	Strong dual of E
(E, F)	Dual pair
$\mathscr{E}'(\Omega)$	Space of distributions with compact support on Ω
Exp	Space of holomorphic functions of exponential type
$\widehat{f}(n)$	nth Fourier coefficient of f
$\widehat{f}(w)$	Fourier transform of a function
$\mathcal{F}$	Fourier transform operator
$\Gamma(A)$	Absolutely convex hull of A
$\mathsf{G}(T)$	Graph of the mapping T
$\mathrm{H}(\Omega)$	Space of holomorphic functions on Ω
$\mathrm{H}(\{0\})$	Space of germs at 0
$H(x)$	Heaviside function
$\lambda_p(A)$	Köthe echelon space of order p
Im	Image of an operator
ker	Kernel of an operator
ℓ_p	Space of p-summable sequences
ℓ_∞	Space of bounded sequences
$L_p(\mu)$	Space of p-integrable functions
$L_\infty(\mu)$	Space of esentially bounded functions
$L_{1,\mathrm{loc}}(\Omega)$	Space of locally integrable functions on Ω
$L(E, F)$	Space of linear continuous operators from E to F
$L(E)$	Space of linear continuous operators from E to E
$L_\beta(E, F)$	Space $L(E, F)$ with the topology of uniform convergence on bounded sets
$L_{co}(E, F)$	Space $L(E, F)$ with the topology of uniform convergence on precompact sets
$L_s(E, F)$	Space $L(E, F)$ with the strong operator topology
$L_w(E, F)$	Space $L(E, F)$ with the weak operator topology
$\mu(E, F)$	Mackey topology on E defined by the dual pair (E, F)
$\widehat{\mu}$	Fourier-Borel transform of a measure μ
m_N	Lebesgue measure on $\mathbb{R}^N$
M_g	Multiplication operator defined by g
$\mathbb{N}_0$	Set of non-negative integer numbers
$\mathrm{Orb}(x, T)$	Orbit of x with respect to T
$\mathscr{P}$	Family of seminorms defining a topology
p_K	Seminorm defined as the supremum on the compact set K
$P(\partial)$	Linear partial differential operator with constant coefficients

$PV\frac{1}{x}$	Cauchy principal value of $\frac{1}{x}$
$\sigma(E, F)$	Weak topology on E defined by the dual pair (E, F)
$\sigma(F, E)$	Weak topology on F defined by the dual pair (E, F)
s	Space of rapidly decreasing sequences
supp	Support of a function or a distribution
$S(\mathbb{R}^N)$	Space of rapidly decreasing functions (Schwartz)
$S'(\mathbb{R}^N)$	Space of tempered distributions
$\tau\|_Y$	Induced topology on Y
τ_d	Topology defined by the distance d
$\tau_{\mathscr{P}}$	Topology defined by the family of seminorms
τ_{co}	Compact-open topology (of uniform convergence on compact sets)
τ_p	Topology of pointwise convergence
$\\|T\\|$	Norm of the operator T
$^*T$	Transpose of the linear mapping T
^tT	Transpose of the continuous linear operator T
T^n	nth iterate of the operator T
T_n	Cesàro mean
T_f	Distribution defined by a function
$\mathbb{T}$	One-dimensional torus
$\mathcal{U}_0(E)$	Set of absolutely convex neighbourhoods of 0
ω	Space of all sequences, with pointwise convergence

Chapter 1
Convergence of Sequences of Functions

1.1 Preliminaries and Notation

Throughout the text, $\mathbb{K}$ will denote either the field of real numbers $\mathbb{R}$ or complex numbers $\mathbb{C}$. On $\mathbb{R}^d$ for some $d \geq 1$, we consider the inner product $x \cdot y = \sum_{n=1}^{d} x_n y_n$ for $x = (x_1, \ldots, x_d)$ and $y = (y_1, \ldots, y_d)$, and the Euclidean norm $|x| = (x \cdot x)^{1/2}$. Then the open and closed balls centred at $a \in \mathbb{R}^d$ with radius $r > 0$ are defined as

$$B(a, r) = \{x \in \mathbb{R}^d : |x - a| < r\} \text{ and } \overline{B(a, r)} = \{x \in \mathbb{R}^d : |x - a| \leq r\}.$$

Given $A \subseteq \mathbb{R}^d$, we write χ_A for the characteristic function, that is, $\chi_A(x)$ is 1 if $x \in A$ and 0 otherwise.

We identify $\mathbb{R}^2$ with $\mathbb{C}$. We recall now some basic notations of complex numbers. Given $z \in \mathbb{C}$, we write $z = x + iy$ with $x = \mathrm{Re}(z)$ (the real part) and $y = \mathrm{Im}(z)$ (the imaginary part). The conjugate is defined as $\overline{z} = x - iy$ and the modulus as $|z|^2 = z\overline{z} = x^2 + y^2$. The (open) disc centred at z_0 with radius $r > 0$ is denoted by

$$\mathbb{D}(z_0, r) = \{z \in \mathbb{C} : |z - z_0| < r\}.$$

Its boundary, the circle, is denoted by

$$C(z_0, r) = \{z \in \mathbb{C} : |z - z_0| = r\};$$

it will always be assumed to be oriented counterclockwise. We write $\mathbb{D} = \mathbb{D}(0, 1)$ and call it the *open unit disc* of $\mathbb{C}$. Throughout this text, we will write $\mathbb{N}_0$ for the set of non-negative integers $\mathbb{N} \cup \{0\}$.

J. Bonet et al., *Function Spaces and Operators between them*,
RSME Springer Series 11, https://doi.org/10.1007/978-3-031-41602-6_1

1.2 Pointwise and Uniform Convergence

We are going to consider functions defined on some subset of $\mathbb{R}^d$ and sequences $(f_n)_n$ of such functions. As we will see there are several possible ways to think of convergence to a certain f of such a sequence. The most elementary one is to ask that the values at each point converge to the value of f at that point. If $\Omega \subseteq \mathbb{R}^d$, a sequence of functions $f_n : \Omega \to \mathbb{K}$ (with $n \in \mathbb{N}$) *converges pointwise* to $f : \Omega \to \mathbb{K}$ if

$$\lim_{n\to\infty} f_n(x) = f(x),$$

for every $x \in \Omega$. This means that for every $x \in \Omega$ and for every $\varepsilon > 0$, there exists some $n_0 = n_0(x, \varepsilon) \in \mathbb{N}$ such that for every $n \geq n_0$,

$$|f_n(x) - f(x)| < \varepsilon . \tag{1.1}$$

Example 1.1 We take $\Omega = [0, 1]$ and define, for each $n \in \mathbb{N}$, the function $f_n : [0, 1] \longrightarrow \mathbb{R}$ by $f_n(x) = x^n$. Note that $\lim_{n\to\infty} f_n(x) = \lim_{n\to\infty} x^n = 0$ for all $0 \leq x < 1$ and $\lim_{x\to\infty} f_n(1) = 1$. This shows that $(f_n)_n$ converges pointwise to

$$f(x) = \begin{cases} 0 & \text{if } x \in [0, 1[\\ 1 & \text{if } x = 1. \end{cases}$$

See Fig. 1.1. □

Example 1.2 Take now $\Omega = \mathbb{R}$ and define, for each $n \in \mathbb{N}$, the function $f_n : \mathbb{R} \to \mathbb{R}$ by $f_n(x) = \dfrac{x^{2n}}{1 + x^{2n}}$. It is easy to check that $(f_n)_n$ converges pointwise to

$$f(x) = \begin{cases} \frac{1}{2} & \text{if } x = \pm 1 \\ 0 & \text{if } |x| < 1, \\ 1 & \text{if } |x| > 1. \end{cases}$$

See Fig. 1.2. □

We need notions of convergence that preserve certain properties of the functions, such as continuity (that is, the limit of continuous functions is again continuous). Examples 1.1 and 1.2 show that pointwise convergence is not good enough for what we need. We need to go a little bit further. Recall that for a sequence to converge pointwise we have some n_0 (depending on ε and x) for which (1.1) is satisfied. What we want now is to have some n_0 that depends only on ε and works for every x.

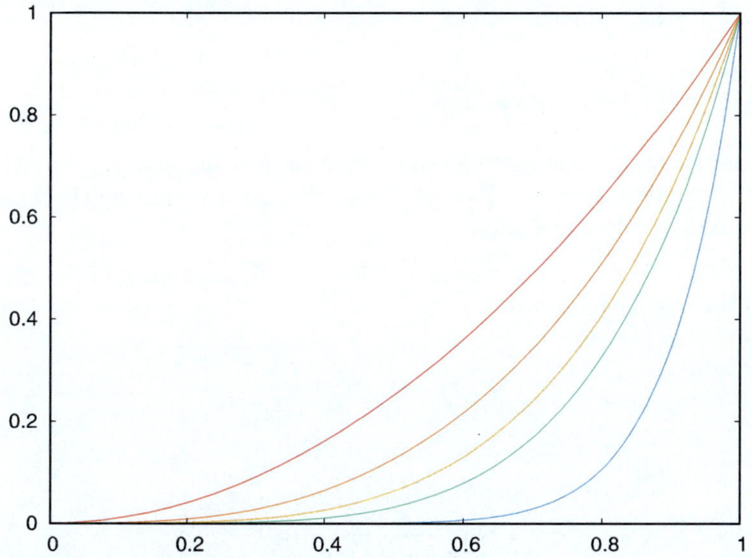

Fig. 1.1 Plot of f_n for $n = 2, 3, 4, 5, 10$

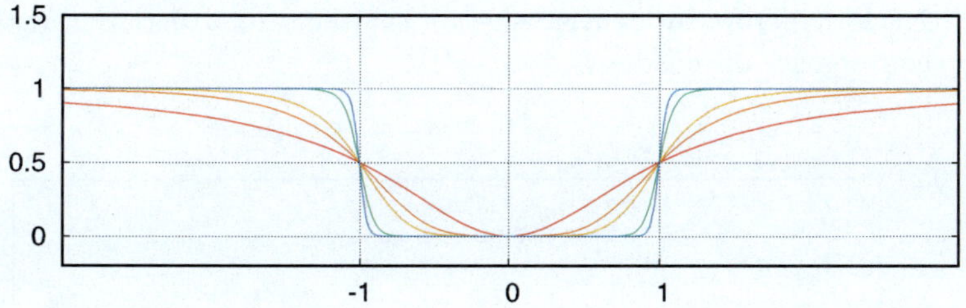

Fig. 1.2 Plot of f_n for $n = 1, 2, 3, 10, 20$

Given $\Omega \subseteq \mathbb{R}^d$, a sequence of functions $f_n : \Omega \to \mathbb{K}$ *converges uniformly* to $f : \Omega \to \mathbb{K}$ if for every $\varepsilon > 0$, there exists $n_0 = n_0(\varepsilon) \in \mathbb{N}$ such that

$$|f_n(x) - f(x)| < \varepsilon ,$$

for every $x \in \Omega$ and every $n \geq n_0$.

Remark 1.1 It is straightforward to check that a sequence $(f_n)_n$ converges uniformly to f if and only if for every $\varepsilon > 0$ there exists some $n_0 \in \mathbb{N}$ such that

$$\sup_{x \in \Omega} |f_n(x) - f(x)| < \varepsilon ,$$

for every $n \geq n_0$, or in other words, if and only if

$$\lim_{n\to\infty} \Big(\sup_{x\in\Omega} |f_n(x) - f(x)| \Big) = 0 \, . \tag{1.2}$$

Clearly, if $(f_n)_n$ converges uniformly to f, then it also converges pointwise to the same function. However, Examples 1.1 and 1.2 show (just by checking (1.2)) that the converse does not hold.

Example 1.3 We take the sequence $f_n : [0, 1] \to \mathbb{R}$ given by $f_n(x) = x^n(1 - x)$. A simple computation shows that

$$\max_{x\in[0,1]} |f_n(x)| = \frac{n^n}{(n+1)^{n+1}} \, .$$

Since

$$\lim_{n\to\infty} \frac{n^n}{(n+1)^{n+1}} = 0 \, ,$$

Remark 1.1 implies that $f_n \to 0$ uniformly on [0, 1] (see Fig. 1.3). □

The Cauchy criterion for convergence of sequences of scalars transfers to the uniform convergence of functions.

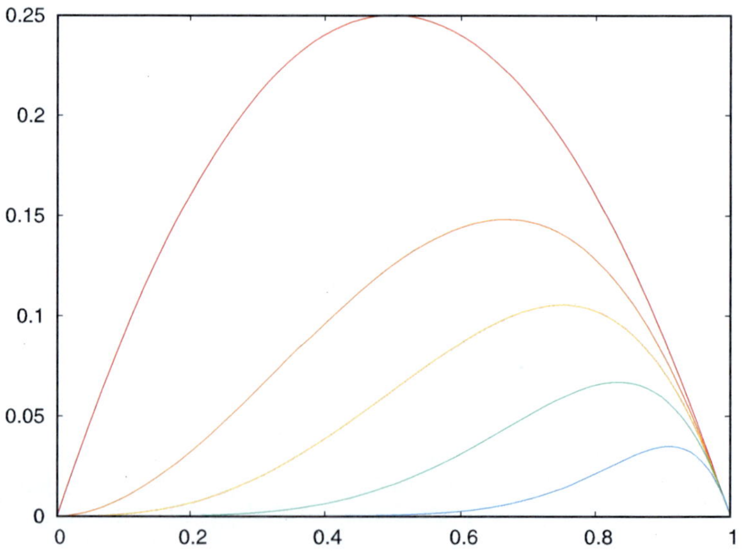

Fig. 1.3 Plot of f_n for $n = 1, 2, 3, 5, 10$

Proposition 1.1 (Cauchy criterion) *A sequence of functions $(f_n)_n$ defined on $\Omega \subseteq \mathbb{R}^d$ converges uniformly if and only if for every $\varepsilon > 0$ there is n_0 such that if $m, n \geq n_0$, then*

$$\sup_{x\in\Omega} |f_n(x) - f_m(x)| < \varepsilon\,. \tag{1.3}$$

Proof Suppose first that the sequence converges uniformly on Ω to some f. Then, given $\varepsilon > 0$, there is n_0 such that $\sup_{x\in\Omega} |f_n(x) - f(x)| < \frac{\varepsilon}{2}$. This immediately implies (1.3).

Let us note that

$$|f_n(x_0) - f_m(x_0)| \leq \sup_{x\in\Omega} |f_n(x) - f_m(x)|\,, \tag{1.4}$$

for every $x_0 \in \Omega$ and all n, m. Then, if (1.3) holds, this implies that the sequence $(f_n(x_0))_n$ is Cauchy in $\mathbb{K}$ for each fixed $x_0 \in \Omega$ and, therefore, converges. We define $f : \Omega \to \mathbb{K}$ by doing $f(x) = \lim_n f_n(x)$. Taking the limit as $m \to \infty$ on the left-hand side of (1.4), we obtain $|f_n(x) - f(x)| \leq \varepsilon$ for all $n \geq n_0$ and $x \in \Omega$. This shows that $(f_n)_n$ converges uniformly to f. □

As we announced, uniform convergence preserves continuity.

Theorem 1.4 *Let $(f_n)_n$ be a sequence of continuous functions on $\Omega \subseteq \mathbb{R}^d$. If $f_n \to f$ uniformly on Ω, then f is continuous on Ω.*

Proof Fix some $x_0 \in \Omega$ and let us show that $\lim_{x\to x_0} f(x) = f(x_0)$. To do that, we fix $\varepsilon > 0$ and choose $n_0 \in \mathbb{N}$ such that $|f_n(x) - f(x)| < \frac{\varepsilon}{3}$ for every $n \geq n_0$ and $x \in \Omega$. Now, since f_{n_0} is continuous at x_0, there is some $\delta > 0$ such that for $x \in \Omega$ with $|x - x_0| < \delta$, we have $|f_{n_0}(x) - f_{n_0}(x_0)| < \frac{\varepsilon}{3}$. Then, if $x \in \Omega$ and $|x - x_0| < \delta$, we have

$$|f(x) - f(x_0)| \leq |f(x) - f_{n_0}(x)| + |f_{n_0}(x) - f_{n_0}(x_0)| + |f_{n_0}(x_0) - f(x_0)| \leq \varepsilon\,.$$

□

Remark 1.2 Suppose that $(f_n)_n$ is a sequence of bounded functions on $\Omega \subseteq \mathbb{R}^d$ that converges uniformly. Let us see that the limit function f is also bounded. To see this, note in first place (recall (1.3)) that there exists $n_0 \in \mathbb{N}$ so that

$$\sup_{x\in\Omega} |f(x) - f_{n_0}(x)| < 1\,.$$

Being f_{n_0} bounded, we have $\sup_{x\in\Omega} |f_{n_0}(x)| = K < \infty$, and if $x_0 \in \Omega$, then

$$|f(x_0)| \leq |f(x) - f_{n_0}(x)| + |f_{n_0}(x)| < 1 + K\,.$$

Therefore f is bounded in Ω.

Proposition 1.2 *Let $(f_n)_n$ be a sequence of continuous functions defined on Ω that converges uniformly to f and $(x_n)_n \subseteq \Omega$ such that $\lim_n x_n = x_0$. Then $\lim_n f_n(x_n) = f(x_0)$.*

Proof We fix $\varepsilon > 0$ and choose $n_1 \in \mathbb{N}$ such that $|f_n(x) - f(x)| < \frac{\varepsilon}{2}$ for every $x \in \Omega$ and $n \geq n_1$. On the other hand, since f is continuous at x_0 (because of Theorem 1.4), we can find $\delta > 0$ such that if $x \in \Omega$ and $|x - x_0| < \delta$ we have $|f(x) - f(x_0)| < \frac{\varepsilon}{2}$. Now we take $n_2 \in \mathbb{N}$ such that $|x_n - x_0| < \delta$ for $n \geq n_2$ and we have that if $n \geq n_0 = \max(n_1, n_2)$, then

$$|f_n(x_n) - f(x_0)| \leq |f_n(x_n) - f(x_n)| + |f(x_n) - f(x_0)| < \varepsilon .$$

This proves the claim. □

This becomes useful to check that a sequence does not converge uniformly.

Example 1.5 For each $n \in \mathbb{N}$, take the function $f_n : [0, 1] \to \mathbb{R}$ given by $f_n(x) = nx^n(1 - x)$. First of all, it is a simple exercise to check that $\lim_n nx^n(1 - x) = 0$ for every $x \in [0, 1]$. Then the sequence $(f_n)_n$ converges pointwise to 0. Take now $x_n = 1 - \frac{1}{n}$, which tends to 1. We have

$$\lim_{n\to\infty} f_n(x_n) = \lim_{n\to\infty} \left(1 - \frac{1}{n}\right)^n = \frac{1}{e} \neq 0 = f(1) .$$

By Proposition 1.2, the sequence $(f_n)_n$ does not converge uniformly (see Fig. 1.4).

□

As we have pointed out, uniform implies pointwise convergence, but the converse is not in general true. The following result shows that if both the set where the functions are defined and the functions themselves are nice enough, then pointwise also implies uniform convergence.

Theorem 1.6 (Dini's Theorem) *Let $K \subseteq \mathbb{R}^d$ be closed and bounded. Let $f_n : K \to \mathbb{R}$ with $n \in \mathbb{N}$ be continuous functions such that $f_{n+1}(x) \leq f_n(x)$ (or $f_{n+1}(x) \geq f_n(x)$) for every $x \in K$ and $n \in \mathbb{N}$ and so that $(f_n)_n$ converges pointwise to a continuous function $f : K \to \mathbb{R}$. Then $(f_n)_n$ converges to f uniformly on K.*

Proof We proceed by contradiction. Assume that f_n does not converge uniformly to f. Then, there exist $\varepsilon_0 > 0$ and a sequence $m_1 < m_2 < \cdots < m_k < \cdots$ such that for each $k \in \mathbb{N}$ there is $x_k \in K$ such that

$$|f_{m_k}(x_k) - f(x_k)| \geq \varepsilon_0 .$$

Since $(x_k)_k \subseteq K$ is a bounded sequence, by the Bolzano–Weierstraß theorem, there is a convergent subsequence (which for the sake of clarity we denote again by $(x_k)_k$), converging to some $x_0 \in K$.

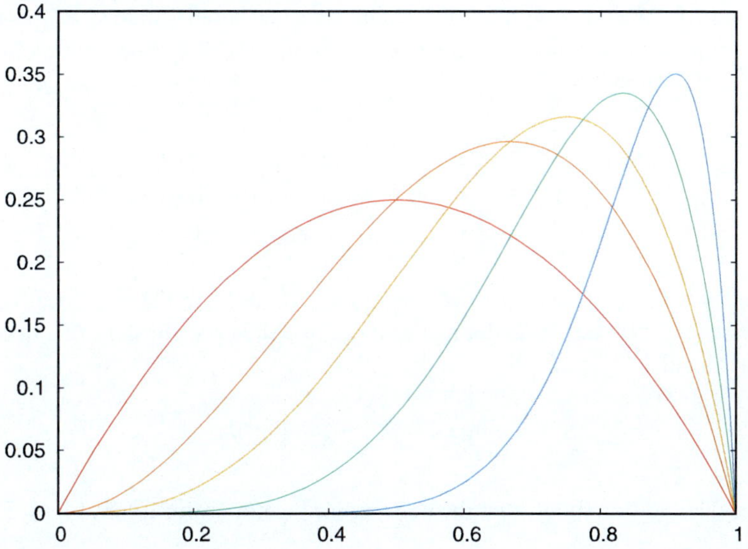

Fig. 1.4 Plot of f_n for $n = 1, 2, 3, 5, 10$

We fix now m_i and we have, for every $k > i$,

$$f_{m_i}(x_k) - f(x_k) \geq f_{m_k}(x_k) - f(x_k) = |f_{m_k}(x_k) - f(x_k)| \geq \varepsilon_0 .$$

Taking limit as k tends to ∞ on the left-hand side and using the fact that both f_{m_i} and f are continuous at x_0, we obtain

$$f_{m_i}(x_0) - f(x_0) \geq \varepsilon_0$$

for every $i \in \mathbb{N}$. This contradicts the fact that $(f_n)_n$ converges pointwise to f (and, in particular, $\lim_n f_n(x_0) = f(x_0)$). □

We have three hypotheses in this theorem: two regarding the set (being closed and bounded) and one on the sequence of functions (being monotone). The following three examples show that each one of them is necessary.

Example 1.7 Consider the sequence $f_n : [0, 1[\to \mathbb{R}$ given by $f_n(x) = x^n$. Note that it is decreasing for each fixed x and converges pointwise to $f : [0, 1[\to \mathbb{R}$ given by $f(x) = 0$. However, the sequence does not converge uniformly in $[0, 1[$ to its pointwise limit since

$$\sup_{x\in[0,1[} |f_n(x) - f(x)| = 1 ,$$

for every n. Here, the set, although being bounded, is not closed. □

Example 1.8 We see now that the hypothesis of K being bounded is also needed. To do that, we consider the sequence $f_n : [0,\infty[\to \mathbb{R}$ (the set is closed but unbounded) defined as

$$f_n(x) = \begin{cases} 0 & \text{if } x \in [0, n-1], \\ x - n + 1 & \text{if } x \in [n-1, n], \\ 1 & \text{if } x \in [n, \infty[\,. \end{cases}$$

We clearly have $f_{n+1}(x) \leq f_n(x)$ for every $x \in [0,\infty[$ and the sequence converges pointwise to $f : [0,\infty[\to \mathbb{R}$ defined as $f(x) = 0$ (i.e., the constant function 0). On the other hand,

$$\sup_{x\in[0,\infty[} |f_n(x) - f(x)| = 1,$$

for every n, and this shows that the convergence is not uniform (see Fig. 1.5). □

Example 1.9 Now, we consider the sequence of continuous functions f_n : $[0, 1/2] \to \mathbb{R}$ (defined on a closed, bounded set) given by

$$f_n(x) = \begin{cases} 2^{n+1}x & \text{if } x \in [0, 2^{-(n+1)}], \\ -2^{n+1}x + 2 & \text{if } x \in]2^{-(n+1)}, 2^{-n}], \\ 0 & \text{if } x \in [2^{-n}, 1/2]\,. \end{cases}$$

This time $(f_n)_n$ is neither decreasing nor increasing. Again, the sequence converges pointwise to 0, but since

$$\sup_{x\in[0,1/2]} |f_n(x) - f(x)| \geq f_n(2^{-(n+1)}) = 1$$

for every n, it does not converge uniformly on $[0, 1/2]$ (see Fig. 1.6). □

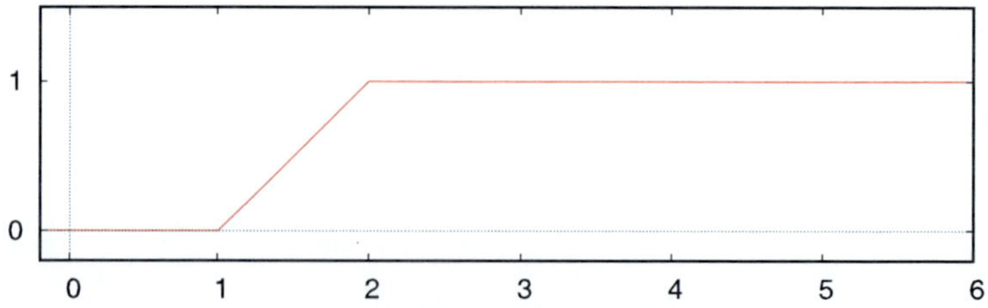

Fig. 1.5 Plot of f_n for $n = 2$. For higher ns, the diagonal slides to the right

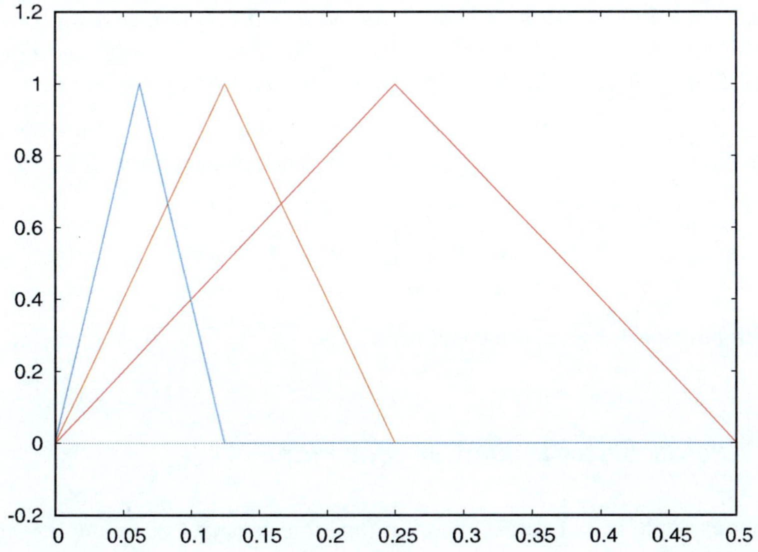

Fig. 1.6 Plot of f_n for $n = 1, 2, 3$

1.3 Series of Functions

Given a sequence of functions $f_n : \Omega \subseteq \mathbb{R}^d \to \mathbb{K}$, the *series of functions* $\sum_{n=1}^{\infty} f_n$ is (if it exists) the limit of the sequence $(S_n)_n$ of partial sums of the sequence $(f_n)_n$, given by $S_n = \sum_{k=1}^{n} f_k$. Depending on what kind of convergence we ask for the sequence of partial sums, we have a different convergence for the series. To be more precise, the series $\sum_{n=1}^{\infty} f_n$ *converges pointwise* on Ω if (S_n) converges pointwise (that is, for each fixed $x \in \Omega$, the scalar series $\sum_{n=1}^{\infty} f_n(x)$ converges in $\mathbb{K}$).

The series *converges uniformly* if the sequence (S_n) converges uniformly on Ω. Finally, we say that the series *converges absolutely* on Ω if the series $\sum_{n=1}^{\infty} |f_n|$ converges pointwise in Ω.

The following criterion for uniform convergence of series is easily deduced from Proposition 1.1.

Proposition 1.3 (Cauchy criterion for series) *The series* $\sum_{n=1}^{\infty} f_n$ *is uniformly convergent on* Ω *if and only if for all* $\varepsilon > 0$ *there is* $n_0 \in \mathbb{N}$ *such that*

$$\sup_{x \in \Omega} \left| \sum_{k=n}^{m} f_k(x) \right| < \varepsilon$$

for every $m > n \geq n_0$.

As a consequence, we obtain the following useful criterion.

Proposition 1.4 (Weierstraß M-test) *Suppose that $(M_n)_n$ is a sequence of non-negative real numbers such that $\sum_{n=1}^{\infty} M_n < \infty$. If for all $x \in \Omega$ and $n \in \mathbb{N}$ we have $|f_n(x)| \leq M_n$, then the series $\sum_{n=1}^{\infty} f_n$ converges uniformly on Ω.*

Proof Given $\varepsilon > 0$, by Cauchy's criterion for scalar series, we can find $n_0 \in \mathbb{N}$ such that for $m > n \geq n_0$, then $\sum_{k=n}^{m} M_n < \varepsilon$. Then, for each $x \in \Omega$,

$$\Big|\sum_{k=n}^{m} f_k(x)\Big| \leq \sum_{k=n}^{m} |f_k(x)| \leq \sum_{k=n}^{m} M_n < \varepsilon .$$

Then Proposition 1.3 gives the conclusion. □

1.3.1 *Power Series in the Complex Plane*

Our first example is a classical object in function theory of one complex variable: *power series*. We take a sequence $(a_n)_n$ of complex numbers and consider the series

$$\sum_{n=0}^{\infty} a_n z^n$$

for $z \in \mathbb{C}$. Note that this can be seen as a series of functions $\sum_n f_n$, where $f_n(z) = a_n z^n$ for each n. We wonder now about the convergence of this series and find out that it converges at discs centred at 0.

Theorem 1.10 *Let $(a_n)_n \subseteq \mathbb{C}$ and $z_0 \in \mathbb{C}$ such that $\sum_{n=0}^{\infty} a_n z_0^n$ converges. Then the power series $\sum_{n=0}^{\infty} a_n z^n$ converges absolutely at every $z \in \mathbb{C}$ with $|z| < |z_0|$ and converges uniformly on $\overline{\mathbb{D}(0,r)}$ for every $0 < r < |z_0|$.*

Proof If $z_0 = 0$, then the result is trivially satisfied. We assume, then, $z_0 \neq 0$. Since the series is convergent, $\lim_{n\to\infty} |a_n z_0^n| = 0$, and we can find $M > 0$ such that $|a_n z_0^n| \leq M$. If $0 < r < |z_0|$ and $|z| \leq r$, then

$$|a_n z^n| = |a_n||z|^n \leq |a_n| r^n = |a_n||z_0|^n \Big(\frac{r}{|z_0|}\Big)^n \leq M\Big(\frac{r}{|z_0|}\Big)^n .$$

Since $\sum_{n=1}^{\infty} M\Big(\frac{r}{|z_0|}\Big)^n$ is a convergent geometric series, we immediately conclude that $\sum_{n=1}^{\infty} |a_n z^n| < \infty$ and that, by Proposition 1.4, the power series converges uniformly on $\{z \in \mathbb{C} : |z| \leq r\}$. □

If we write

$$R = \sup\{|z| : \sum_{n=0}^{\infty} a_n z^n \text{ is convergent}\} \in [0, , +\infty] ,$$

then the power series converges (even absolutely) for every $|z| < R$ and diverges for every $|z| > R$. Even more, the power series converges uniformly on $\overline{\mathbb{D}(0,r)}$ for every $r < R$. The number R is called the *radius of convergence* of the power series. We have no information on what does the series do for zs with $|z| = R$. For example, the power series $\sum_{n=1}^{\infty} \frac{1}{n} z^n$ has the radius of convergence $R = 1$, but for $z = 1$, the series diverges, and for $z = -1$ it converges.

The radius of convergence of a power series can be computed using the Hadamard formula

$$R = \frac{1}{\limsup_n \sqrt[n]{|a_n|}}\,.$$

With all this in mind, the function $f : \mathbb{D}(0, R) \to \mathbb{C}$ given by $f(z) = \sum_{n=0}^{\infty} a_n z^n$ is well defined and, as we will see later (in Remark 4.1), holomorphic.

We see, then, that the natural domains where power series are defined are open discs. Even more, if we consider the numbers

$$R_c = \sup\{|z| : \sum_{n=0}^{\infty} a_n z^n \text{ is convergent}\}$$

$$R_u = \sup\{r : \sum_{n=0}^{\infty} a_n z^n \text{ converges uniformly on } \mathbb{D}(0,r)\}$$

$$R_a = \sup\{|z| : \sum_{n=0}^{\infty} a_n z^n \text{ is absolutely convergent}\},$$

then

$$R_c = R_u = R_a\,. \tag{1.5}$$

1.3.2 Fourier Series

We consider now periodic functions and ask to what point can these be represented by some series expansion. We start with the most elementary functions of this type. A *trigonometric polynomial* is a function $p : \mathbb{R} \to \mathbb{C}$ of the form

$$p(x) = \sum_{|j| \le n} c_j e^{ijx},$$

where $c_j \in \mathbb{C}$ and $|c_n| + |c_{-n}| \neq 0$. The number n is called the *degree of* p. Such a function p is clearly continuous and 2π-periodic, so that the polynomial is determined by its values in the interval $]-\pi, \pi]$. Since

$$\int_{-\pi}^{\pi} e^{ikx}\,dx = \begin{cases} 2\pi & \text{if } k = 0 \\ 0 & \text{if } k \neq 0, \end{cases}$$

a straightforward computation yields

$$c_j = \frac{1}{2\pi}\int_{-\pi}^{\pi} p(x)e^{-ijx}\,dx$$

for every $|j| \leq n$. Starting from this idea, if $f : \mathbb{R} \to \mathbb{C}$ is 2π-periodic and $|f|$ is integrable in $]-\pi, \pi]$ (that is, f belongs to $L_1(-\pi, \pi)$), then for each $j \in \mathbb{Z}$, the jth *Fourier coefficient* of f is defined as

$$c_j = \frac{1}{2\pi}\int_{-\pi}^{\pi} f(x)e^{-ijx}\,dx\,. \tag{1.6}$$

We will also denote these by $\widehat{f}(j)$. The series $\sum_{j\in\mathbb{Z}} c_j e^{ijx}$ is called the *Fourier series* of f, and we write $f \sim \sum_{j\in\mathbb{Z}} c_j e^{ijx}$ (or $f \sim \sum_{j\in\mathbb{Z}} \widehat{f}(j)e^{ijx}$). The two questions that arise naturally are

1. Is f determined by its Fourier coefficients?
2. If this is the case, to what extent does the Fourier series represent the function?

The first question can be reformulated as whether or not the fact that $\widehat{f}(j) = \widehat{g}(j)$ for every j implies $f = g$ or, equivalently, if $\widehat{f}(j) = 0$ for every j gives that necessarily $f = 0$. Another way to look at the second question is to understand under which conditions do the trigonometric polynomials defined by the partial sums of the Fourier series

$$s_n(f, x) = \sum_{|j|\leq n} \widehat{f}(j)e^{ijx} \quad \text{for } x \in \mathbb{R} \tag{1.7}$$

converge (and in what sense) to f. So, our aim now is to give some answers to these questions.

In many cases, we restrict our attention to 2π-periodic real functions. In this case, the description of the Fourier series with sine and cosine might be more convenient. Note that

$$c_j e^{ijx} + c_{-j}e^{-ijx} = (c_j + c_{-j})\cos(jx) + i(c_j - c_{-j})\sin(jx)$$

for every $j \in \mathbb{N}$. Then we can define $a_0 = 2c_0$ and

$$a_j = c_j + c_{-j} \quad \text{and} \quad b_j = i(c_j - c_{-j}) \tag{1.8}$$

for $j \in \mathbb{N}$. Note that, since f is real-valued, we have $c_{-j} = \overline{c}_j$, and then $a_j, b_j \in \mathbb{R}$ for every j. With all this, we can rewrite the Fourier series of f as

$$f \sim \sum_{j\in\mathbb{Z}} c_j e^{ijx} = \frac{a_0}{2} + \sum_{j=1}^{\infty} a_j \cos(jx) + \sum_{j=1}^{\infty} b_j \sin(jx)\,.$$

These are called the *exponential and trigonometric Fourier series* of f (respectively). We have

$$a_j = \frac{1}{\pi}\int_{-\pi}^{\pi} f(x)\cos(jx)dx \text{ and } b_j = \frac{1}{\pi}\int_{-\pi}^{\pi} f(x)\sin(jx)dx\,.$$

We can also recover the c_j coefficients from these by doing

$$c_j = \frac{1}{2}(a_j - ib_j)$$

for $j \in \mathbb{Z}$ (here we take $b_0 = 0$).

From now on and for the rest of this section, we identify $L_1(-\pi, \pi)$ with the space of 2π-periodic functions $f : \mathbb{R} \to \mathbb{R}$ for which $|f|$ is integrable in $]-\pi, \pi]$. Also, if a function is defined in $]-\pi, \pi]$, we assume that it is extended 2π-periodically to $\mathbb{R}$.

Example 1.11 Fix some $a > 0$, and let us consider the function defined as

$$f(x) = \begin{cases} 0 & \text{if } -\pi < x < 0 \\ a & \text{if } 0 \le x \le \pi \end{cases}$$

(extended periodically to $\mathbb{R}$). For each $j \in \mathbb{N}$, we have

$$a_j = \frac{1}{\pi}\int_0^{\pi} a\cos(jx)dx = 0\,,$$

and

$$b_j = \frac{1}{\pi}\int_0^{\pi} a\sin(jx)dx = \begin{cases} 0 & \text{if } j \text{ is even} \\ \frac{2a}{j\pi} & \text{if } j \text{ is odd}\,. \end{cases}$$

Clearly $a_0 = a$, and therefore

$$f \sim \frac{a}{2} + \frac{2a}{\pi}\sum_{k=1}^{\infty}\frac{1}{2k-1}\sin\left((2k-1)x\right).$$

□

Example 1.12 Consider the function $f(x) = e^x$ for $x \in]-\pi, \pi]$ (extended periodically to the whole $\mathbb{R}$). For each $j \in \mathbb{Z}$, we have

$$c_j = \frac{1}{2\pi}\int_{-\pi}^{\pi} e^{(1-ij)x}dx = \frac{e^{\pi} - e^{-\pi}}{2\pi}(-1)^j\frac{1}{1-ij},$$

so that

$$f \sim \frac{e^{\pi} - e^{-\pi}}{2\pi}\sum_{j\in\mathbb{Z}}\frac{(-1)^j}{1-ij}e^{ijx}.$$

From (1.8), we immediately have

$$a_j = \frac{e^{\pi} - e^{-\pi}}{2\pi}2\frac{(-1)^j}{j^2+1} \text{ and } b_j = \frac{e^{\pi} - e^{-\pi}}{2\pi}2\frac{(-1)^{j+1}j}{j^2+1}$$

for $j \in \mathbb{N}$, and $a_0 = \frac{e^{\pi}-e^{-\pi}}{\pi}$. This altogether yields

$$f \sim \frac{e^{\pi} - e^{-\pi}}{2\pi}\left(1 + \sum_{j=1}^{\infty}\frac{2(-1)^j}{j^2+1}\cos(jx) + \sum_{j=1}^{\infty}\frac{2j(-1)^{j+1}}{j^2+1}\sin(jx)\right).$$

□

Example 1.13 Take now the function given by $f(x) = x^2$ for $x \in]-\pi, \pi]$ (again, extended to $\mathbb{R}$ periodically). Straightforward computations show that $a_0 = \frac{2\pi^2}{3}$ and

$$a_j = \frac{1}{\pi}\int_{-\pi}^{\pi} x^2\cos(jx)dx = 4\frac{(-1)^j}{j^2} \text{ and } b_j = \frac{1}{\pi}\int_{-\pi}^{\pi} x^2\sin(jx)dx = 0.$$

With this,

$$f \sim \frac{\pi^2}{3} + 4\sum_{j=1}^{\infty}\frac{(-1)^j}{j^2}\cos(jx).$$

□

Example 1.14 Take now the function given by $f(x) = |x|$ for $x \in]-\pi, \pi]$ (once more, extended periodically to $\mathbb{R}$). We have

$$a_0 = \frac{1}{\pi}\int_{-\pi}^{\pi}|x|dx = \frac{2}{\pi}\int_0^{\pi} xdx = \pi,$$

and

$$a_j = \frac{1}{\pi}\int_{-\pi}^{\pi} |x|\cos(jx)dx = \frac{2}{\pi}\int_0^{\pi} x\cos(jx)dx = \begin{cases} 0 & \text{for } j \text{ even} \\ -\frac{4}{\pi j^2} & \text{for } j \text{ odd} \end{cases}$$

as well as

$$b_j = \frac{1}{\pi}\int_{-\pi}^{\pi} |x|\sin(jx)dx = \frac{1}{\pi}\Big(\int_0^{\pi} x\sin(jx)dx - \int_{-\pi}^{0} x\sin(jx)dx\Big) = 0$$

for $j \in \mathbb{N}$. This yields

$$f \sim \frac{\pi}{2} - \frac{4}{\pi}\sum_{k=1}^{\infty} \frac{1}{(2k-1)^2}\cos\big((2k-1)x\big)\,. \tag{1.9}$$

□

Let us point out that if $f : \mathbb{R} \to \mathbb{R}$ is 2π-periodic and

- Even, then $b_j = 0$ and

$$a_j = \frac{2}{\pi}\int_0^{\pi} f(x)\cos(jx)dx\,,$$

  for every $j \in \mathbb{N}_0$.
- Odd, then $a_j = 0$ and

$$b_j = \frac{2}{\pi}\int_0^{\pi} f(x)\sin(jx)dx\,,$$

  for every $j \in \mathbb{N}_0$.

Let us give now a first look at the behaviour of the sequence $\big(\widehat{f}(j)\big)_{j\in\mathbb{Z}}$ of Fourier coefficients of a function f. We start with a result that we state without proof. There are a number of proofs available; one of them can be found in [11, Theorem 2.2].

Theorem 1.15 (Riemann–Lebesgue Lemma) *Let $f \in L_1(\mathbb{R})$, then*

$$\lim_{|t|\to\infty} \int_{-\infty}^{\infty} f(x)e^{-ixt}dx = 0\,.$$

As an immediate consequence, we have the following on the behaviour of the Fourier coefficients.

Corollary 1.1 *Let $f \in L_1(-\pi, \pi)$, then*

1. $|\widehat{f}(j)| \le \int_{-\pi}^{\pi} |f(x)| dx$ *for every* $j \in \mathbb{Z}$.
2. $\lim_{|j|\to\infty} |\widehat{f}(j)| = 0$.

Proof Statement 1 is a direct consequence of the definition (1.6), whereas 2 follows from an application of Theorem 1.15 to $f\chi_{]-\pi,\pi]}$. □

1.3.2.1 Dirichlet Kernel

For each $n \in \mathbb{N}$, we define the *nth Dirichlet kernel* D_n as the function defined on $\mathbb{R}$ by

$$D_n(x) = \sum_{|j|\le n} e^{ijx} , \tag{1.10}$$

which is clearly continuous and 2π-periodic. Moreover we have (see [11, Lemma 4.2])

$$D_n(x) = 1 + 2\sum_{j=1}^{n} \cos(jx) = \begin{cases} \frac{\sin((n+1)/2)x)}{\sin(x/2)} & \text{if } 0 < |x| < \pi \\ 2n+1 & \text{if } x = 0 \,. \end{cases} \tag{1.11}$$

In particular, D_n is real-valued and even. It is a convenient tool to describe the partial sums of a Fourier series of a function.

Proposition 1.5 *Let $f \in L_1(-\pi, \pi)$, then*

$$\begin{aligned} s_n(f, x) = \frac{1}{2\pi}\int_{-\pi}^{\pi} f(t)D_n(x-t)dt &= \frac{1}{2\pi}\int_{-\pi}^{\pi} f(x-t)D_n(t)dt \\ &= \frac{1}{2\pi}\int_{0}^{\pi} (f(x+t) - f(x-t))D_n(t)dt \,, \end{aligned}$$

for every $x \in]-\pi, \pi]$ and $n \in \mathbb{N}$. In particular,

$$\int_0^{\pi} D_n(x)dx = \pi \,,$$

for every $n \in \mathbb{N}$.

Proof The result follows by direct computation, using the definition of the Fourier coefficients (1.6) and the periodicity of f. □

We will see later (see (6.2)) that this is the convolution of f with the Dirichlet kernel. With this, we can already give a first result on the pointwise convergence of the Fourier series of a function (providing a first answer for our second question). Given $f : \mathbb{R} \to \mathbb{R}$ and $a \in \mathbb{R}$, we define

$$f(a^+) = \lim_{h\to 0^+} f(a+h) \text{ and } f(a^-) = \lim_{h\to 0^-} f(a+h)\,. \tag{1.12}$$

Whenever these limits exist, we also define

$$f'_+(a) = \lim_{h\to 0^+} \frac{f(a+h) - f(a^+)}{h}$$

and

$$f'_-(a) = \lim_{h\to 0^-} \frac{f(a+h) - f(a^-)}{h}\,.$$

Theorem 1.16 (Dini–Dirichlet) *Let $f \in L_1(-\pi,\pi)$ be continuous, except for a finite number of points, in which the limits in (1.12) exist. Let $c \in \mathbb{R}$ so that both $f'_+(c)$ and $f'_-(c)$, then*

$$\lim_{n\to\infty} s_n(f,c) = \frac{1}{2}\big(f(c^+) + f(c^-)\big)\,.$$

In particular, if f is differentiable in $\mathbb{R}$, then

$$\lim_{n\to\infty} s_n(f,x) = f(x)\,,$$

for every $x \in \mathbb{R}$

Proof See [2, Theorem 11.9], [10, Theorem 8.45] or [11, Theorem 4.5]. □

Remark 1.3 All the functions in Examples 1.11–1.14 clearly satisfy the hypothesis of Theorem 1.16, so that the corresponding Fourier series converge (pointwise) at every $x \in \mathbb{R}$. Note that, as a consequence of the Weierstraß M-test (Proposition 1.4), the Fourier series of e^x, x^2, and $|x|$ converge uniformly on $[-\pi,\pi]$. We can use the expansion as a Fourier series of $|x|$ to compute the value of $\sum_{k=1}^{\infty} \frac{1}{k^2}$. To begin with, evaluating the function at $x = 0$ and using that the series in (1.9) converges, we have

$$0 = \frac{\pi}{2} - \frac{4}{\pi}\sum_{j=1}^{\infty} \frac{1}{(2j-1)^2}\,,$$

from which we get

$$\sum_{j=1}^{\infty} \frac{1}{(2j-1)^2} = \frac{\pi^2}{8}\,.$$

So, we have the sum of the odd terms in the series. Let us compute now the sum of the even terms.

$$\sum_{j=1}^{\infty} \frac{1}{(2j)^2} = \frac{1}{4} \sum_{j=1}^{\infty} \frac{1}{j^2} = \frac{1}{4} \Big(\sum_{j=1}^{\infty} \frac{1}{(2j)^2} + \sum_{j=1}^{\infty} \frac{1}{(2j-1)^2} \Big) = \frac{1}{4} \sum_{j=1}^{\infty} \frac{1}{(2j)^2} + \frac{\pi^2}{32},$$

and

$$\sum_{j=1}^{\infty} \frac{1}{(2j)^2} = \frac{\pi^2}{24}.$$

Joining these two finally yields

$$\sum_{k=1}^{\infty} \frac{1}{k^2} = \frac{\pi^2}{8} + \frac{\pi^2}{24} = \frac{\pi^2}{6}.$$

Remark 1.4 Theorem 1.16 has two major drawbacks. First of all, the conditions on the function are rather restrictive and may not be accomplished even by continuous functions. As a matter of fact, we will see later (see Theorem 3.20) that there are continuous functions for which the Fourier series does not converge at 0. Second, even when the series converges, we can only ensure pointwise convergence, and we would like to have a more powerful (such as, for example, uniform) convergence. In order to overcome these two problems, we need a different approach.

1.3.2.2 Cesàro Means: Féjer Kernel

Given a function, the idea now is to take, instead of the partial sums, the arithmetic means of them. To be more precise, for $f \in L_1(-\pi, \pi)$ and $n \in \mathbb{N}_0$, the *nth Cesàro mean* of f is defined as the arithmetic mean of the partial sums up to n, that is,

$$\sigma_n(f, x) = \frac{1}{n+1} \sum_{k=0}^{n} s_k(f, x),$$

for $x \in \mathbb{R}$. Note that, for each f and n, the Cesàro mean is again a trigonometric polynomial. Just as the Dirichlet kernel was a helpful tool when describing the partial sum of the Fourier series, we consider a new kernel that allows us to deal with the Cesàro means. For each $n \in \mathbb{N}$, the *nth Féjer kernel* is defined as the arithmetic mean of the first Dirichlet kernels, that is,

$$K_n(x) = \frac{1}{n+1} \sum_{k=0}^{n} D_k(x),$$

for $x \in \mathbb{R}$. As a straightforward consequence of Proposition 1.5, given $f \in L_1(-\pi, \pi)$, we have

$$\sigma_n(f, x) = \frac{1}{2\pi} \int_{-\pi}^{\pi} f(t) K_n(x-t) dt = \frac{1}{2\pi} \int_{-\pi}^{\pi} f(x-t) K_n(t) dt\,, \tag{1.13}$$

for every $n \in \mathbb{N}_0$ and $x \in \mathbb{R}$. Once again, this is the convolution (see (6.2)) of f and K_n. We establish now some basic properties of the Féjer kernel. A proof can be found, for example, in [11, Lemmas 4.3 and 4.4].

Proposition 1.6 *For each $n \in \mathbb{N}_0$, we have*

1. $K_n(0) = n + 1$.
2. $K_n(x) = \dfrac{1}{n+1}\left(\dfrac{\sin\left(\frac{n+1}{2}x\right)}{\sin\left(\frac{x}{2}\right)}\right)^2$ *for every* $x \in]-\pi, \pi] \setminus \{0\}$.
3. K_n *is non-negative and even.*
4. $\displaystyle\int_{-\pi}^{\pi} K_n(x) dx = 2\pi$.
5. $\displaystyle\lim_{n\to\infty} \int_{\delta}^{\pi} K_n(x) dx = 0$ *for each fixed* $0 < \delta < \pi$.

With this, we can get what we aimed at uniform convergence, giving another answer to our second question.

Theorem 1.17 (Féjer) *Let $f : \mathbb{R} \to \mathbb{R}$ be continuous and 2π-periodic. Then $\sigma_n(f, x) \to f$ uniformly on $\mathbb{R}$ as $n \to \infty$.*

Proof Since f is continuous and periodic, it is bounded, so that we can find $M > 0$ with $|f(x)| \leq M$ for every $x \in \mathbb{R}$. Then, the function is also uniformly continuous- So, for each fixed $\varepsilon > 0$ there is $\delta > 0$ so that $|f(x+t) - f(x)| < \varepsilon$ for every $x \in \mathbb{R}$ and $|t| < \delta$.
Given any $x \in \mathbb{R}$ and $n \in \mathbb{N}$, we have from (1.13) and Proposition 1.6

$$\begin{aligned}
|\sigma_n(f, x) - f(x)| &\leq \frac{1}{2\pi} \int_{-\pi}^{\pi} |f(x+t) - f(x)| K_n(t) dt \\
&= \frac{1}{2\pi}\Bigg(\int_{-\pi}^{-\delta} |f(x+t) - f(x)| K_n(t) dt + \int_{-\delta}^{\delta} |f(x+t) - f(x)| K_n(t) dt \\
&\qquad + \int_{\delta}^{\pi} |f(x+t) - f(x)| K_n(t) dt \Bigg).
\end{aligned}$$

Let us bound now each one of these three integrals. For the first and third one, let us observe in first place that $\left(\sin\left(\frac{u}{2}\right)\right)^2 \geq \frac{u^2}{\pi^2}$ for every $u \in [0, \pi]$, which, if $\delta < |u| < \pi$, implies

$$0 \leq K_n(u) = \frac{1}{n+1}\left(\frac{\sin\left(\frac{n+1}{2}u\right)}{\sin\left(\frac{u}{2}\right)}\right)^2 \leq \frac{\pi^2}{(n+1)u^2} \sin\left(\tfrac{n+1}{2}u\right)^2 \leq \frac{\pi^2}{(n+1)\delta^2}\,.$$

Hence

$$\int_{-\pi}^{-\delta} |f(x+t) - f(x)| K_n(t)dt \leq 2M \frac{\pi^3}{(n+1)\delta^2},$$

and

$$\int_{\delta}^{\pi} |f(x+t) - f(x)| K_n(t)dt \leq 2M \frac{\pi^3}{(n+1)\delta^2}.$$

For the remaining integral, just observe that

$$\int_{-\delta}^{\delta} |f(x+t) - f(x)| K_n(t)dt \leq \varepsilon \int_{-\delta}^{\delta} K_n(t)dt \leq 2\pi\varepsilon.$$

Joining all these together, we obtain

$$|\sigma_n(f,x) - f(x)| \leq \varepsilon + \frac{2M\pi^2}{(n+1)\delta^2}$$

for every $x \in \mathbb{R}$. Given ε and δ (recall that these are fixed), it is enough to choose $n_0 \in \mathbb{N}$ with $\frac{1}{n_0+1} < \varepsilon \frac{\delta^2}{2M\pi^2}$ to get

$$|\sigma_n(f,x) - f(x)| \leq 2\varepsilon$$

for every $x \in \mathbb{R}$ and $n \geq n_0$. This completes the proof. □

Remark 1.5 This result also answers our first question on whether or not the coefficients determine the function whenever it is continuous. Indeed, suppose that f is continuous and $\widehat{f}(j) = 0$ for every $j \in \mathbb{Z}$. Then $s_n(f,x) = 0$ for every n (and x) and so $\sigma_n(f,x) = 0$. Theorem 1.17 implies that $f = 0$.

As a consequence of Féjer's theorem, we obtain that the continuous functions can be uniformly approximated by polynomials.

Corollary 1.2 *Let $f : \mathbb{R} \to \mathbb{R}$ be continuous and 2π-periodic. For each $\varepsilon > 0$, there exists a trigonometric polynomial T so that*

$$\sup_{x \in \mathbb{R}} |f(x) - T(x)| < \varepsilon.$$

Corollary 1.3 *Let f be continuous in $[a,b]$. For each $\varepsilon > 0$, there exists a polynomial p so that*

$$\sup_{x \in [a,b]} |f(x) - p(x)| < \varepsilon.$$

1.3.2.3 Poisson Kernel

Given a function $f :]-\pi,\pi] \to \mathbb{C}$ (or a 2π-periodic function on $\mathbb{R}$), we can think of it as defined on $\mathbb{T} = \partial\mathbb{D} = \{z \in \mathbb{C} : |z| = 1\}$ just by identifying $]-\pi,\pi]$ with $\mathbb{T}$ by means of $t \leftrightsquigarrow e^{it}$. The question then arises naturally of whether or not can this function be extended to the disc $\mathbb{D}$. The tool to achieve this is the *Poisson kernel*, defined as

$$P_r(t) = \sum_{j\in\mathbb{Z}} r^{|j|} e^{ijt} , \tag{1.14}$$

for $t \in]-\pi,\pi]$ and $0 \le r < 1$. Note that since $0 \le r < 1$ the series converges (absolutely) for every t and r, and the function is well defined. The kth partial sum in (1.14) gives

$$\sum_{j=-k}^{k} r^{|j|} e^{ijt} = \sum_{j=0}^{k} r^j e^{ijt} + \sum_{j=1}^{k} r^j e^{-ijt} ,$$

which, summing up the geometric series, yields

$$P_r(t) = \frac{1}{1-re^{it}} + \frac{re^{-it}}{1-re^{it}} = \frac{1-r^2}{1-2r\cos t + r^2} . \tag{1.15}$$

Proposition 1.7 *For each $0 \le r < 1$, the function P_r is 2π-periodic, positive, and even. It satisfies*

$$\int_{-\pi}^{\pi} P_r(t)dt = 2\pi$$

and

$$\frac{1-r}{1+r} \le P_r(t) \le \frac{1+r}{1-r}$$

for every r and t.

Proof The series in (1.14) converges uniformly on $[-\pi,\pi]$. This gives the statement on the integral. All the rest follows easily from (1.15). □

By doing $z = re^{it}$ (for $z \in \mathbb{D}$), we may look at it as a function defined on the disc

$$P(z) = P(re^{it}) = P_r(t) = \sum_{j\in\mathbb{Z}} r^{|j|} e^{ijt} = \frac{1-|z|^2}{|1-z|^2} .$$

Proposition 1.8 *Fix $0 < \delta < \pi$, and then for each $0 < r < 1$, we have $P_r(t) \leq P_r(\delta)$ for every $\delta < |t| < \pi$. Also*

$$\lim_{r \to 1^-} \Big[\sup\{P_r(t) \colon \delta < |t| < \pi\} \Big] = 0 .$$

Proof Let us define the function $g : [\delta, \pi] \to \mathbb{R}$ by $g(t) = P_r(t)$. Since $g'(t) < 0$ for every $\delta < t < \pi$ (this follows by direct computation from (1.15)), this gives the conclusion. Also,

$$\sup\{P_r(t) \colon \delta < |t| < \pi\} = P_r(\delta) = \frac{1 - r^2}{1 - 2r\cos\delta + r^2} ,$$

which clearly tends to 0 as $r \to 1^-$. □

This is the right tool to extend our functions from $\mathbb{T}$ to the disc $\mathbb{D}$. Given $f \in L_1(-\pi, \pi)$ (recall that we are identifying this with 2π-periodic functions on $\mathbb{R}$) for each $0 \leq r < 1$ and $t \in \mathbb{R}$, the integral

$$P[f](re^{it}) = \frac{1}{2\pi} \int_{-\pi}^{\pi} P_r(x - t) f(t) dt = \frac{1}{2\pi} \int_{-\pi}^{\pi} f(x - t) P_r(t) dt$$

is finite (note that, once again, this is a convolution). So, in this way, we may define a function $P[f] : \mathbb{D} \to \mathbb{C}$ (called the *Poisson integral* of f). If f has Fourier coefficients $\big(\widehat{f}(j)\big)_j$, then we can rewrite the previous as

$$P[f](re^{it}) = \sum_{j \in \mathbb{Z}} r^{|j|} \widehat{f}(j) e^{ijt} .$$

So, starting from a function f defined on $\mathbb{T}$, we get a function $P[f]$ defined on $\mathbb{D}$. Let us see now that (at least in some cases) this actually extends f.

Theorem 1.18 *Let $f : \mathbb{R} \to \mathbb{C}$ be continuous and 2π-periodic. Let us define a function $F : \overline{\mathbb{D}} \to \mathbb{C}$ by*

$$F(e^{ix}) = f(x) \text{ for } x \in \mathbb{R}$$

and

$$F(re^{ix}) = P[f](re^{ix}) \text{ for } 0 \leq r < 1 \text{ and } x \in \mathbb{R} .$$

Then F is continuous on $\overline{\mathbb{D}}$.

Proof The function F is clearly well defined. If $z = re^{ix}$ with $|z| < 1$, we have

$$F(z) = \sum_{j\in\mathbb{Z}} r^{|j|}\widehat{f}(j)e^{ijt} = \sum_{j=0}^{\infty}\widehat{f}(j)z^j + \sum_{j=1}^{\infty}\widehat{f}(-j)\overline{z}^j .$$

Since $\left(\widehat{f}(j)\right)_{j\in\mathbb{Z}}$ is bounded (recall Corollary 1.1), each one of these two series converges absolutely for each $z \in \mathbb{D}$ and, then, uniformly on each disc of radius strictly smaller than 1. Hence, F is continuous in $\mathbb{D}$.

It is only left to see that F is continuous at each $e^{ix} \in \mathbb{T}$. To do this, we are going to see that $F(re^{ix}) \to F(e^{ix})$ uniformly on $[-\pi, \pi]$ as $r \to 1^-$, following a similar idea as in the proof of Theorem 1.17. To begin with, since f is continuous and 2π-periodic, it is bounded (by, say, $M > 0$) and uniformly continuous. Then, for a fixed $\varepsilon > 0$, we can find $\delta > 0$ so that $|f(x-t) - f(x)| < \varepsilon$ for every $x \in \mathbb{R}$ and $|t| < \delta$. Then, given $x \in \mathbb{R}$, we have

$$\begin{aligned}
|F(re^{ix}) - F(e^{ix})| &= \frac{1}{2\pi}\left|\int_{-\pi}^{\pi} f(x-t)P_r(t)dt - \int_{-\pi}^{\pi} f(x)P_r(t)\right| \\
&\leq \frac{1}{2\pi}\int_{-\pi}^{\pi} |f(x-t) - f(x)|P_r(t)dt \\
&= \frac{1}{2\pi}\left[\int_{-\pi}^{-\delta} |f(x-t) - f(x)|P_r(t)dt + \int_{-\delta}^{\delta} |f(x-t) - f(x)|P_r(t)dt \right. \\
&\quad \left. + \int_{\delta}^{\pi} |f(x-t) - f(x)|P_r(t)dt\right].
\end{aligned}$$

We bound now each one of these three integrals. On the one hand,

$$\int_{-\delta}^{\delta} |f(x-t) - f(x)|P_r(t)dt \leq \varepsilon \int_{-\delta}^{\delta} P_r(t)dt \leq 2\pi\varepsilon ,$$

and, on the other hand,

$$\int_{\delta}^{\pi} |f(x-t) - f(x)|P_r(t)dt \leq 2M \int_{\delta}^{\pi} P_r(t)dt \leq 2M\pi P_r(\delta) .$$

The same bound holds for the remaining integral. This altogether gives

$$|F(re^{ix}) - F(e^{ix})| \leq \varepsilon + 2MP_r(\delta) .$$

Recall that $\delta > 0$ is fixed and then, taking Proposition 1.8 into account, we can find $0 < r_0 < 1$ so that $P_r(\delta) < \frac{\varepsilon}{2M}$ for every $r_0 < r < 1$. Then,

$$|F(re^{ix}) - F(e^{ix})| < \varepsilon$$

for every $x \in [-\pi, \pi]$ and $r_0 < r < 1$. Since the choice of r_0 does not depend on x, this proves our claim and completes the proof. □

1.3.3 Dirichlet Series

Each sequence $(a_n)_n$ of complex numbers defines a *Dirichlet series*

$$\sum_{n=1}^{\infty} a_n n^{-s} ,$$

where the variable $s \in \mathbb{C}$ (usually we would write z for a complex variable, but for historical reasons we write s). Note that, once again, this is a series of functions where $f_n(s) = a_n n^{-s}$.

We have seen in Theorem 1.10 that if a power series converges at some point, then it converges absolutely at every point with strictly smaller modulus. In this way, there is a maximal disc on which a power series converges, and it converges uniformly on every disc of strictly smaller radius (recall (1.5)). There are certain parallelism between Dirichlet and power series, replacing 'modulus' by 'real part' and 'disc' by half plane'. To be more precise, if a Dirichlet series converges at some point, then it converges at every point with strictly bigger real part (see Theorem 1.19). Then, the natural domains where Dirichlet series converge are half-planes $[\mathrm{Re} > \sigma] = \{s \in \mathbb{C} \colon \mathrm{Re}\, s > \sigma\}$. But there are also big differences: while the radius of convergence and the absolute convergence of a given power series are the same (recall (1.5)), in the case of Dirichlet series, the maximal half-planes of convergence and absolute convergence may not coincide (see (1.17)).

We start by analysing the convergence of Dirichlet series. To begin with, we need a basic fact, known as Abel summation: given $a_1, \dots, a_N \in \mathbb{C}$ and $b_1, \dots, b_N \in \mathbb{C}$, we write $A_n = \sum_{k=1}^{n} a_k$ for $2 \leq n \leq N$, then

$$\sum_{n=1}^{N} a_n b_n = A_N b_N + \sum_{n=1}^{N-1} A_n (b_n - b_{n+1}) . \tag{1.16}$$

If we observe that $A_1 = a_1$ and $a_n = A_n - A_{n-1}$, we have

$$\sum_{n=1}^{N} a_n b_n = A_1 b_1 + \sum_{n=2}^{N} (A_n - A_{n-1}) b_n = A_N b_N + \sum_{n=1}^{N-1} A_n b_n - \sum_{n=2}^{N} A_{n-1} b_n$$

$$= A_N b_N + \sum_{n=1}^{N-1} A_n b_n - \sum_{n=1}^{N-1} A_n b_{n+1} = A_N b_N + \sum_{n=1}^{N-1} A_n (b_n - b_{n+1}) .$$

Theorem 1.19 *Let $\sum_n a_n \frac{1}{n^s}$ be a Dirichlet series that converges at $s_0 = \sigma_0 + it_0 \in \mathbb{C}$, then it converges at s for all $\operatorname{Re} s > \sigma_0$. Even more, the series converges uniformly on every compact set in $[\operatorname{Re} > \sigma_0]$.*

Proof Given $\operatorname{Re} s > \operatorname{Re} s_0$, we write $u = s_0 - s$, and by (1.16), we have, for $M > N$,

$$\sum_{n=N}^{M} a_n \frac{1}{n^s} = \sum_{n=N}^{M} \frac{a_n}{n^{s_0}} \frac{1}{n^u} = \Big(\sum_{n=N}^{M} \frac{a_n}{n^{s_0}} \Big) \frac{1}{M^u} + \sum_{n=N}^{M-1} \Big(\sum_{k=N}^{n} \frac{a_k}{k^{s_0}} \Big) \Big(\frac{1}{n^u} - \frac{1}{(n+1)^u} \Big) .$$

Take now some $K \subseteq [\operatorname{Re} > \sigma_0]$ and fix $\varepsilon > 0$. Since $K_1 = \{ s - s_0 : s \in K \}$ is again compact, we may define $A = \sup_{u \in K_1} \frac{|u|}{\operatorname{Re} u} < \infty$ (which clearly satisfies $A > 0$). Since the series $\sum_n a_n \frac{1}{n^{s_0}}$ is convergent, we can find N_0 so that

$$\Big| \sum_{n=N}^{M} \frac{a_n}{n^{s_0}} \Big| < \frac{\varepsilon}{A}$$

for every $M > N \geq N_0$. Also

$$\begin{aligned} \Big| \frac{1}{n^u} - \frac{1}{(n+1)^u} \Big| &= \Big| \int_n^{n+1} \frac{u}{x^{u+1}} dx \Big| \leq \int_n^{n+1} \frac{|u|}{x^{\operatorname{Re} u + 1}} dx \\ &= \frac{|u|}{\operatorname{Re} u} \Big(\frac{1}{n^{\operatorname{Re} u}} - \frac{1}{(n+1)^{\operatorname{Re} u}} \Big) . \end{aligned}$$

Taking all this into account, if $M > N \geq N_0$ and $s \in K$, we have

$$\begin{aligned} \Big| \sum_{n=N}^{M} a_n \frac{1}{n^s} \Big| &\leq \Big| \sum_{n=N}^{M} \frac{a_n}{n^{s_0}} \Big| \frac{1}{M^{\operatorname{Re} u}} + \sum_{n=N}^{M-1} \Big| \sum_{k=N}^{n} \frac{a_k}{k^{s_0}} \Big| \frac{|u|}{\operatorname{Re} u} \Big(\frac{1}{n^{\operatorname{Re} u}} - \frac{1}{(n+1)^{\operatorname{Re} u}} \Big) \\ &\leq \frac{\varepsilon}{A} \frac{|u|}{\operatorname{Re} u} \Big(\frac{1}{M^{\operatorname{Re} u}} + \sum_{n=N}^{M-1} \Big(\frac{1}{n^{\operatorname{Re} u}} - \frac{1}{(n+1)^{\operatorname{Re} u}} \Big) \Big) < \frac{\varepsilon}{A} A \frac{1}{N^{\operatorname{Re} u}} < \varepsilon . \end{aligned}$$

Proposition 1.3 gives the conclusion. □

This shows that if a Dirichlet series converges at some point s_0, then it converges on the open half-plane defined by the abscissa $\operatorname{Re} s_0$. In other words, the natural regions of convergence of Dirichlet series are half-planes. As we did for power series, we may consider

$$\sigma_c = \inf \Big\{ \sigma \in \mathbb{R} \colon \sum a_n \tfrac{1}{n^s} \text{ converges in } [\operatorname{Re} > \sigma] \Big\} \in [-\infty, \infty] ,$$

the abscissa that defines the biggest half-plane on which a given Dirichlet series converges (this would be in some sense the analogue to the radius of convergence of a power series). Then the Dirichlet series converges pointwise for every $\mathrm{Re}\, s > \sigma_c$ and diverges for every $\mathrm{Re}\, s < \sigma_c$. In this way, the function $f : [\mathrm{Re}\, s > \sigma_c] \to \mathbb{C}$ given by

$$f(s) = \sum_{n=1}^{\infty} a_n \frac{1}{n^s}$$

is well defined. Moreover, the fact that the series converges uniformly on compact sets implies that the function defined by the Dirichlet series is holomorphic (check Theorem 4.4).

We may also consider the abscissa of absolute convergence

$$\sigma_a = \inf \left\{ \sigma \in \mathbb{R} \colon \sum a_n \tfrac{1}{n^s} \text{ converges absolutely in } [\mathrm{Re} > \sigma] \right\} \in [-\infty, \infty]$$

and wonder if, as in the case of power series (recall (1.5)), this equals σ_c. Of course, for certain Dirichlet series, these two abscissas may coincide (just take a Dirichlet series with positive coefficients), but this need not be the case. Take for example the series $\sum_n (-1)^n \frac{1}{n^s}$. This converges for every $\mathrm{Re}\, s > 0$ (just apply Leibniz criterion) but converges absolutely only on the plane $[\mathrm{Re} > 1]$. In other terms,

$$\sigma_c = 0 \text{ and } \sigma_a = 1 .$$

Now the question is 'how big can it get?' or 'how far apart can σ_a and σ_c be to each other?' Actually not really too far away. Assume that $\sum a_n \frac{1}{n^s}$ converges at some $s_0 = \sigma_0 + it$. Then the sequence $(\frac{|a_n|}{n^{\sigma_0}})_n$ is bounded by, say, K, and we have, for every $\varepsilon > 0$,

$$\sum_{n=1}^{\infty} \left| \frac{a_n}{n^{s_0+1+\varepsilon}} \right| = \sum_{n=1}^{\infty} \frac{|a_n|}{n^{\sigma_0+1+\varepsilon}} = \sum_{n=1}^{\infty} \frac{|a_n|}{n^{\sigma_0}} \frac{1}{n^{1+\varepsilon}} \leq K \sum_{n=1}^{\infty} \frac{1}{n^{1+\varepsilon}} < \infty .$$

Then $\sigma_a \leq \sigma_0 + 1 + \varepsilon$. This then shows that for every Dirichlet series $\sigma_a \leq \sigma_c + 1$ and, together with the previous example, we obtain

$$\sup \left\{ \sigma_a - \sigma_c \colon D \text{ Dirichlet series } \right\} = 1 . \tag{1.17}$$

We focus now on a third abscissa: that of uniform convergence. Let

$$\sigma_u = \inf \left\{ \sigma \in \mathbb{R} \colon \sum_{n=1}^{\infty} a_n \tfrac{1}{n^s} \text{ converges uniformly on } [\mathrm{Re} > \sigma] \right\} \in [-\infty, \infty] .$$

Again, the question arises of whether or not this coincides with any of the other two abscissas. As a matter of fact, it does not. Taking the series $\sum_n (-1)^n p_n^{-s}$ (where p_n stands for the nth prime number), it is easy to check using Leibniz criterion for alternate series that $\sigma_c = 0$. On the other hand, it can be shown that $\sigma_u = 1$ (this is less trivial, a proof can be found in [4, page 16]). Taking (1.17) into account (and that $\sigma_u \leq \sigma_a$), this shows

$$\sup \left\{ \sigma_u - \sigma_c : D \text{ Dirichlet series} \right\} = 1 .$$

A much more intricate problem is to separate σ_a from σ_u. It can be shown (see [4, Chapter 4]) that there is a Dirichlet series with $\sigma_a = 1$ and $\sigma_u = \frac{1}{2}$ and that, in fact,

$$\sup \left\{ \sigma_a - \sigma_u : D \text{ Dirichlet series} \right\} = \frac{1}{2} .$$

The proofs of these facts exceed the scope of this text. We can, however, show an interesting and important fact about the abscissa of uniform convergence: it determines the biggest half-plane on which the holomorphic function defined by the series is bounded. Let us introduce two more abscissas. First of all, we define

$$\begin{aligned}\sigma_b = \inf \Big\{ \sigma \in \mathbb{R} \colon \sum_{n=1}^{\infty} a_n \tfrac{1}{n^s} & \text{ converges and defines a} \\ & \text{bounded holomorphic function on } [\mathrm{Re} > \sigma] \Big\} \in [-\infty, \infty] .\end{aligned}$$

Second, if $\sum a_n \frac{1}{n^s}$ is a somewhere convergent Dirichlet series (and defines a function f), we define σ_b^{ext} as the infimum of all $\sigma \in \mathbb{R}$ so that f can be extended to $[\mathrm{Re} > \sigma]$ as a bounded holomorphic function. Obviously $\sigma_b^{\text{ext}} \leq \sigma_b$. On the other hand, it is easy to see that finite sums are bounded on half-planes and, then, Remark 1.2 gives $\sigma_b \leq \sigma_u$. A remarkable result, due to Harald Bohr (known today as *Bohr's theorem*), shows that all three coincide.

Theorem 1.20 (Bohr) *For every Dirichlet series, we have* $\sigma_b^{\text{ext}} = \sigma_b = \sigma_u$.

This follows from the following result.

Lemma 1.1 *Let* $\sum a_n \frac{1}{n^s}$ *be a Dirichlet series that converges absolutely in* $[\mathrm{Re} > a]$ *(and then defines a holomorphic function* $f : [\mathrm{Re} > a] \to \mathbb{C}$*). Suppose that the function* f *can be extended as a bounded, holomorphic function to* $[\mathrm{Re} > b]$*. Then the Dirichlet series* $\sum a_n \frac{1}{n^s}$ *converges uniformly on* $[\mathrm{Re} > b + \varepsilon]$ *for every* $\varepsilon > 0$.

Let us note that as a straightforward consequence of this result $\sigma_u \leq \sigma_b^{\text{ext}}$, which gives Theorem 1.20. The proof relies heavily on a careful use of Cauchy's theorem (see Theorem 4.1 ahead),

Proof Let us denote $K = \sup_{\mathrm{Re}\, s > b} |f(s)|$ and fix $0 < \varepsilon < 1$. Our aim is to show that there is a constant $C = C(\varepsilon, K) > 0$ such that

$$\Big| f(s_0) - \sum_{n=0}^{M} a_n \frac{1}{n^{s_0}} \Big| \leq C \frac{1}{M^{\varepsilon}} \log M \tag{1.18}$$

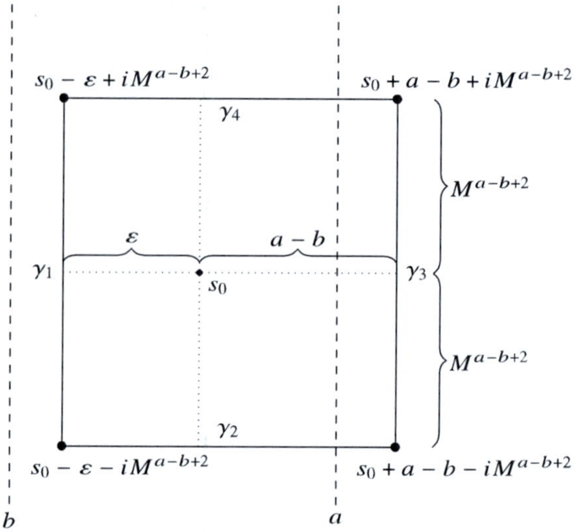

Fig. 1.7 Contour of integration R

for every $M \geq 2$ and all $\operatorname{Re} s_0 > b+\varepsilon$. This clearly will give our claim.

Fix, then, s_0 with $\operatorname{Re} s_0 > b+\varepsilon$ and $M \geq 2$. We consider the rectangle of vertices $s_0 - \varepsilon \pm iM^{a-b+2}$, $s_0 + a - b \pm iM^{a-b+2}$ (we denote it by R, see Fig. 1.7). Note that $\operatorname{Re} s_0 - \varepsilon > b$ and $\operatorname{Re} s_0 + a - b > a + \varepsilon$.

The function $g(z) = f(z)\big(M + \frac{1}{2}\big)^{z-s_0}$ is holomorphic on a connected open set containing the rectangle (actually, on $[\operatorname{Re} > b]$). Then, by Cauchy's Integral Formula (see Theorem 4.1),

$$\int_R \frac{g(z)}{z-s_0}dz = 2\pi i g(s_0) = 2\pi i f(s_0)$$

and

$$\begin{aligned}\left|f(s_0) - \sum_{n=0}^{M} a_n \frac{1}{n^{s_0}}\right| &= \left|\frac{1}{2\pi i}\int_R \frac{g(z)}{z-s_0}dz - \sum_{n=0}^{M} a_n \frac{1}{n^{s_0}}\right| \\ &\leq \left|\frac{1}{2\pi i}\int_{\gamma_3} \frac{g(z)}{z-s_0}dz - \sum_{n=0}^{M} a_n \frac{1}{n^{s_0}}\right| + \left|\frac{1}{2\pi i}\int_{\gamma_1} \frac{g(z)}{z-s_0}dz\right| \\ &\quad + \left|\frac{1}{2\pi i}\int_{\gamma_2} \frac{g(z)}{z-s_0}dz\right| + \left|\frac{1}{2\pi i}\int_{\gamma_4} \frac{g(z)}{z-s_0}dz\right|. \end{aligned} \tag{1.19}$$

Now we have to bound each one of these integrals. In order to keep the notation as neat as possible, we set $c = a - b$ and $T = M^{a-b+2}$. We begin by bounding the integral on the vertical left side. We have

$$\left|\int_{\gamma_1} \frac{g(z)}{z - s_0} dz\right| = \left|\int_{-T}^{T} \frac{f(s_0 - \varepsilon + it)(M + 1/2)^{-\varepsilon+it}}{-\varepsilon + it} i dt\right|$$
$$\leq \int_{-T}^{T} \frac{|f(s_0 - \varepsilon + it)|}{\sqrt{\varepsilon^2 + t^2}} \left(M + \tfrac{1}{2}\right)^{-\varepsilon} dt \leq 2K \frac{1}{(M + 1/2)^{\varepsilon}} \int_0^T \frac{1}{\sqrt{\varepsilon^2 + t^2}} dt .$$

A change of variable $u = t/\varepsilon$ gives

$$\int_0^T \frac{1}{\sqrt{\varepsilon^2 + t^2}} dt = \int_0^{T/\varepsilon} \frac{1}{\sqrt{1 + u^2}} du = \int_0^1 \frac{1}{\sqrt{1 + u^2}} du + \int_1^{T/\varepsilon} \frac{1}{\sqrt{1 + u^2}} du$$
$$\leq 1 + \int_1^{T/\varepsilon} \frac{1}{u} du = 1 + \log T - \log \varepsilon = 1 + (a - b + 2) \log M - \log \varepsilon .$$

Then, if $M > e/\varepsilon$ (so that $1 - \log \varepsilon < \log M$), we have

$$\left|\int_{\gamma_1} \frac{g(z)}{z - s_0} dz\right| \leq 2K \frac{1}{(M + 1/2)^{\varepsilon}} (a - b + 1) \log M ,$$

and hence there is $C_1 > 0$ so that

$$\left|\frac{1}{2\pi i} \int_{\gamma_1} \frac{g(z)}{z - s_0} dz\right| \leq C_1 \frac{1}{M^{\varepsilon}} \log M , \tag{1.20}$$

for every $M \geq 2$.

We now bound the integrals along the horizontal sides. We have

$$\left|\int_{\gamma_2} \frac{g(z)}{z - s_0} dz\right| = \left|\int_{-\varepsilon}^{c} \frac{f(s_0 + t + iT)}{t + iT} \left(M + \tfrac{1}{2}\right)^{t+iT} dt\right|$$
$$\leq \int_{-\varepsilon}^{c} \frac{|f(s_0 + t + iT)|}{\sqrt{t^2 + T^2}} \left(M + \tfrac{1}{2}\right)^{t} dt$$
$$\leq K \int_{-\varepsilon}^{c} \frac{1}{\sqrt{t^2 + T^2}} \left(M + \tfrac{1}{2}\right)^{t} dt \leq \frac{K}{T} \int_{-\varepsilon}^{c} \left(M + \tfrac{1}{2}\right)^{t} dt$$
$$= \frac{K}{M^{a-b+2}} \frac{1}{\log(M + 1/2)} \left(\left(M + \tfrac{1}{2}\right)^{a-b} - \left(M + \tfrac{1}{2}\right)^{-\varepsilon}\right)$$
$$\leq \frac{K}{M^{a-b+2}} (2M)^{a-b} .$$

Then, there is $C_2 > 0$ so that

$$\left| \frac{1}{2\pi i} \int_{\gamma_2} \frac{g(z)}{z - s_0} dz \right| \leq C_2 \frac{1}{M^2}, \tag{1.21}$$

for every $M \geq 2$.

Exactly the same argument shows that we can find $C_4 > 0$ with

$$\left| \frac{1}{2\pi i} \int_{\gamma_4} \frac{g(z)}{z - s_0} dz \right| \leq C_4 \frac{1}{M^2}, \tag{1.22}$$

for every $M \geq 2$.

The remaining term in (1.19) requires some more work. Let us note first that the segment is within the half-plane of absolute convergence of the series. Hence the series converges uniformly on the segment, and we may replace f by the Dirichlet series and then interchange the sum and the integral to get

$$\int_{\gamma_3} \frac{g(z)}{z - s_0} dz = \int_{\gamma_3} \frac{f(z)}{z - s_0} (M + \tfrac{1}{2})^{z - s_0} dz$$

$$= \sum_{n=1}^{\infty} \int_{\gamma_3} a_n \frac{1}{n^z} \frac{(M + 1/2)^{z - s_0}}{z - s_0} dz = \sum_{n=1}^{\infty} \frac{a_n}{n^{s_0}} \int_{\gamma_3} \Big(\frac{M + 1/2}{n} \Big)^{z - s_0} \frac{1}{z - s_0} dz .$$

Therefore,

$$\begin{aligned} \Bigg| \sum_{n=1}^{M} a_n \frac{1}{n^{s_0}} - \frac{1}{2\pi i} \int_{\gamma_3} \frac{g(z)}{z - s_0} dz \Bigg| & \\ \leq \Bigg| \sum_{n=1}^{M} \frac{a_n}{n^{s_0}} \Big(1 - \frac{1}{2\pi i} \int_{\gamma_3} \Big(\frac{M + 1/2}{n} \Big)^{z - s_0} \frac{1}{z - s_0} dz \Big) \Bigg| & \\ + \Bigg| \frac{1}{2\pi i} \sum_{n=M+1}^{\infty} \frac{a_n}{n^{s_0}} \int_{\gamma_3} \Big(\frac{M + 1/2}{n} \Big)^{z - s_0} \frac{1}{z - s_0} dz \Bigg| . & \end{aligned} \tag{1.23}$$

So, to complete the proof, we have to bound each one of these terms, which correspond to sums with $1 \leq n \leq M$ and with $n \geq M + 1$. From now on, we write

$$g_n(z) = \Big(\frac{M + 1/2}{n} \Big)^{z - s_0} \frac{1}{z - s_0} .$$

We begin by dealing with the case $n \geq M + 1$. For a given $\sigma > 0$ (which we later will make tend to ∞), we consider a new rectangle (which we denote by R_1, see Fig. 1.8) with vertices $s_0 + c \pm iT$ and $s_0 + c + \sigma \pm iT$.

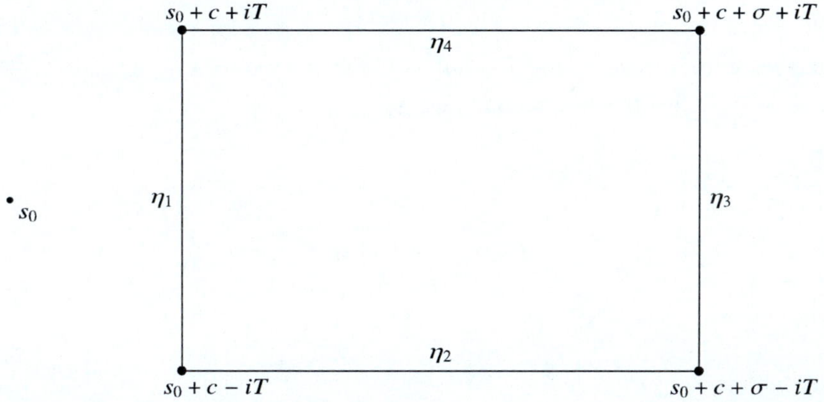

Fig. 1.8 New contour of integration R_1

Note that the vertical left side here (η_1) is the same as the vertical right side of R (γ_3). Now, since $s_0 \notin R_1$, the function g_n is holomorphic on an open set containing R_1 and, by Cauchy's Theorem 4.1, $\int_{R_1} g_n(z)dz = 0$. Hence

$$\left|\int_{\gamma_3} g_n(z)dz\right| \leq \left|\int_{\eta_2} g_n(z)dz\right| + \left|\int_{\eta_3} g_n(z)dz\right| + \left|\int_{\eta_4} g_n(z)dz\right|. \tag{1.24}$$

Once again, we bound each one of these integrals. For the horizontal sides, we have

$$\begin{aligned}
\left|\int_{\eta_2} g_n(z)dz\right| &= \left|\int_c^{c+\sigma} \left(\frac{M+1/2}{n}\right)^{t+iT} \frac{1}{t+iT}dt\right| \\
&\leq \int_c^{c+\sigma} \left(\frac{M+1/2}{n}\right)^{t} \frac{1}{\sqrt{t^2+T^2}}dt \leq \frac{1}{T}\int_c^{c+\sigma} \left(\frac{M+1/2}{n}\right)^{t} dt \\
&= \frac{1}{T}\frac{1}{\log\left(\frac{M+1/2}{n}\right)}\left[\left(\frac{M+1/2}{n}\right)^{c+\sigma} - \left(\frac{M+1/2}{n}\right)^{c}\right] \\
&= \frac{1}{T}\frac{1}{\left|\log\left(\frac{M+1/2}{n}\right)\right|}\left[\left(\frac{M+1/2}{n}\right)^{c} - \left(\frac{M+1/2}{n}\right)^{c+\sigma}\right] \\
&\leq \frac{1}{T}\frac{1}{\left|\log\left(\frac{M+1/2}{n}\right)\right|}\left(\frac{M+1/2}{n}\right)^{c}.
\end{aligned} \tag{1.25}$$

Note now that $\left|\log\left(\frac{M+1/2}{n}\right)\right| \geq \left|\log\left(\frac{M+1/2}{M+1}\right)\right| > \frac{1}{2M+2}$ for every $n \geq M+1$ (the latter being true because $|\log(1-x)| > x$ for every $0 < x < 1$, and $\frac{M+1/2}{M+1} = 1 - \frac{1}{2M+2}$). Taking this into account, we have

$$\begin{aligned}\left|\int_{\eta_2} g_n(z)dz\right| & \\ &\leq \frac{1}{M^{a-b+2}}(2M+2)\left(\frac{M+1/2}{n}\right)^{a-b}\left(\frac{2M+1}{2M}\right)^{a-b}(2M+2)\frac{1}{M^2}\frac{1}{n^{a-b}} \\ &\leq 3^{a-b}\frac{1}{M^2}\frac{1}{n^{a-b}}.\end{aligned}$$

The same estimate holds for $\left|\int_{\eta_4} g_n(z)dz\right|$.

We bound now the integral along the vertical right side.

$$\begin{aligned}\left|\int_{\eta_3} g_n(z)dz\right| &= \left|\int_{\eta_3}\left(\frac{M+1/2}{n}\right)^{c+\sigma+it}\frac{1}{c+\sigma+it}dt\right| \\ &\leq \int_{\eta_3}\left(\frac{M+1/2}{n}\right)^{c+\sigma}\frac{1}{|c+\sigma+it|}dt \\ &= \left(\frac{M+1/2}{n}\right)^{c+\sigma}\int_{\eta_3}\left(\frac{M+1/2}{n}\right)^{c+\sigma}\frac{1}{|c+\sigma+it|}dt \\ &\leq \left(\frac{M+1/2}{M+1}\right)^{c+\sigma}\frac{2T}{c+\sigma}.\end{aligned}$$

Plugging all these in (1.24), we have

$$\left|\int_{\gamma_3} g_n(z)dz\right| \leq 2\times 3^{a-b}\frac{1}{M^2}\frac{1}{n^{a-b}} + \left(\frac{M+1/2}{M+1}\right)^{c+\sigma}\frac{2T}{c+\sigma}$$

and, since this holds for every $\sigma > 0$,

$$\left|\int_{\gamma_3} g_n(z)dz\right| \leq 2\times 3^{a-b}\frac{1}{M^2}\frac{1}{n^{a-b}},$$

for every $n \geq M+1$. Whit this,

$$\left|\sum_{n=M+1}^{\infty}\frac{a_n}{n^{s_0}}\int_{\gamma_3}\left(\frac{M+1/2}{n}\right)^{z-s_0}\frac{1}{z-s_0}dz\right| \leq 2\times 3^{a-b}\sum_{n=M+1}^{\infty}\frac{|a_n|}{n^{\operatorname{Re}s_0+a-b}}\frac{1}{M}.$$

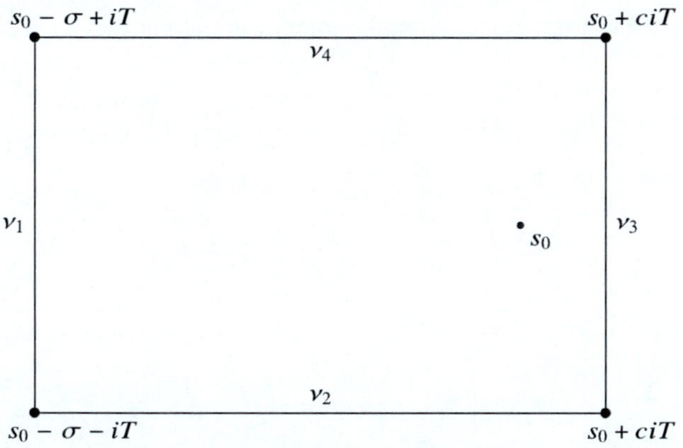

Fig. 1.9 Another contour of integration R_2

But this is the tail of a convergent series (because $\operatorname{Re} s_0 + a - b > a + \varepsilon$) and, so, it is bounded. Then we can find $K_1 > 0$ so that

$$\left| \sum_{n=M+1}^{\infty} \frac{a_n}{n^{s_0}} \int_{\gamma_3} \left(\frac{M+1/2}{n}\right)^{z-s_0} \frac{1}{z-s_0} dz \right| \leq K_1 \frac{1}{M} . \tag{1.26}$$

We now consider the case $1 \leq n \leq M$ in (1.23). In this case, we have to estimate

$$\left| 1 - \frac{1}{2\pi i} \int_{\gamma_3} \left(\frac{M+1/2}{n}\right)^{z-s_0} \frac{1}{z-s_0} dz \right| .$$

Again, for each $\sigma > 0$, we take the rectangle (which we now call R_2, see Fig. 1.9) of vertices $s_0 - \sigma \pm iT$ and $s_0 + c \pm iT$

Note that now the vertical right sides of R (γ_3) and of R_2 (ν_3) coincide. Now, by the Cauchy Integral Formula (Theorem 4.2),

$$\int_{R_2} g_n(z) dz = \int_{R_2} \left(\frac{M+1/2}{n}\right)^{z-s_0} \frac{1}{z-s_0} dz = 2\pi i ,$$

and then

$$\left| 1 - \frac{1}{2\pi i} \int_{\gamma_3} g_n(z) dz \right| \leq \frac{1}{2\pi} \left(\left| \int_{\nu_1} g_n(z) dz \right| + \left| \int_{\nu_2} g_n(z) dz \right| + \left| \int_{\nu_4} g_n(z) dz \right| \right) .$$

Essentially the same computations as in (1.25) (using now that $\log(x+1) > x/2$ for every $0 < x < 1$) yield

$$\left| \int_{\nu_2} g_n(z) dz \right| \leq \frac{1}{M^{a-b+2}} \frac{1}{\log\left(\frac{M+1/2}{M}\right)} \left(\frac{M+1/2}{n}\right)^{a-b} \leq 4 \left(\frac{M+1/2}{M}\right)^{a-b} \frac{1}{M} \frac{1}{n^{a-b}} .$$

The same estimate holds for ν_4. Finally, integrating along the vertical left side, we have

$$\left|\int_{\nu_1} g_n(z)dz\right| = \left|\int_{-T}^{T}\left(\frac{M+1/2}{n}\right)^{-\sigma+it}\frac{i}{-\sigma+it}dt\right|$$
$$\leq \int_{-T}^{T}\left(\frac{M+1/2}{n}\right)^{-\sigma}\frac{1}{\sqrt{\sigma^2+t^2}}dt \leq \left(M+\tfrac{1}{2}\right)^{-\sigma}\frac{2T}{\sigma}.$$

Hence

$$\left|1-\frac{1}{2\pi i}\int_{\gamma_3} g_n(z)dz\right| \leq \frac{1}{2\pi}\left(8\times 2^{a-b}\frac{1}{M}\frac{1}{n^{a-b}}+\frac{2M^{a-b+2}}{\sigma(M+1/2)^{\sigma}}\right)$$

for every $\sigma > 0$. Using this and proceeding as in (1.26), we find some $K_2 > 0$ so that

$$\left|\sum_{n=1}^{M}\frac{a_n}{n^{s_0}}\left(1-\frac{1}{2\pi i}\int_{\gamma_3}\left(\frac{M+1/2}{n}\right)^{z-s_0}\frac{1}{z-s_0}dz\right)\right| \leq K_2\frac{1}{M},$$

and this, together with (1.26) and (1.23), gives $C_3 > 0$ such that

$$\left|\sum_{n=1}^{M}a_n\frac{1}{n^{s_0}}-\frac{1}{2\pi i}\int_{\gamma_3}\frac{g(z)}{z-s_0}dz\right| \leq C_3\frac{1}{M}. \tag{1.27}$$

We have now at hand everything we need. Using (1.20), (1.21), (1.22), and (1.27) in (1.19), we get

$$\left|f(s_0)-\sum_{n=0}^{M}a_n\frac{1}{n^{s_0}}\right| \leq C_1\frac{1}{M^{\varepsilon}}\log M + C_4\frac{1}{M^2}+C_3\frac{1}{M}+C_4\frac{1}{M^2}.$$

Noting that $\frac{1}{M^2} \leq \frac{1}{M} \leq \frac{1}{M^{\varepsilon}}\log M$ for every $M \geq 3$ finally gives (1.18) and completes the proof. □

1.4 Exercises

1.1 Calculate the pointwise limit of each sequence of functions $f_n : \Omega \to \mathbb{R}$

1. $f_n(x) = \dfrac{nx}{1+nx}$ for $\Omega = [0, \infty[$.
2. $f_n(x) = x^2e^{-nx}$ for $\Omega = [0, \infty[$
3. $f_n(x) = n^2x^2e^{-nx}$ for $\Omega = [0, \infty[$.

4. $f_n(x) = \dfrac{x^n}{1+x^n}$ for $\Omega = [0, \infty[$.
5. $f_n(x) = \cos^{2n} \pi x$ for $\Omega = \mathbb{R}$.

1.2 Consider the sequence $f_n(x) = \dfrac{x}{x+n}$. Show that $(f_n)_n$ converges uniformly on $[0, a]$ for every $a > 0$ and pointwise but not uniformly on $[0, \infty[$.

1.3 Consider the sequence $f_n(x) = \dfrac{nx}{1+n^2x^2}$. Show that $(f_n)_n$ converges uniformly on $[a, \infty[$ for every $a > 0$ and pointwise but not uniformly on $[0, \infty[$.

1.4 Show that if $f_n(x) = x + 1/n$ and $f(x) = x$ for $x \in \mathbb{R}$, then (f_n) converges uniformly on $\mathbb{R}$ to f, but the sequence (f_n^2) does not converge uniformly on $\mathbb{R}$. *Thus the product of uniformly convergent sequences may not converge uniformly.*

1.5 Let (f_n) and (g_n) be sequences of bounded functions on Ω that converge uniformly on Ω to f and g, respectively. Show that $(f_n g_n)$ converges uniformly on Ω to fg.

1.6 Let (f_n) be a sequence of functions that converges uniformly to f on Ω and that satisfies $|f_n(x)| \leq M$ for all $n \in \mathbb{N}$ and all $x \in \Omega$. If g is continuous on the interval $[-M, M]$, show that the sequence $(g \circ f_n)$ converges uniformly to $g \circ f$ on Ω.

1.7 Let $f_n : [0, 1] \to \mathbb{R}$ be given by $f_n(x) = \dfrac{1}{(1+x)^n}$. Find the pointwise limit f of the sequence (f_n) on $[0, 1]$. Does (f_n) converge uniformly to f on $[0, 1]$?

1.8 Discuss the convergence and the uniform convergence of the series $\sum f_n$, where $f_n(x)$ is given in each case by

1. $f_n(x) = \dfrac{1}{x^2+n^2}$.
2. $f_n(x) = \dfrac{1}{(nx)^2}$ for $x \neq 0$.
3. $f_n(x) = \sin\left(\frac{x}{n^2}\right)$.
4. $f_n(x) = \dfrac{1}{x^n+1}$ for $x \neq 0$.
5. $f_n(x) = \dfrac{x^n}{x^n+1}$ for $x \geq 0$.
6. $f_n(x) = \dfrac{(-1)^n}{n+x}$ for $x \geq 0$.

1.9 Determine the radius of convergence of the series $\sum a_n x^n$, where a_n in each case is given by

1. $a_n = \dfrac{1}{n^n}$.
2. $a_n = \dfrac{n^\alpha}{n!}$.

3. $a_n = \dfrac{n^n}{n!}$.
4. $a_n = \dfrac{1}{\log n}$.
5. $a_n = \dfrac{(n!)^2}{(2n)!}$.
6. $a_n = \dfrac{1}{n\sqrt{n}}$.

1.10 For the following 2π-periodic functions, compute the Fourier series:

1. $f(x) = 0,\ -\pi < x < 0,\ f(x) = a > 0,\ 0 < x < \pi$.
2. $f(x) = 0,\ -\pi < x < 0,\ f(x) = x,\ 0 < x < \pi$.
3. $f(x) = -1,\ -\pi < x < 0,\ f(x) = 1,\ 0 < x < \pi$.
4. $f(x) = \pi - x,\ -\pi < x < \pi$.
5. $f(x) = x\cos 2x,\ -\pi < x < \pi$.

1.11 Give the Fourier series and study its convergence of the following functions (extended periodically):

1. $f(x) = x^4 - 2\pi^2x^2$, $|x| < \pi$. Compute $\sum_{n=1}^{\infty} \frac{1}{n^4}$.
2. $f(x) = |\cos x|$, $|x| \le \pi$. Compute $\sum_{n=1}^{\infty} \frac{(-1)^n}{4n^2-1}$.
3. $f(x) = x\sin x$, $|x| < \pi$.
4. $f(x) = x+\pi$, $-\pi < x < 0$, and $f(x) = 0$, $0 \le x \le \pi$. Compute $\sum_{n=1}^{\infty} \frac{1}{(2n-1)^2}$.

1.12 Show that the series $\zeta(s) = \sum_{n=1}^{\infty} n^{-s}$ converges uniformly in each semi-infinite interval of the form $[1+h, +\infty[$ with $h > 0$.

1.13 Show that for every Dirichlet series $\sum a_n \frac{1}{n^s}$, we have

$$\sigma_c \le \limsup_N \frac{\left|\sum_{n=1}^{N} a_n\right|}{\log N}$$

and

$$\sigma_a \le \limsup_N \frac{\sum_{n=1}^{N} |a_n|}{\log N}.$$

Show also that, in each case, if the abscissa is not negative, then we have equality.
Hint: Use Abel's summation.

1.14 Let $f(s) = \sum_{n=1}^{\infty} a_n \frac{1}{n^s}$ be a Dirichlet series which converges absolutely in $[\mathrm{Re} > \sigma_a]$. Prove that

$$\lim_{\sigma \to +\infty} f(\sigma + it) = f(1) = a_1$$

uniformly for $t \in \mathbb{R}$.

1.15 Deduce from the previous exercise that the function $f(s) = \frac{s-1}{s+1}$, which is analytic and bounded in the half-plane [Re > 0], cannot be represented by a Dirichlet series on any half-plane [Re $> \sigma$] (for any $\sigma > 0$).

1.16 The sum $f(s) \sum_{n=1}^{\infty} a_n \frac{1}{n^s}$ of a Dirichlet series is holomorphic in the half-plane of convergence [Re $> \sigma_c$] (recall Theorem 1.19). Show that the derivative $f'(s)$ is represented in this half-plane by the series

$$f'(s) = -\sum_{n=1}^{\infty} a_n \log n \frac{1}{n^s}.$$

1.17 Prove that if $\sum_{n=1}^{\infty} a_n \frac{1}{n^s}$ converges on some half-plane [Re $> \sigma_0$], then the series

$$-\sum_{n=2}^{\infty} \frac{a_n}{\log n} \frac{1}{n^s}$$

also converges in the same half-plane. As a consequence, if $a_1 = 0$, then the integral term by term of a Dirichlet series on a half-plane is represented by a Dirichlet series on the same half-plane.

Further Reading
[1–12]

References

1. Ahlfors, L.: Complex Analysis—An Introduction to the Theory of Analytic Functions of One Complex Variable, 3rd edn. AMS Chelsea Publishing, Providence, RI (2021). Reprint of the 1978 original
2. Apostol, T.M.: Mathematical Analysis, 2nd edn. Addison-Wesley Publishing Co., Reading, Mass.-London-Don Mills, Ont (1974)
3. Conway, J.B.: Functions of one complex variable. In: Graduate Texts in Mathematics, vol. 11, 2nd edn. Springer, New York/Berlin (1978)
4. Defant, A., García, D., Maestre, M., Sevilla-Peris, P.: Dirichlet series and holomorphic functions in high dimensions. In: New Mathematical Monographs, vol. 37. Cambridge University Press, Cambridge (2019). https://doi.org/10.1017/9781108691611
5. Duoandikoetxea, J.: Fourier analysis. In: Graduate Studies in Mathematics, vol. 29. American Mathematical Society, Providence, RI (2001). Translated and revised from the 1995 Spanish original by David Cruz-Uribe
6. Gamelin, T.W.: Complex analysis. Undergraduate Texts in Mathematics. Springer, New York (2001). https://doi.org/10.1007/978-0-387-21607-2
7. Queffelec, H., Queffelec, M.: Diophantine approximation and Dirichlet series. In: Texts and Readings in Mathematics, vol. 80, 2nd edn. Hindustan Book Agency, New Delhi; Springer, Singapore (2020). https://doi.org/10.1007/978-981-15-9351-2
8. Rudin, W.: Principles of mathematical analysis, 3rd edn. In: International Series in Pure and Applied Mathematics. McGraw-Hill Book Co., New York-Auckland-Düsseldorf (1976)

9. Rudin, W.: Real and Complex Analysis, 3rd edn. McGraw-Hill Book Co., New York (1987)
10. Stromberg, K.R.: An Introduction to Classical Real Analysis. AMS Chelsea Publishing, Providence, RI (1981). https://doi.org/10.1090/chel/376. (2015, corrected reprint of the 1981 original)
11. Vretblad, A.: Fourier analysis and its applications. In: Graduate Texts in Mathematics, vol. 223. Springer, New York (2003). https://doi.org/10.1007/b97452
12. Werner, D.: Einführung in die höhere Analysis, corrected edn. Springer-Lehrbuch. [Springer Textbook]. Springer, Berlin (2009). https://doi.org/10.1007/978-3-540-79696-1

Chapter 2
Locally Convex Spaces

2.1 Topological Preliminaries

2.1.1 Basic Definitions

We collect in this section some very basic definitions and facts on topology that will be used from now on. A *topological space* is a pair (X, τ), where X is a set and τ (or τ_X, when we need to stress the set) is a family of subsets of X (called *open sets*) satisfying

(T1) $\varnothing, X \in \tau$.
(T2) If $G_1, G_2 \in \tau$, then $G_1 \cap G_2 \in \tau$.
(T3) If I is an (finite or infinite) indexing set and $G_i \in \tau$ for every $i \in I$, then $\bigcup_{i \in I} G_i \in \tau$.

A set $C \subseteq X$ is *closed* if $X \setminus C$ is open. A set $U \subseteq X$ in a topological space (X, τ) is a *neighbourhood* of $x \in X$ if there exists an open set G such that $x \in G \subseteq U$. A *basis of neighbourhoods* of x is a family of sets $(U_i)_{i \in I}$ such that

(BN1) U_i is a neighbourhood of x for every $i \in I$.
(BN2) For every neighbourhood V of x, there is $i_0 \in I$ such that $U_{i_0} \subseteq V$.

Remark 2.1 It is an easy exercise to see that a set in X is open if and only it is a neighbourhood of all its elements. If each point has a basis of neighbourhoods $(U_i(x))_{i \in I}$, then G is open if and only if for all $x \in G$ there exists $i \in I$ such that $U_i(x) \subseteq G$.

Remark 2.2 Let X be a set with two topologies τ_1 and τ_2. If every τ_1-open set is τ_2-open (that is, $\tau_1 \subseteq \tau_2$), then τ_1 are said to be *coarser* (also *smaller* or *weaker*) than τ_2 and τ_2 is said to be *finer* (also *bigger* or *stronger*) than τ_1.

J. Bonet et al., *Function Spaces and Operators between them*,
RSME Springer Series 11, https://doi.org/10.1007/978-3-031-41602-6_2

If each $x \in X$ has basis of neighbourhoods $(U_i(x))_{i\in I}$ of τ_1 and $(V_j(x))_{j\in J}$ of τ_2, then it is an easy exercise to show that τ_1 is coarser than τ_2 if and only if for every x, given $i \in I$ there is $j \in J$ such that $V_j(x) \subseteq U_i(x)$.

A topological space (X, τ) is said to be *Hausdorff* if for every $x, y \in X$, $x \neq y$, there are open sets G_1 and G_2 with $x \in G_1$ and $y \in G_2$, such that $G_1 \cap G_2 = \varnothing$. It is clear from what we have explained so far that the definition is equivalent if we replace 'open sets' by 'neighbourhoods'.

A function between two topological spaces $f : (X, \tau_X) \to (Y, \tau_Y)$ is *continuous* at $x \in X$ if for every $G \in \tau_Y$ with $f(x) \in G$ we have $f^{-1}(G) \in \tau_X$ (that is, the pre-image of every open set in Y containing $f(x)$ is an open set in X).

Remark 2.3 Suppose that X and Y are topological spaces, $f : X \to Y$, $(U_i)_{i\in I}$ is a basis of neighbourhoods of x, and $(V_j)_{j\in J}$ is a basis of neighbourhoods of $f(x)$. Then f is continuous at x if and only if for every $j \in J$ there is $i \in I$ such that $U_i \subseteq f^{-1}(V_j)$.
Also, f is continuous at x if and only if for every $j \in J$ there is $i \in I$ such that $f(U_i) \subseteq V_j$.

Remark 2.4 A sequence $(x_n)_n$ in a topological space X *converges* to some $x \in X$ (we write $\lim x_n = x$ or $x_n \to x$ and call x the *limit* of the sequence) if for every neighbourhood U of x there exists $n_0 \in \mathbb{N}$ such that $x_n \in U$ for all $n \geq n_0$. If $(U_j)_{j\in J}$ is a basis of neighbourhoods of x, then $x_n \to x$ if and only if for every $j \in J$ there exists $n_j \in \mathbb{N}$ such that $x_n \in U_j$ for every $n \geq n_j$.

Remark 2.5 The notion of sequence is generalized by nets. Let us briefly recall the definition. First of all, a set I is *partially ordered* if there is a relationship $i \leq j$, defined for certain pairs of elements $i, j \in I$, which is reflexive ($i \leq i$), transitive ($i \leq j$ and $j \leq k$ imply $i \leq k$), and antisymmetric ($i \leq j$ and $j \leq i$ imply $i = j$). A partially ordered set is *totally ordered* if for every $i, j \in I$ either $i \leq j$ or $j \leq i$. A set I is said to be *directed* if it is partially ordered and for every pair $i, j \in I$ there is some $k \in I$ so that $i \leq k$ and $j \leq k$.
If X is a topological space, a *net* in X is any function $I \to X$, where I is a directed set, and is denoted by $(x_i)_{i\in I}$ (meaning that the index i is mapped to the element $x_i \in X$). Note that a sequence is just a net indexed on the natural numbers. A net $(x_i)_{i\in I}$ in a topological space X *converges* to some $x \in X$ (we write $x_i \to x$) if for every neighbourhood U of x there exists $i_0 \in I$ such that $x_i \in U$ for all $i \geq i_0$. If $(U_j)_{j\in J}$ is a basis of neighbourhoods of x, then $x_i \to x$ if and only if for every $j \in J$ there exists $i_j \in I$ such that $x_i \in U_j$ for every $i \geq i_j$.

Take $A \subseteq X$, and then $x \in X$ is a *closure point* of A if $A \cap U \neq \varnothing$ for every neighbourhood U of x. The set of all closure points of A is called the *closure* of A and is denoted by $\overline{A}$. It is the smallest closed set that includes A. The point $x \in X$ is *interior* to A if there is an open set G so that $x \in G \subseteq A$. We denote $\mathring{A}$ for the set of all interior points of A (called the *interior* of A). This is the biggest open set contained in A.

A set $A \subseteq X$ is *dense* in X if $\overline{A} = X$. A straightforward computation shows that a set A is dense in X if and only if $A \cap U \neq \varnothing$ for every non-empty open set $U \subseteq X$. A space X is *separable* if it has a countable dense subset.

A set $K \subseteq X$ is *compact* (we usually denote this by $K \Subset X$) if whenever there are open sets $(G_i)_{i \in I}$ (being I some infinite indexing set) with $K \subseteq \bigcup_{i \in I} G_i$ there are $i_1, \dots, i_n \in I$ such that

$$K \subseteq G_{i_1} \cup \ldots \cup G_{i_n} .$$

A set is *relatively compact* if its closure is compact.

If (X, τ) is a topological space, then each subset $Y \subseteq X$ inherits a topology, defined as

$$\tau|_Y = \{G \cap Y : G \in \tau\} .$$

This is often called the *induced topology* on Y. The inclusion $i : (Y, \tau|_Y) \to (X, \tau)$ is clearly continuous.

If (X_1, τ_1) and (X_2, τ_2) are topological spaces, then the set $X_1 \times X_2$ is a topological space with the topology defined by

$$\tau_1 \times \tau_2 = \{G_1 \times G_2 : G_1 \in \tau_1, \ G_2 \in \tau_2\} .$$

This definition carries over in an obvious way for the product of finitely many topological spaces. More generally, if (X_i, τ_i) is an arbitrary family of topological spaces, the *product topology* on the product space $X = \prod_{i \in I} X_i$ is defined in the following way. Given $x = (x_i) \in X$, a set $U \subseteq X$ is a neighbourhood of x if (and only if) is of the form $U = \prod_{i \in I} U_i$, in such a way that there is a finite $J \subseteq I$ so that $U_i = X_i$ for every $i \in I \setminus J$ and U_i is a neighbourhood of x_i in X_i for every $i \in J$. This is the smallest topology that makes all projections $\pi_i : X \to X_i$ on the i-component continuous.

Theorem 2.1 (Tikhonov) *If K_i for $i \in I$ is a family of compact spaces, then the product space $K = \prod_{i \in I} K_i$ is also compact.*

Proof [19, Chapter 3], [20, Chapter 1, § 3], [35, Theorem 7.4.1] □

2.1.2 Metric and Normed Spaces

Given a non-empty set X, a *metric* (also *distance*) on X is any mapping $d : X \times X \to [0, +\infty[$ satisfying

(D1) $d(x, y) \geq 0$, for all $x, y \in X$, and if $d(x, y) = 0$, then $x = y$.
(D2) $d(x, y) = d(y, x)$ for all $x, y \in X$.
(D3) $d(x, y) \leq d(x, z) + d(z, y)$ for all $x, y, z \in X$.

The pair (X, d) is called a *metric space*.

Example 2.2 There are plenty of examples of such metrics. Probably the most classical one is the Euclidean distance on $X = \mathbb{K}^n$

$$d_2(x, y) = \Big(\sum_{i=1}^{n} |x_i - y_i|^2\Big)^{\frac{1}{2}}, \tag{2.1}$$

for $x = (x_1, \dots, x_n)$ and $y = (y_1, \dots, y_n)$ two vectors in $\mathbb{K}^n$. But there are other examples. Already on $\mathbb{K}^n$, we can consider

$$d_p(x, y) = \Big(\sum_{i=1}^{n} |x_i - y_i|^p\Big)^{\frac{1}{p}} \tag{2.2}$$

for $1 \leq p < \infty$ and

$$d_\infty(x, y) = \max_{1 \leq i \leq n} |x_i - y_i|. \tag{2.3}$$

Another example of a metric space is to consider $X =]0, \infty[$ and

$$d(x, y) = \log \left|\tfrac{y}{x}\right|. \tag{2.4}$$

□

Remark 2.6 A metric on a space defines in a natural way a topology (which we denote by τ_d) as follows:

- $\varnothing$ is open.
- $\varnothing \neq G \subseteq X$ is open if for each $x \in G$, there is $\varepsilon > 0$ such that

$$B_d(x, \varepsilon) = \{y \in X : d(y, x) < \varepsilon\} \subseteq G.$$

To begin with, (T1) is clearly satisfied. Take now two open sets G_1 and G_2 and fix $x \in G_1 \cap G_2$. Then there are $\varepsilon_1, \varepsilon_2 > 0$ so that $B_d(x, \varepsilon_i) \subseteq G_i$ (for $i = 1, 2$). Taking $\varepsilon = \min\{\varepsilon_1, \varepsilon_2\}$ gives $B_d(x, \varepsilon) \subseteq G_1 \cap G_2$ and shows that (T2) is also satisfied. Finally, choose open sets G_i with $i \in I$ and $x \in \bigcup_{i \in I} G_i$. Then there exists $i_0 \in I$ such that $x \in G_{i_0}$. So there is $\varepsilon_0 > 0$ such that $B_d(x, \varepsilon_0) \subseteq G_{i_0} \subseteq \bigcup_{i \in I} G_i$, and (T3) is also satisfied. Clearly every metric space is Haussdorff.

The set $B_d(x, \varepsilon)$ is called the *ball* in the metric d, centred at x of radius ε. It is fairly easy to check that for each $x \in X$ the families

$$\big(B_d(x, \varepsilon)\big)_{\varepsilon > 0} \quad \text{and} \quad \big(B_d(x, 1/n)\big)_{n \in \mathbb{N}}$$

define bases of neighbourhoods of x for the topology τ_d. Observe that in both cases they consist of open sets.

In view of Remark 2.4 and with this at hand, a sequence $(x_n)_n$ in a metric space X converges to $x \in X$ if and only if for all $\varepsilon > 0$ there exists $n_0 \in \mathbb{N}$ such that $d(x_n, x) < \varepsilon$ for all $n \geq n_0$ or, in other words, if $d(x_n, x) \overset{n}{\to} 0$ (as a sequence of real numbers).

Being closed can be described in metric spaces through convergence of sequences. The proof is easy and left as an exercise to the reader.

Proposition 2.1 *Let C be a subset of a metric space X. The following are equivalent:*

1. *C is closed.*
2. *If $(x_n)_n \subseteq C$ converges to some x, then $x \in C$.*

In a metric space X, the *distance* between a point $x \in X$ and a set $A \subseteq X$ is defined as

$$d(x, A) = \inf\{d(x, a) \colon a \in A\} .$$

We consider now a vector space X (over $\mathbb{R}$ or $\mathbb{C}$). A *norm* on X is any mapping $\| \cdot \| : X \to [0, +\infty[$ which satisfies

(N1) $\|x\| \geq 0$, for all $x \in X$ and $\|x\| = 0$ if and only if $x = 0$.
(N2) $\|\alpha x\| = |\alpha| \|x\|$, for all $\alpha \in \mathbb{K}$, and all $x \in X$.
(N3) $\|x + y\| \leq \|x\| + \|y\|$ for all $x, y \in X$.

A pair $(X, \| \cdot \|)$ is called a *normed space*. A norm defines a metric (and then a topology) on X in a natural way by doing

$$d(x, y) = \|x - y\| \text{ for } x, y \in X . \tag{2.5}$$

The ball $B_X = B_d(0, 1)$ is often called the *open unit ball* of X.

Example 2.3 Given $x = (x_1, \ldots, x_n) \in \mathbb{K}^n$, we can define

$$\|x\|_\infty = \max_{k=1,\ldots,n} |x_k| ,$$

which clearly satisfies (N1)–(N3) and, then, defines a norm on $\mathbb{K}^n$ (note that the distance defined in (2.3) is associated with this norm). For $1 \leq p < \infty$, we may define

$$\|x\|_p = \Big(\sum_{k=1}^{n} |x_k|^p \Big)^{\frac{1}{p}} .$$

This clearly satisfies (N1) and (N2). The inequality in (N3) is known as *Minkowski inequality*:

$$\Big(\sum_{k=1}^{n} |x_k + y_k|^p\Big)^{\frac{1}{p}} \leq \Big(\sum_{k=1}^{n} |x_k|^p\Big)^{\frac{1}{p}} + \Big(\sum_{k=1}^{n} |y_k|^p\Big)^{\frac{1}{p}}, \tag{2.6}$$

a proof of which can be found in [13, (2.11.4)], [24, Proposition 7.1], or [9, Theorem 1.12]. So, $\|\cdot\|_p$ also defines a norm on $\mathbb{K}^n$, with which the distances defined on (2.1) and (2.2) are associated.
For $x = (x_1, \ldots, x_n)$, $y = (y_1, \ldots, y_n) \in \mathbb{K}^n$, we obviously have

$$\sum_{k=1}^{n} |x_k y_k| \leq \|x\|_1 \|y\|_\infty .$$

This simple remark extends to other pairs of norms. If $1 < p < \infty$ and $\frac{1}{p} + \frac{1}{p'} = 1$ (we say that p and p' are conjugate to each other, from now on we use the convention that 1 and ∞ are conjugated exponents), then

$$\sum_{k=1}^{n} |x_k y_k| \leq \|x\|_p \|y\|_{p'} . \tag{2.7}$$

This is known as *Hölder's inequality*, and a proof can be found in [13, (2.8.3)], [24, Proposition 7.6] or [9, Theorem 1.12]. □

Example 2.4 There are metric spaces whose distance cannot be obtained from a norm as in (2.5). One example is the distance defined in (2.4), but in this case, the metric space is $]0, +\infty[$, which is not a vector space (hence no norm can be defined on it). There are, however, vector spaces on which a metric can be defined that cannot be obtained from a norm as in (2.5). For $x, y \in \mathbb{R}$, consider

$$d(x, y) = \frac{|x - y|}{1 + |x - y|} .$$

It is not difficult to see that this defines a metric on $\mathbb{R}$. To begin with, (D1) and (D2) are obviously satisfied. To see that (D3) also holds, note that the function $f(s) = \frac{s}{1+s}$ is increasing and $|x - y| \leq |x - z| + |z - y|$ for all x, y, z. Then

$$\begin{aligned} d(x, y) = f\big(|x - y|\big) \leq f\big(|x - z| + |z - y|\big) &= \frac{|x - z| + |z - y|}{1 + |x - z| + |z - y|} \\ &= \frac{|x - z|}{1 + |x - z| + |z - y|} + \frac{|z - y|}{1 + |x - z| + |z - y|} \\ &\leq \frac{|x - z|}{1 + |x - z|} + \frac{|z - y|}{1 + |z - y|} = d(x, z) + d(z, y) . \end{aligned}$$

This shows that $(\mathbb{R}, d)$ is a metric space (and $\mathbb{R}$ is obviously a vector space). If there were a norm satisfying (2.5), we would have $\|x\| = d(x, 0)$ for every $x \in \mathbb{R}$. Taking $x_0 = 1$, we have $\|x_0\| = \frac{1}{2}$, but $\|2x_0\| = \frac{2}{3} \neq 2\|x_0\|$, and (N2) is not satisfied. □

Example 2.5 The norms defined for finite tuples in Example 2.3 can be easily extended to sequences. Given a sequence $x = (x_n)_{n=1}^{\infty} \in \mathbb{K}^{\mathbb{N}}$, we can consider

$$\sum_{n=1}^{\infty} |x_n|^p \text{ for } 1 \leq p < \infty \text{ and } \sup_{n \in \mathbb{N}} |x_n| .$$

Note that these may be finite or infinite. For $1 \leq p < \infty$, we consider the set

$$\ell_p = \Big\{x = (x_n)_{n=1}^{\infty} \in \mathbb{K}^{\mathbb{N}} \colon \sum_{n=1}^{\infty} |x_n|^p < +\infty\Big\} \text{ for } 1 \leq p < \infty ,$$

and for $x \in \ell_p$, we define

$$\|x\|_p = \Big(\sum_{n=1}^{\infty} |x_n|^p\Big)^{\frac{1}{p}} . \tag{2.8}$$

Also, for $p = \infty$, we define the space

$$\ell_\infty = \Big\{x \in \mathbb{K}^{\mathbb{N}} \colon \|x\|_\infty = \sup_{i \in \mathbb{N}} |x_i| < +\infty\Big\} .$$

Note that for $1 \leq p \leq \infty$, the mapping $\|\cdot\|_p$ clearly satisfies (N1) and (N2). Minkowski's inequality (2.6) extends to series, so that if $x, y \in \ell_p$ then $x + y \in \ell_p$ and

$$\|x + y\|_p \leq \|x\|_p + \|y\|_p .$$

Hence, they also satisfy (N3), and ℓ_p is a normed space for every $1 \leq p \leq \infty$. Hölder's inequality also extends to sequences in the following way: if $x \in \ell_p$ and $y \in \ell_{p'}$ (where p and p' are conjugated exponents), then $(x_n y_n)_n \in \ell_1$ and

$$\|(x_n y_n)_n\|_1 \leq \|x\|_p \|y\|_{p'} .$$

Another important example is the space of null sequences

$$c_0 = \Big\{x \in \mathbb{K}^{\mathbb{N}} \colon \lim_{n \to \infty} x_n = 0\Big\} . \tag{2.9}$$

This is clearly a vector subspace of ℓ_∞, and the topology defined by $\|\cdot\|_\infty$ on c_0 is the induced topology.

For each $n \in \mathbb{N}$, we define the sequence e_n, which has 1 in the nth coordinate and 0 elsewhere. All these clearly belong to ℓ_p for every $1 \leq p \leq \infty$, and $\|e_n\|_p = 1$ for every n and p. These sequences are commonly known as the *canonical vectors*. □

Example 2.6 The spaces defined in the previous example can be seen as a particular case of a wider class of spaces of functions. Let Ω be an arbitrary measure space and μ a positive measure. Let us briefly recall that in this context we say that a given property is satisfied *for almost every* (a.e. for short) $x \in \Omega$ if there is $\Omega_1 \subseteq \Omega$ with $\mu(\Omega_1) = 0$ such that the property holds for every $x \in \Omega \setminus \Omega_1$. For $1 \leq p < \infty$, we write $L_p(\mu)$ for the space of all measurable functions $f : \Omega \to \mathbb{K}$ (where $\mathbb{K}$ can be either $\mathbb{R}$ or $\mathbb{C}$) so that

$$\int_\Omega |f|^p d\mu < \infty$$

and define, for each such function,

$$\|f\|_p = \left(\int_\Omega |f|^p d\mu \right)^{\frac{1}{p}}.$$

For $p = \infty$, we define $L_\infty(\mu)$ as the space of measurable functions f for which

$$\|f\|_\infty = \inf\{M : |f(x)| \leq M \text{ a.e. } x \in \Omega\} < \infty.$$

Again we have a version of Minkowski's inequality (2.6): if $1 \leq p \leq \infty$ and $f, g \in L_p(\mu)$, then $f + g \in L_p(\mu)$ and

$$\|f + g\|_p \leq \|f\|_p + \|g\|_p,$$

as well as Hölder's inequality (2.7): if $f \in L_p(\mu)$ and $g \in L_{p'}(\mu)$ (where p and p' are conjugated), then $fg \in L_1(\mu)$ and

$$\|fg\|_1 \leq \|f\|_p \|g\|_{p'}. \tag{2.10}$$

We clearly have that $\|f\|_p = 0$ for $f = 0$ and $\|\alpha f\|_p = |\alpha| \, \|f\|_p$. The only feature left to be a norm is to have that $\|f\|_p = 0$ implies $f = 0$, but this is not true. To overcome this, we define a relation of equivalence in $L_p(\mu)$ by saying that $f \sim g$ if $\mu\big(\{x \in \Omega : f(x) \neq g(x)\}\big) = 0$ and defining $L_p(\mu)$ as the space of all equivalence classes. If f and g are in the same class of equivalence, then $\|f\|_p = \|g\|_p$. So, on $L_p(\mu)$, we may define $\| \cdot \|_p$ as $\|f\|_p$ for any representative of the equivalence class. With this, $L_p(\mu)$ is a normed space for every $1 \leq p \leq \infty$.

A complete account on these spaces can be found, for example, in [28, Chapter 3]. In particular, Minkowski's and Hölder's inequalities are [28, Theorems 3.8 and 3.9]. These can also be found in [24, Theorem 13.4 and Corollary 13.3] or [9, Theorems 1.19 and 1.20].

Let us note that the spaces defined in Example 2.5 are a particular case of this construction, taking as measure space $\mathbb{N}$ with the counting measure.
When Ω is some subset of $\mathbb{R}^d$ and $\mu = m_d$ is the Lebesgue measure, we will write $L_p(\Omega)$ instead of $L_p(m_d)$. □

Example 2.7 Another classical example (within our setting) of normed space is that of continuous functions on a closed, bounded interval. Let $-\infty < a < b < \infty$ and consider

$$C[a,b] = \{f : [a,b] \to \mathbb{K} : \ \text{such that} \ f \ \text{is continuous}\}.$$

This is a vector space and

$$\|f\| = \sup_{x\in[a,b]} |f(x)|$$

(that as a matter of fact, since $[a,b]$ is compact, can be taken as a maximum) defines a norm on $C[a,b]$.

Let us note that a sequence $(f_n)_n \subseteq C[a,b]$ converges to $f \in C[a,b]$ if

$$\lim_{n\to\infty} \|f_n - f\| = 0.$$

In other words, the convergence in the norm is the uniform convergence (see Remark 1.1). □

A sequence in a normed space is called *Cauchy* if for all $\varepsilon > 0$ there exists $n_0 \in \mathbb{N}$ such that $\|x_n - x_m\| < \varepsilon$ for every $n, m \geq n_0$. A normed space is said to be *complete* if every Cauchy sequence is convergent.

Theorem 2.8 *The space* $(C[a,b], \|\cdot\|)$ *is complete.*

Proof Proposition 1.1 implies that every Cauchy sequence $(f_n)_n \subseteq C[a,b]$ converges uniformly to some function that, by Theorem 1.4, is continuous because the f_ns are so. □

With exactly the same idea, one can show that if $K \subseteq \mathbb{R}^d$ is compact, then $C(K)$ (the space of continuous functions on K) with the norm $\|f\| = \sup_{x\in K} |f(x)|$ is a complete normed space.

Complete normed spaces are known as *Banach spaces*. All normed spaces given in Examples 2.3, 2.5, and 2.6 are Banach spaces.

2.2 Seminorms

We have seen in Theorem 2.8 that $C[a,b]$ (or, more generally $C(K)$ for some compact $K \subseteq \mathbb{R}^d$) with the supremum norm is a Banach space. Our next step is to give a structure to the space of continuous functions on an open set $\Omega \subseteq \mathbb{R}^d$. Here we cannot define a norm, and we need to develop different tools.

If E is a vector space, a *seminorm* is a mapping $p : E \to [0, +\infty[$ satisfying

(S1) $p(x) \geq 0$ for all $x \in E$.
(S2) $p(\lambda x) = |\lambda| p(x)$ for all $\lambda \in \mathbb{K}$ and all $x \in E$.
(S3) $p(x + y) \leq p(x) + p(y)$ for all $x, y \in E$.

2.2.1 Locally Convex Topology

With a similar idea as that for a metric (recall Remark 2.6), if we have a family of seminorms that behaves well enough, we can define a topology on E. To begin with, we need a family of seminorms on E (which we call $\mathscr{P}$) satisfying

$$\forall p_1, p_2 \in \mathscr{P}, \exists p \in \mathscr{P} : \max\{p_1(x), p_2(x)\} \leq p(x) \;\; \forall x \in E \,. \tag{2.11}$$

This family induces a topology on E (which we denote by $\tau_{\mathscr{P}}$) by doing

- $\varnothing$ is open.
- $\varnothing \neq G \subseteq X$ is open if for every $x \in G$ there is some $p \in \mathscr{P}$ and $\varepsilon > 0$ so that

$$B_p(x, \varepsilon) = \{y \in X : p(y - x) < \varepsilon\} \subseteq G \,.$$

It is easy to see that $\tau_{\mathscr{P}}$ is a topology and that, for each $x \in E$,

$$\{B_p(x, \varepsilon) \colon p \in \mathscr{P}, \; \varepsilon > 0\} \tag{2.12}$$

is a basis of neighbourhoods of x consisting of open sets. As a matter of fact, the basis on each point can be expressed in terms of a basis at 0 since

$$B_p(x, \varepsilon) = x + B_p(0, \varepsilon) \tag{2.13}$$

for all $p \in \mathscr{P}$ and all $\varepsilon > 0$.

A set A in some vector space E is called *convex* if

$$\{\lambda x + (1 - \lambda) y \colon 0 \leq \lambda \leq 1\} \,, \tag{2.14}$$

for every $x, y \in A$ and *balanced* if $\lambda x \in A$ for every $x \in A$ and $\lambda \in \mathbb{K}$ with $|\lambda| \leq 1$. Sets that are balanced and convex are called *absolutely convex*. As a straightforward consequence of (S2) and (S3), the set $B_p(x_0, \varepsilon)$ is convex for every p, x_0, and ε, and $B_p(0, \varepsilon)$ is absolutely convex for every p and ε. We denote by $\mathcal{U}_0(E)$ the set of absolutely convex neighbourhoods of 0.

The property of being absolutely convex is clearly preserved by taking intersections, and, for a given set A we can define the *absolutely convex hull* as

$$\Gamma(A) = \bigcap \{B \subseteq E \colon B \text{ absolutely convex and } A \subseteq B\}.$$

It is a simple exercise to show that a set B is absolutely convex if and only if $\lambda x + \mu y \in B$ for every $x, y \in B$ and $\lambda, \mu \in \mathbb{K}$ with $|\lambda| + |\mu| \leq 1$. Then the absolutely convex hull of a set A can be reformulated as

$$\Gamma(A) = \Big\{ \sum_{i=1}^{n} \lambda_i x_i \colon x_1, \dots, x_n \in A,\ \lambda_1, \dots, \lambda_n \in \mathbb{K},\ \sum_{i=1}^{n} |\lambda_i| \leq 1,\ n \in \mathbb{N} \Big\}. \tag{2.15}$$

A set A in a vector space E is *absorbing* if for every $x \in E$ there is $\lambda > 0$ such that $x \in \mu A$ for every $\mu \in \mathbb{K}$ with $|\mu| \geq \lambda$. If A is balanced, then this is equivalent to

$$\bigcup_{\lambda > 0} \lambda A = \bigcup_{\lambda > 0} \{\lambda x \colon x \in A\} = E.$$

Given a seminorm p and $x \in E$ so that $p(x) > 0$, defining $\lambda = p(x) + 1$, we have that $z = \frac{1}{\lambda} x \in B_p(0, 1)$ and $x = \lambda z$. Hence $B_p(0, 1)$ is absorbing.

Proposition 2.2 *Let E be a vector space with a family of seminorms $\mathscr{P}$ satisfying (2.11). Then $(E, \tau_{\mathscr{P}})$ is Hausdorff if and only if for all $x \in E$, $x \neq 0$ there is $p \in \mathscr{P}$ such that $p(x) > 0$.*

Proof We suppose first that $(E, \tau_{\mathscr{P}})$ is Hausdorff and choose $x \in E$ with $x \neq 0$. Then we can find $G_1, G_2 \in \tau$ with $0 \in G_1$, $x \in G_2$, and $G_1 \cap G_2 = \varnothing$. By the definition of the topology, there is some $\varepsilon_0 > 0$ such that $B_p(0, \varepsilon_0) \subseteq G_1$. This implies that $x \notin B_p(0, \varepsilon_0)$ and then $p(x) \geq \varepsilon_0 > 0$. This gives one implication.

To prove the converse, take $x, y \in E$ with $x \neq y$. Then $y - x \neq 0$ and we can find $p \in \mathscr{P}$ such that $p(y - x) = \varepsilon > 0$. We define the open sets $G_1 = B_p(x, \varepsilon/2)$ and $G_2 = B_p(y, \varepsilon/2)$. Obviously $G_1 \cap G_2 = \varnothing$, $x \in G_1$, and $y \in G_2$. Hence $(E, \tau_{\mathscr{P}})$ is Hausdorff. □

With all this at hand, we call (Hausdorff) *locally convex space* (in short, l.c.s.) to every vector space E provided with a family of seminorms $\mathscr{P}$ satisfying the following properties:

(LC1) For every $p_1, p_2 \in \mathscr{P}$, there exists $p \in \mathscr{P}$ such that $\max\{p_1, p_2\} \leq p$.
(LC2) For all $x \in E$, $x \neq 0$, there is $p \in \mathscr{P}$ such that $p(x) > 0$.

Note that by condition (LC1) the family of seminorms defines a topology $\tau_{\mathscr{P}}$. On the other hand, condition (LC2) and Proposition 2.2 give that $(E, \tau_{\mathscr{P}})$ (which we also denote by $(E, \mathscr{P})$) is Hausdorff. So from now on and for convenience, **we assume that every locally convex space is already a Hausdorff topological space without**

mentioning it explicitly. Note also that, by (2.12) and (2.14), every point in a locally convex space has a basis of neighbourhoods consisting of convex sets. Also, 0 has a basis of neighbourhoods consisting of absolutely convex sets. Note also that every point x_0 in a locally convex space E has a basis of closed neighbourhoods, since, for example, $B_p(x_0, \varepsilon/2) \subseteq \{x \in E \colon p(x - x_0) \leq \varepsilon/2\} \subseteq B_p(x_0, \varepsilon)$ for every seminorm p and $\varepsilon > 0$.

Example 2.9 Let us consider the space $\mathbb{K}^{\mathbb{N}}$ of all sequences $x = (x_n)_{n\in\mathbb{N}}$ in $\mathbb{K}$. For each $k \in \mathbb{N}$ and $x \in \mathbb{K}^{\mathbb{N}}$, we set

$$p_k(x) = \max\{|x_1|, \ldots, |x_k|\} .$$

Each one of these is a seminorm, and they clearly satisfy (LC1) and (LC2) (for the latter just take k with $x_k \neq 0$). The space $\mathbb{K}^{\mathbb{N}}$ endowed with the locally convex Hausdorff topology defined by this family of seminorms is denoted by ω. □

Remark 2.7 A straightforward application of Remark 2.4 allows to reformulate the convergence of sequences in terms of seminorms. Then, a sequence $(x_k)_k \subseteq E$ in a locally convex space $(E, \mathscr{P})$ converges to x_0 if and only if for all $p \in \mathscr{P}$ and $\varepsilon > 0$ there is k_0 such that $p(x_k - x_0) < \varepsilon$ for every $k \geq k_0$.

A subset A of an l.c.s. $(E, \mathscr{P})$ is said to be *bounded* if

$$\sup_{x\in A} p(x) < +\infty$$

for every $p \in \mathscr{P}$ or, equivalently, if for each $U \in \mathcal{U}_0(E)$ there is $\lambda > 0$ so that $A \subseteq \lambda U$ (that is, $x/\lambda \in U$ for every $x \in A$). A straightforward use of the definition yields the following.

Proposition 2.3 *Let E be a locally convex space and $A, B \subseteq E$. Then*

1. *If both are bounded, then so also is $\overline{\Gamma(A \cup B)}$.*
2. *If A is bounded and $B \subseteq A$, then B is bounded.*
3. *If A is finite, then it is bounded.*

Example 2.10 If $(E, \|\cdot\|)$ is a normed space, then clearly $A \subseteq E$ is bounded if and only if there is $M > 0$ such that $\|x\| \leq M$ for every $x \in A$. □

2.2.2 *Continuity*

We see now how some of the basic properties on continuity that we formulated in terms of neighbourhoods can be stated using the seminorms. If $(E, \mathscr{P})$ and $(F, \mathscr{Q})$ are two locally convex spaces, a straightforward computation from Remark 2.3 shows that a mapping $f : E \to F$ is continuous at $x \in E$ if and only if for every $q \in \mathscr{Q}$ and $\varepsilon > 0$ there are $p \in \mathscr{P}$ and $\delta > 0$ such that, if $p(x - y) < \delta$, then $q(f(x) - f(y)) < \varepsilon$. An important property is that the sum and the product by scalars are continuous.

Proposition 2.4 *Let $(E, \mathscr{P})$ be a locally convex space, then the mappings*

1. *$S : (E, \mathscr{P}) \times (E, \mathscr{P}) \to (E, \mathscr{P})$ given by $S(x, y) = x + y$*
2. *$P : \mathbb{K} \times (E, \mathscr{P}) \to (E, \mathscr{P})$ given by $P(\lambda, x) = \lambda x$*

are continuous.

Proof First of all, given $x_0, y_0 \in E$, $\varepsilon > 0$, and $p \in \mathscr{P}$, it is easy to check that if $p(x - x_0) < \varepsilon/2$ and $p(y - y_0) < \varepsilon/2$, then $p((x + y) - (x_0 - y_0)) < \varepsilon$. This gives the continuity of S.

On the other hand, given $x_0 \in E$, $\lambda_0 \in \mathbb{K}$, $\varepsilon > 0$, and $p \in \mathscr{P}$, if

$$|\lambda - \lambda_0| < \frac{\varepsilon}{2(p(x_0) + 1)} \text{ and } p(x - x_0) < \frac{\varepsilon}{2(|\lambda_0| + 1)},$$

then $p(\lambda x - \lambda_0 x_0) < \varepsilon$. This gives that P is continuous. □

Lemma 2.1 *Let p and q be seminorms on E and $\delta, \varepsilon > 0$. Then*

$$B_p(0, \delta) = \{x : p(x) < \delta\} \subseteq B_q(0, \varepsilon) = \{x : q(x) < \varepsilon\}$$

if and only if $q \leq \frac{\varepsilon}{\delta} p$.

Proof Let us suppose first that $B_p(0, \delta) \subseteq B_q(0, \varepsilon)$ and choose $x \in E$. First of all, if $p(x) = 0$, then $p(\lambda x) = \lambda p(x) = 0 < \delta$ for all $\lambda > 0$, and this gives $\lambda q(x) = q(\lambda x) < \varepsilon$ for every $\lambda > 0$. Hence $q(x) = 0$.
On the other hand, if $p(x) > 0$, we fix $0 < r < 1$ to have

$$p\left(\frac{r\delta x}{p(x)}\right) = r\delta < \delta .$$

Then $q\left(\frac{r\delta x}{p(x)}\right) < \varepsilon$ and this gives

$$q(x) < \frac{\varepsilon}{\delta r} p(x) .$$

Letting $r \to 1^-$ yields

$$q(x) \leq \frac{\varepsilon}{\delta} p(x) .$$

Let us suppose now that $q \leq \frac{\varepsilon}{\delta} p$. If $p(x) < \delta$, then $q(x) \leq \frac{\varepsilon}{\delta} p(x) < \varepsilon$ and this gives the conclusion. □

Remark 2.8 Suppose that $\mathscr{P}$ and $\mathscr{Q}$ are two families of seminorms on E, each one satisfying (2.11). Suppose that every $q \in \mathscr{Q}$ is continuous with respect to $\tau_{\mathscr{P}}$, and then, by Remark 2.3, for each $x \in X$ and $q \in \mathcal{Q}$, given $\varepsilon > 0$ we can find $\delta > 0$ so that

$$B_p(x, \delta) \subseteq B_q(x, \varepsilon) .$$

But by Remark 2.2, this implies that $\tau_{\mathcal{Q}}$ is coarser that $\tau_{\mathcal{P}}$ (that is, $\tau_{\mathcal{Q}} \subseteq \tau_{\mathcal{P}}$).

Proposition 2.5 *Let $(E, \mathcal{P})$ be a locally convex space. A seminorm q defined on E is continuous if and only if there exist $p \in \mathcal{P}$ and $\lambda > 0$ so that*

$$q(x) \leq \lambda p(x) \text{ for all } x \in E . \tag{2.16}$$

Proof The result is a straightforward consequence of Remark 2.3 and Lemma 2.1. □

If the mapping is linear, then the situation is nice: if it is continuous at 0, then it is continuous everywhere, and moreover this is much easier to check.

Theorem 2.11 *Let $(E, \mathcal{P})$ and $(F, \mathcal{Q})$ be locally convex spaces and $T : E \to F$ linear. The following are equivalent:*

1. *T is continuous.*
2. *T is continuous at 0.*
3. *For every $q \in \mathcal{Q}$, there exist $p \in \mathcal{P}$ and $C > 0$ so that $q(Tx) \leq Cp(x)$ for all $x \in E$.*
4. *If $V \in \mathcal{U}_0(F)$, then $T^{-1}(V) \in \mathcal{U}_0(E)$.*
5. *For each $V \in \mathcal{U}_0(F)$, there exists $U \in \mathcal{U}_0(E)$ so that $T(U) \subseteq V$.*

Proof Note first that 1 obviously implies 2. Suppose now that T is continuous at 0 and take some $q \in \mathcal{Q}$. We can find $\delta > 0$ and $p \in \mathcal{P}$ so that $q(Tx) < 1$ whenever $p(x) < \delta$. We want to show that

$$q(Tx) \leq \frac{2}{\delta} p(x)$$

for every x. We consider two cases. If $p(x) = 0$, then $p(\lambda x) = 0$ for every $\lambda > 0$, and then $\lambda q(Tx) = q(T(\lambda x)) < 1$. Since this holds for every $\lambda > 0$, we have $q(Tx) = 0$ and the claim holds.
Now, if $p(x) > 0$, then $p\big(\frac{\delta x}{2p(x)}\big) = \frac{\delta}{2} < \delta$. Therefore, $q\big(T\big(\frac{\delta x}{2p(x)}\big)\big) < 1$, and we conclude $q(Tx) < \frac{2}{\delta} p(x)$, and this finally yields the claim.

Let us assume that 3 holds, and let us see that T is continuous at an arbitrary $x_0 \in X$. We take, then, $q \in \mathcal{Q}$ and $\varepsilon > 0$. Since 3 holds, we can find a seminorm p and some $C > 0$ satisfying the inequality. If $p(x - x_0) < \varepsilon / C$, then

$$q(Tx - Tx_0) = q(T(x - x_0)) \leq Cp(x - x_0) < \varepsilon ,$$

and T is continuous at x_0.

This shows the equivalence between 1, 2, and 3. The fact that 4 and 5 are equivalent to 2 is a simple exercise using the definition of continuity and the linearity of T. □

A linear map $T : E \to F$ is an *isomorphism* if it is one to one, surjective, and continuous and $T^{-1} : F \to E$ is also continuous. In this case, X and Y are called *isomorphic* and we write $E \cong F$.

Remark 2.9 A particular case of Theorem 2.11 is when we have normed spaces. Then the system of seminorms reduced to one in each space, and the condition in Theorem 2.11–3 is simply to have a constant $C > 0$ so that

$$\|Tx\|_F \leq C\|x\|_E \tag{2.17}$$

for every $x \in E$. It is an easy exercise to check that, in this case,

$$\|T\| = \inf\{C : \text{ satisfying (2.17)}\}$$

defines a norm for the continuous linear operators between E and F, which can also be computed as $\|T\| = \sup_{\|x\|_E \leq 1} \|Tx\|_F$.

If E and F are locally convex spaces, we define the space

$$L(E, F) = \{T : E \to F \mid \text{linear and continuous}\}. \tag{2.18}$$

If $E = F$, we simply write $L(E) = L(E, E)$. Note that if $A \subseteq E$ is bounded and $T : E \to F$ is linear and continuous, then $T(A)$ is bounded in F.

Example 2.12 Given a sequence $(\alpha_i)_{i\in\mathbb{N}} \in \mathbb{K}^{\mathbb{N}}$, we define $T_\alpha : \ell_1 \longrightarrow \mathbb{K}^{\mathbb{N}}$ by $T_\alpha((x_i)_i) = (\alpha_i x_i)_i$. If we consider $T_\alpha : \ell_1 \to \omega$ (recall Example 2.9), then it is obviously well defined and continuous (since the topology is given by the pointwise convergence) for every $(\alpha_i)_i$.
If we take a different space as target, then the situation changes. The following statements are equivalent:

1. $T : \ell_1 \to \ell_1$ is well defined and continuous.
2. $T\ell_1 \subseteq \ell_1$.
3. $\alpha \in \ell_\infty$.

That 1 implies 2 is trivial. To see that 2 implies 3, take some $\alpha \notin \ell_\infty$. For each $k \in \mathbb{N}$, we can find n_k so that $|\alpha_{n_k}| \geq k$. With this in mind, we define the sequence

$$x_n = \begin{cases} \frac{1}{k^2} & \text{if } n = n_k \\ 0 & \text{otherwise}. \end{cases}$$

Clearly $x = (x_n)_n \in \ell_1$ but $T_\alpha(x) \notin \ell_1$, and this completes the argument. Finally, if $\alpha \in \ell_\infty$, then

$$\|Tx\|_1 = \sum_{n=1}^{\infty} |\alpha_n x_n| \leq \Big(\sup_{k \in \mathbb{N}} |\alpha_k| \Big) \sum_{n=1}^{\infty} |x_n| = \|\alpha\|_\infty \|x\|_1$$

for every $x \in \ell_1$, which shows that T_α is well defined and continuous. $\square$

Suppose that H is a family of continuous functions $f : X \to F$ (where X is a Hausdorff topological space and $(F, \mathscr{Q})$ is an l.c.s.). Given $x_0 \in X$, $q \in \mathscr{Q}$, and $\varepsilon > 0$, the continuity of each $f \in H$ implies that there is a neighbourhood $V = V_f$ so that $q(f(x) - f(x_0)) < \varepsilon$ for every $x \in V_f$. Of course, the neighbourhood depends not only on f but also on q and ε, but we want to stress here the dependence on f. Now, H is said to be *equicontinuous* at x_0 if the neighbourhood can be taken independent of f (that is, the same V works for all fs in H); more precisely, if for every $q \in \mathscr{Q}$ and $\varepsilon > 0$, there is a neighbourhood V of x such that

$$q(f(x) - f(x_0)) < \varepsilon \tag{2.19}$$

for every $x \in V$ and every $f \in H$. The family H is said to be equicontinuous if it is equicontinuous at x_0 for every $x_0 \in X$.
The following result, dealing with equicontinuous families of linear mappings between locally convex spaces, follows with essentially the same proofs as its analogues for single linear mappings (Theorem 2.11).

Proposition 2.6 *Let $(E, \mathscr{P})$ and $(F, \mathscr{Q})$ be two l.c.s. and $H \subseteq L(E, F)$. The following conditions are equivalent:*

1. *H is equicontinuous at* 0.
2. *H is equicontinuous at every $x \in E$.*
3. *For each $q \in \mathscr{Q}$, there is $p \in \mathscr{P}$ and a constant $C > 0$ such that $q(T(x)) \leq C\,p(x)$ for every $x \in E$ and every $T \in H$.*
4. *If $V \in \mathcal{U}_0(F)$ then $\bigcap_{T \in H} T^{-1}(V) \in \mathcal{U}_0(E)$.*
5. *For each $V \in \mathcal{U}_0(F)$ there exists $U \in \mathcal{U}_0(E)$ so that $T(U) \subseteq V$ for every $T \in H$.*

Remark 2.10 Given a sequence $(T_n)_n \subseteq L(E, F)$ such that $\lim_n T_n x$ exists (in F) for every $x \in E$, then the mapping $T : E \to F$ defined as $Tx = \lim_n T_n x$ is clearly linear. If $(T_n)_n$ is equicontinuous, then it is an easy exercise to show that T is also continuous. Compare this with Proposition 2.16.

Remark 2.11 Suppose that $(T_n)_n \subseteq L(E, F)$ is equicontinuous. If B is dense in A and $T_n y \to 0$ for every $y \in B$, then it is easy to see that $T_n x \to 0$ for every $x \in A$. In order to see this, take some continuous seminorm q and fix $\varepsilon > 0$. Being $(T_n)_n$ equicontinuous, we can find a seminorm p so that $p(T_n z - z) \leq p(z)$ for every $z \in E$. Now, by density, there is $y \in B$ with $p(y - x) < \varepsilon/2$. Finally, since $T_n y \to 0$, there is n_0 so that $p(T_n y) < \varepsilon/2$ for every $n \geq n_0$. With all this, if

$n \geq n_0$, then

$$q(T_n x) \leq q(T_n(x - y)) + q(T_n y) \leq p(x - y) + q(T_n y) < \varepsilon ,$$

which yields our claim.

2.2.3 Metrizable Locally Convex Spaces

An l.c.s. E is *metrizable* if there is a metric on E that defines the same topology. Our next step is to find conditions that ensure that a given l.c.s. is metrizable.

Theorem 2.13 *A locally convex space* $(E, \mathscr{P})$ *is metrizable if and only if there exists a sequence* $(p_n)_n$ *of continuous seminorms on* $(E, \mathscr{P})$ *such that*

1. $p_n \leq p_{n+1}$ *for all* $n \in \mathbb{N}$.
2. *For each* $p \in \mathscr{P}$, *there is* m *such that* $p \leq p_m$.

In this case, the metric can be taken invariant by translations.

Before we go to the proof, let us note that if every p_n is continuous on $(E, \mathscr{P})$, then by Remark 2.8, $\tau_{(p_n)} \subseteq \tau_{\mathscr{P}}$. On the other hand, if 2 is satisfied, then again Remark 2.8 gives $\tau_{\mathscr{P}} \subseteq \tau_{(p_n)}$. All in all, if the family $(p_n)_n$ satisfies the conditions of the theorem, then the topologies $\tau_{(p_n)}$ and $\tau_{\mathscr{P}}$ coincide.

Proof Suppose that E is metrizable. Then, there exists a distance d such that τ_d and $\tau_{\mathscr{P}}$ coincide. Then, by Remark 2.2, for each n, we can find $\tilde{q}_n \in \mathscr{P}$ and $\tilde{\varepsilon}_n$ so that $B_{\tilde{q}_n}(0, \tilde{\varepsilon}_n) \subseteq B_d(0, 1/n)$. Now we define

$$q_n(x) = \max\{\tilde{q}_1(x), \dots, \tilde{q}_n(x)\} \quad \text{for } x \in X$$

$$\varepsilon_n = \min\{\tilde{\varepsilon}_1, \dots, \tilde{\varepsilon}_n\} .$$

We obviously have

$$B_{q_n}(0, \varepsilon_n) \subseteq B_{\tilde{q}_n}(0, \varepsilon_n) \subseteq B_d(0, 1/n) \tag{2.20}$$

for all $n \in \mathbb{N}$. Moreover, $B_{q_{n+1}}(0, \varepsilon_{n+1}) \subseteq B_{q_n}(0, \varepsilon_n)$, and this, by Lemma 2.1, gives $q_n \leq \frac{\varepsilon_n}{\varepsilon_{n+1}} q_{n+1}$ for every n.

Now, given $p \in \mathscr{P}$, there is n such that $B_d(0, 1/n) \subseteq B_p(0, 1)$. This, together with (2.20) and Lemma 2.1, gives $p \leq \frac{1}{\varepsilon_n} q_n$. Defining $p_n = \frac{1}{\varepsilon_n} q_n$, we have the desired sequence of seminorms.

Conversely let us assume that $(p_n)_n$ is a sequence of continuous seminorms satisfying the two conditions in the statement. We define $d : E \times E \longrightarrow [0, \infty[$ by

$$d(x, y) = \sum_{n=1}^{\infty} \frac{1}{2^n} \min\{1, p_n(x - y)\} ,$$

which is a translation-invariant metric since it satisfies the following conditions:

1. $0 \leq d(x, y) \leq 1$
2. $d(x, y) = d(y, x)$
3. $d(x, y) \leq d(x, z) + d(z, y)$ (because $p_n(x - y) \leq p_n(x - z) + p_n(z - y)$)
4. $d(x + z, y + z) = d(x, y)$

Now, we see that the topology τ_d given by the metric and the topology $\tau_{(p_n)}$ (which, as we have pointed out, is the same as $\tau_{\mathscr{P}}$) coincide. We begin by seeing that $\tau_{(p_n)} \subseteq \tau_d$ using Remark 2.2. Note first that for each $x \in E$ we can write

$$B_d(x, \varepsilon) = x + B_d(0, \varepsilon) .$$

With this and (2.13) in mind, we can deal only with neighbourhoods of 0. Take n and $1 > \varepsilon > 0$. If $d(x, 0) < \varepsilon/2^n$, then $\frac{\min\{1, p_n(x)\}}{2^n} < \frac{\varepsilon}{2^n}$ and this gives $p_n(x) < \varepsilon$. Hence

$$B_d(0, \varepsilon/2^n) \subseteq B_{p_n}(0, \varepsilon) ,$$

which gives the conclusion.

On the other hand, to see that $\tau_d \subseteq \tau_{(p_n)}$, we fix $\delta > 0$ and take $m \in \mathbb{N}$ such that $\sum_{n=m+1}^{\infty} 1/2^n < \delta/2$, and for $p_m(x) < \delta/2$, we have (recall that since the sequence of seminorms is increasing, we have $p_k(x) < \delta/2$ for $k = 1, \dots, m$)

$$\begin{aligned} d(0, x) = \sum_{k=1}^{m} 2^{-k} \min\{1, p_k(x)\} + \sum_{k=m+1}^{\infty} 2^{-k} \min\{1, p_k(x)\} \\ < \sum_{k=1}^{m} 2^{-k} p_k(x) + \sum_{k=m+1}^{\infty} 2^{-k} < \frac{\delta}{2} \sum_{k=1}^{m} 2^{-k} + \frac{\delta}{2} < \delta . \end{aligned}$$

This proves $B_{p_m}(0, \delta/2) \subseteq B_d(0, \delta)$ and completes the proof. □

Note that if E is a metrizable locally convex space, the translation invariance of the metric d defining the topology gives

$$d(x, 0) = d(0, -x) = d(-x, 0)$$

and

$$d(x + y, 0) = d(x, -y) \leq d(x, 0) + d(0, -y) = d(x, 0) + d(y, 0) , \tag{2.21}$$

for every $x, y \in E$.

If the space is metrizable, then **we may (and from now on will always do) assume that the topology is defined by an increasing sequence of seminorms.**

Then we can just deal with this sequence of seminorms, and we immediately have the following.

Proposition 2.7 *Let* $(E, \mathscr{P})$ *be a metrizable l.c.s. Let* $(p_n)_n$ *be the sequence of seminorms coming from Theorem 2.13. Then a sequence* $(x_k)_k \subseteq E$

1. *Converges to* x_0 *if and only if for all* $n \in \mathbb{N}$ *and* $\varepsilon > 0$ *there is* k_0 *such that* $p_n(x_k - x_0) < \varepsilon$ *for every* $k \geq k_0$.
2. *Is a Cauchy sequence if and only if for all* $n \in \mathbb{N}$ *and* $\varepsilon > 0$ *there is* k_0 *such that* $p_n(x_k - x_l) < \varepsilon$ *for every* $l, k \geq k_0$.

Remark 2.12 If E is a locally convex space (not necessarily complete), a sequence $(x_n)_n$ in E is said to be *Cauchy* if for every U, convex neighbourhood of 0, there is $n_0 \in \mathbb{N}$ such that $x_n - x_m \in U$ for all $n, m \geq n_0$ or, equivalently, if for all $p \in \mathscr{P}$ and $\varepsilon > 0$ there is k_0 such that $p(x_k - x_l) < \varepsilon$ for every $l, k \geq k_0$. Then E is said to be *sequentially complete* if every Cauchy sequence in E is convergent. If E is metrizable, we will omit 'sequentially' and say that the space is *complete*. Notice that this coincides with the notion of completeness that we introduced above for normed spaces.

A metrizable and complete locally convex space is called a *Fréchet space*. We have some immediate examples of Fréchet spaces.

Example 2.14

1. Every Banach space is a Fréchet space.
2. Every closed subspace of a Fréchet space is also a Fréchet space.
3. The family of seminorms $(p_k)_k$ defining the topology in ω (recall Example 2.9) clearly satisfies that $p_k \leq p_{k+1}$ for all $k \in \mathbb{N}$. Then, ω is a Fréchet space.

□

More examples are given in Sect. 2.4.

2.3 The Dual of a Locally Convex Space

We have dealt so far with linear operators between locally convex spaces. A particularly interesting case is when the target space is the scalar field. Let us look at this with some detail. In general, if E is a vector space, then the *algebraic dual* (denoted by E^*) is the space of all linear mappings $u : E \to \mathbb{K}$. If E has more structure, then we may also consider the *topological dual* (or simply *dual space*, denoted by E'), consisting of all $u : E \to \mathbb{K}$ that are linear and continuous. If E is a locally convex space, then using the notation that introduced above we have $E' = L(E, \mathbb{K})$. The elements of E' are called *functionals*. As a direct consequence of Theorem 2.11, we have the following.

Proposition 2.8 *Let $(E, \mathscr{P})$ be a locally convex space and $u \in E^*$. Then the following are equivalent:*

1. *$u \in E'$.*
2. *u is continuous at 0.*
3. *There exist $p \in \mathscr{P}$ and $C > 0$ so that*

$$|u(x)| \leq Cp(x) \tag{2.22}$$

   for every $x \in E$.
4. *There exists a 0-neighbourhood U in E such that $\sup_{x \in U} |u(x)| < +\infty$.*

Remark 2.13 As a consequence of Proposition 2.6, if $(E, \mathscr{P})$ is an l.c.s., then $B \subseteq E'$ is equicontinuous if and only if there are $p \in \mathscr{P}$ and $C > 0$ such that $|u(x)| \leq Cp(x)$ for all $x \in E$ and $u \in B$.

Let us describe now the duals of some of the classical spaces that we introduced before.

Example 2.15 Let us see that the dual of ℓ_1 can be identified with ℓ_∞. Note in first place that if $u \in \ell_1'$, then by (2.22) there is some $C > 0$ so that

$$|u(x)| \leq C\|x\|_{\ell_1},$$

for every $x \in \ell_1$. This in particular means that the sequence $\big(u(e_n)\big)_n$ is bounded, and the mapping $T : \ell_1' \to \ell_\infty$ given by $T(u) = \big(u(e_n)\big)_n$ is well defined. It is clearly linear. Let us see that it is also surjective. Given $a = (a_i)_{i=1}^\infty \in \ell_\infty$, we define $u_a : \ell_1 \to \mathbb{K}$ by

$$u_a(x) = \sum_{i=1}^\infty a_i x_i, \tag{2.23}$$

for $x = (x_i)_i \in \ell_1$. Note that, for each $n \in \mathbb{N}$, we have

$$\sum_{i=1}^n |a_i x_i| \leq \big(\sup_{1 \leq i \leq n} |a_i|\big) \sum_{i=1}^n |x_i| \leq \|a\|_{\ell_\infty} \|x\|_{\ell_1}.$$

Then the series in (2.23) converges, u_a is well defined, and $|u_a| \leq \|a\|_{\ell_\infty} \|x\|_{\ell_1}$ for every $x \in \ell_1$. Since it is linear, this shows that it is also continuous, and $u_a \in \ell_1'$. It clearly satisfies $u_a(e_n) = a_n$ for every n, which shows that T is onto. Moreover, this gives for each $u \in \ell_1'$ that

$$u(x) = \sum_{i=1}^\infty u(e_i) x_i,$$

for every $x = (x_i)_i \in \ell_1$. Finally, if $T(u) = 0$, then $u(e_i) = 0$ for every i and $u(x) = 0$ for every x. This shows that T is injective and yields $\ell_1' = \ell_\infty$. □

Example 2.16 We describe now the dual of ℓ_p for $1 < p < \infty$. Fix some $u \in \ell_p'$, and, for each $i \in \mathbb{N}$ take $\omega_i \in \mathbb{K}$ with $|\omega_i| = 1$ so that $u(e_i)\omega_i = |u(e_i)|$. Now we consider p' so that $\frac{1}{p} + \frac{1}{p'} = 1$, and, for each $n \in \mathbb{N}$, we take

$$x^n = \big(|u(e_1)|^{p'-1}\omega_1, \dots, |u(e_n)|^{p'-1}\omega_n, 0, 0, \dots\big) = \sum_{i=1}^{n} |u(e_i)|^{p'-1}\omega_i \,.$$

Since u is continuous, there is some $C > 0$ so that $|u(x)| \le C\|x\|_{\ell_p}$, which in particular implies

$$\sum_{i=1}^{n} |u(e_i)|^{p'} = |u(x^n)| \le C\Big(\sum_{i=1}^{n} |u(e_i)|^{(p'-1)p}\Big)^{1/p} = C\Big(\sum_{i=1}^{n} |u(e_i)|^{p'}\Big)^{1/p},$$

and

$$\Big(\sum_{i=1}^{n} |u(e_i)|^{p'}\Big)^{1/p'} \le C\,.$$

Since this holds for every n, it implies that $\big(u(e_i)\big)_i \in \ell_{p'}$, and the mapping $T : \ell_p' \to \ell_{p'}$ given by $T(u) = \big(u(e_n)\big)_n$ is well defined (and linear). Now, recall that for $a \in \ell_{p'}$ and $x \in \ell_p$, Hölder's inequality (2.7) gives

$$\sum_{i=1}^{n} |a_i x_i| \le \|a\|_{\ell_{p'}} \|x\|_{\ell_p}\,,$$

for every n. Thus, the operator u_a defined on ℓ_p as in (2.23) belongs to ℓ_p'. Proceeding as in Example 2.15, we conclude that T is a linear bijection that gives the identification $\ell_p' = \ell_{p'}$. □

Example 2.17 With an up to some point similar idea, one has that for $1 \le p < \infty$, the dual of $L_p(\Omega)$ (recall Example 2.6) can be identified with $L_{p'}(\Omega)$. A proof of this fact can be found in [9, Theorems 2.18 and 2.19]. □

Remark 2.14 It is an easy exercise to show that if E is finite dimensional, then every linear operator defined on E is continuous (and, in particular, the algebraic and the topological duals coincide). However E is infinite dimensional and this may not be the case. Let us show a couple of examples.

Take E to be the space of all real polynomials of one variable and endow it with the norm

$$\|P\| = \sup_{x\in[0,1]} |P(x)|\,.$$

Consider now the operator $T : E \to E$ given by $T(P) = P'$. This is clearly linear, but it is not continuous. To see this, consider, for each $n \in \mathbb{N}$, the polynomial $P_n(x) = x^n$. These satisfy $\|P_n\| = 1$ but $\|T P_n\| = n$, and this shows that (2.17) does not hold, and hence T is not continuous.

Let us consider now E the space of functions $f : [-1, 1] \to \mathbb{R}$ that are continuous on $[-1, 1]$ and differentiable on $]-1, 1[$ and endow it with the norm

$$\|f\| = \sup_{x\in[-1,1]} |f(x)| .$$

Define $\varphi : E \to \mathbb{R}$ as $\varphi(f) = f'(0)$. Again this is linear but not continuous. Take $f_n(x) = \frac{\sin(n^2 x)}{n}$. Clearly $\|f_n\| \leq \frac{1}{n}$ but $|\varphi(f_n)_n| = n$. This shows that (2.22) does not hold and φ is not continuous. Hence E' and E^* are in general not equal.

What these two examples have in common is that the spaces are not complete. This allows to construct the non-continuous linear mappings. As a matter of fact in every infinite dimensional l.c.s. (also if the space is complete), there exist linear non-continuous operators. This is more complicated and requires the Axiom of Choice.

2.4 Examples of Spaces

2.4.1 *Space of Continuous Functions*

We have now at hand all tools that we need to define a topology on the space

$$C(\Omega) = \{f : \Omega \to \mathbb{K}: \text{ continuous}\},$$

where $\Omega \subseteq \mathbb{R}^d$ is an open set. What we have to do is to define a family of seminorms, and we do it in the following way. For each compact set $K \subseteq \Omega$, we define

$$p_K(f) = \sup_{x\in K} |f(x)| . \tag{2.24}$$

It is easy to check that each one of these is a seminorm. Now we consider the family

$$\mathscr{P} = \{p_K : K \Subset \Omega \text{ compact}\} .$$

Since the union of two compact sets is again compact and $\{x\}$ is compact for every x, this family satisfies (LC1) and (LC2) and, then, defines a locally convex topology (denoted τ_{co}) on $C(\Omega)$. Our aim is to show that, in fact, it is a Fréchet space (metrizable and complete). But before we get into that, we see that the convergence in this topology has a very concrete meaning.

As a straightforward application of Remark 2.7, we have that a sequence of functions $(f_n)_n \subseteq C(\Omega)$ converges in τ_{co} to some f if and only if

$$\lim_n \Big(\sup_{x \in K} |f_n(x) - f(x)| \Big) = 0 ,$$

for every compact set $K \subseteq \Omega$. In other words, $f_n \xrightarrow{\tau_{co}} f$ if and only if $(f_n)_n$ converges to f uniformly on every compact subset of K. For this reason, the topology τ_{co} is known as the *topology of uniform convergence over the compact sets* or *compact-open topology*.

As we have said, we want to prove that the space $(C(\Omega), \tau_{co})$ is Fréchet. In view of Theorem 2.13, we have to find a countable family of increasing seminorms that generates the same topology. Given $\Omega \subsetneq \mathbb{R}^d$ open, for each $n \in \mathbb{N}$, we consider the set

$$K_n = \{x \in \Omega : \|x\| \leq n, d(x, \mathbb{R}^d \setminus \Omega) \geq 1/n\} . \tag{2.25}$$

It is a simple exercise to check that the family $(K_n)_n$ satisfies

- Every $K_n \subseteq \Omega$ is compact.
- $K_n \subseteq \mathring{K}_{n+1}$ for all $n \in \mathbb{N}$.
- $\Omega = \bigcup_{k=1}^{\infty} \mathring{K}_n$.

In particular, for every compact $K \Subset \Omega$, there exists $n = n(K) \in \mathbb{N}$ such that $K \subseteq K_n$. A family of sets satisfying these three conditions is called a *fundamental sequence of compact sets* of Ω. A fundamental sequence of compact sets of $\mathbb{R}^d$ is given by

$$K_n = \{x \in \Omega : \|x\| \leq n\} .$$

Theorem 2.18 *If $\Omega \subseteq \mathbb{R}^d$ is an open set, then $(C(\Omega), \tau_{co})$ is a Fréchet space.*

Proof As we already pointed out, the fact that the union of two compact sets is again compact gives (LC1). On the other hand, if $f \neq 0$, then there exists $x_0 \in \Omega$ for which $f(x_0) \neq 0$. Taking $K = \{x_0\}$, we have $p_K(f) > 0$, and the seminorms satisfy (LC2).

We see now that the topology is metrizable. To do that, we take $(K_n)_n$ a fundamental sequence of compact sets of Ω and denote, for the sake of clarity, $p_n = p_{K_n}$ for $n \in \mathbb{N}$. Clearly these seminorms are all continuous and the sequence is increasing. Also, given any compact set $K \Subset \Omega$, since we have a fundamental sequence of compact sets, there is some n such that $p_K \leq p_n$. With all this, Theorem 2.13 gives that the topology is metrizable.

In order to see that it is complete, take $(f_n)_n$ a Cauchy sequence in $(C(\Omega), \tau_{co})$. By Remark 2.12, for each compact set K and each $\varepsilon > 0$, there is n_0 such that

$$\sup_{x \in K} |f_n(x) - f_m(x)| < \varepsilon \text{ for all } m > n \geq n_0. \tag{2.26}$$

In particular, given any $x \in \Omega$ and taking $K = \{x\}$, the sequence $(f_n(x))_n$ is Cauchy in $\mathbb{K}$ and hence convergent. Let $f : \Omega \to \mathbb{K}$ be the pointwise limit, i.e., $f(x) = \lim_n f_n(x)$ for each $x \in \Omega$. To finish, we have to see that $f \in C(\Omega)$ and $f_n \to f$ in τ_{co}.

Let $K \Subset \Omega$ be a compact set. Then, from (2.26), we have that for all $\varepsilon > 0$ there is some $n_0 \in \mathbb{N}$ such that

$$|f_n(x) - f_m(x)| < \varepsilon \text{ for all } m > n \geq n_0 \text{ and all } x \in K .$$

Letting $m \to \infty$, this yields

$$|f_n(x) - f(x)| \leq \varepsilon \text{ for all } n \geq n_0 \text{ and all } x \in K .$$

Hence $(f_n)_n$ converges uniformly on K for every compact set K of Ω or, in other words, $f_n \to f$ in τ_{co}. Moreover, this implies that $f \in C(K)$ for every compact $K \subseteq \Omega$ and this, by Theorem 1.4, gives $f \in C(\Omega)$. □

Example 2.19 We consider $(C(\Omega), \tau_{co})$. If $v : \Omega \to [0, +\infty[$ is continuous, then the set $A_v = \{f \in C(\Omega) : |f| \leq v\}$ is bounded in $(C(\Omega), \tau_{co})$. To see this, take a compact set K in Ω and denote $M = \sup_{x \in K} v(x)$. Then,

$$\sup_{f \in A_v} p_K(f) = \sup_{f \in A_v} \Big(\sup_{x \in K} |f(x)| \Big) \leq \sup_{x \in K} v(x) = M .$$

□

2.4.2 *Köthe Echelon Spaces*

A *Köthe matrix* is a sequence $(a_n)_{n \in \mathbb{N}}$ of functions $a_n : \mathbb{N} \to [0, +\infty[$ such that

(K1) $a_n(i) \leq a_{n+1}(i)$ for all $i \in \mathbb{N}$ and $n \in \mathbb{N}$.
(K2) For all $i \in \mathbb{N}$, there exists $n \in \mathbb{N}$ such that $a_n(i) > 0$.

Example 2.20 Define $a_n(i) = n^i$. Then A can be represented by a matrix

		i				
		1	2	3	4	
n	1	1	1	1	1	$\cdots$
	2	2	2^2	2^3	2^4	$\cdots$
	3	3	3^2	3^3	3^4	$\cdots$
	4	4	4^2	4^3	4^4	$\cdots$
		$\vdots$	$\vdots$	$\vdots$	$\vdots$	$\ddots$

□

Example 2.21 Take $a_n(i) = (1 - \frac{1}{n})^i$, and then A is

	i: 1	2	3	4	
n 1	0	0	0	0	$\cdots$
2	1/2	$1/2^2$	$1/2^3$	$1/2^4$	$\cdots$
3	2/3	$(2/3)^2$	$(2/3)^3$	$(2/3)^4$	$\cdots$
4	3/4	$(3/4)^2$	$(3/4)^3$	$(3/4)^4$	$\cdots$
	$\vdots$	$\vdots$	$\vdots$	$\vdots$	$\ddots$

□

Example 2.22 Let $a_n(i) = i^n$. Then A is the matrix

	i: 1	2	3	4	
n 1	1	2	3	4	$\cdots$
2	1	2^2	3^2	4^2	$\cdots$
3	1	2^3	3^3	4^3	$\cdots$
4	1	2^4	3^4	4^4	$\cdots$
	$\vdots$	$\vdots$	$\vdots$	$\vdots$	$\ddots$

□

We define now a scale of spaces associated with a Köthe matrix that, in some sense, are an extension of the ℓ_p-spaces that we introduced in Example 2.5. Given $1 \leq p < \infty$ and a Köthe matrix $A = (a_n)_n$, the *Köthe echelon space of order p* is defined as

$$\lambda_p(A) = \{x = (x_i)_{i=1}^\infty \in \mathbb{K}^\mathbb{N} \colon q_n(x) = \Big(\sum_{i=1}^\infty a_n(i)|x_i|^p\Big)^{\frac{1}{p}} < +\infty, \ \forall n \in \mathbb{N}\},$$

and, for $p = \infty$,

$$\lambda_\infty(A) = \{x = (x_i)_{i=1}^\infty \in \mathbb{K}^\mathbb{N} \colon q_n(x) = \sup_{i \in \mathbb{N}} a_n(i)|x_i| < +\infty \ \forall n \in \mathbb{N}\}. \tag{2.27}$$

Proposition 2.9 *For each $1 \leq p \leq \infty$, the space $\lambda_p(A)$ with the topology generated by the corresponding family of seminorms is Fréchet.*

Proof Fix p and consider the system of seminorms $(q_n)_n$ corresponding to $\lambda_p(A)$. Note first that (K1) immediately implies that the system of seminorms is increasing. On the other hand, if $x \neq 0$, then $x_i \neq 0$ for some i and, by (K2), there is some n

for which $a_n(i) > 0$. Hence $p_n(x) > q_n(x) > a_n(i)|x_i| > 0$ and the topology is Hausdorff (thus, by Theorem 2.13, metrizable).
We only sketch the proof of the completeness (which finally gives the conclusion) for $p = 1$. If $(x^k)_k \subseteq \lambda_1(A)$ is a Cauchy sequence in $\lambda_1(A)$, for all $n \in \mathbb{N}$ and $\varepsilon > 0$, there is $k_0 \in \mathbb{N}$ such that if $k, l \geq k_0$, then

$$\sum_{i=1}^{\infty} a_n(i)|x_i^k - x_i^l| = p_n(x^k - x^l) < \varepsilon \,.$$

By (K2), for each $i \in I$, the limit $x_i = \lim_k x_i^k$ exists. We denote $x = (x_i)_i$. Then the proof will be completed by showing first that $x \in \lambda_1(A)$ (use that every Cauchy sequence is bounded) and then that $x^k \to x$ in the topology of $\lambda_1(A)$. The completeness of $\lambda_p(A)$ with $1 < p \leq \infty$ follows by a similar argument. □

As we said above, these Köthe spaces are some sort of generalization of the ℓ_p-spaces. So, we can adapt some of the ideas in Example 2.15 to describe the dual of $\lambda_1(A)$ in terms of some sort of analogue of ℓ_∞. Let A be a Köthe matrix so that $a_1(i) > 0$ for every $i \in \mathbb{N}$ (and then $a_n(i) > 0$ for all n, i). Define, for each n and i, the number

$$v_n(i) = \frac{1}{a_n(i)}\,,$$

and consider the spaces

$$\ell_\infty(v_n) = \big\{u = (u_i)_{i=1}^{\infty} \in \mathbb{K}^{\mathbb{N}} \colon \sup_i v_n(i)|u_i| < \infty\big\}\,,$$

for each $n \in \mathbb{N}$ and

$$k_\infty(A) = \bigcup_{n=1}^{\infty} \ell_\infty(v_n)\,.$$

Proposition 2.10 *Let A be a Köthe matrix so that $a_1(i) > 0$ for every $i \in \mathbb{N}$. Then $\lambda_1(A)'$ can be identified with $k_\infty(A)$, with the duality given by*

$$\langle x, u\rangle = \sum_{i=1}^{\infty} x_i u_i\,,$$

for $x = (x_i)_i \in \lambda_1(A)$ and $u = (u_i)_i \in k_\infty(A)$.

Proof To begin with, if $L \in \lambda_1(A)'$, then there exist $n_0 \in \mathbb{N}$ and $c > 0$ so that

$$|L(x)| \leq c \sum_{j=1}^{\infty} a_{n_0}(j)|x_j|$$

for every $x \in \lambda_1(A)$. This, in particular, yields $|L(e_i)| \leq c a_{n_0}(i)$ for every $i \in \mathbb{N}$, and this immediately gives $\big(L(e_i)\big)_i \in k_\infty(A)$. Then the mapping $T : \lambda_1(A)' \to k_\infty(A)$ given by $\varphi(L) = \big(L(e_i)\big)_i$ is well defined and clearly linear. Let us see that it is surjective. Given any $u = (u_i)_i \in k_\infty(A)$, let us consider $L_u : \lambda_1(A) \to \mathbb{K}$ given by $L_u(x) = \sum_{i=1}^\infty x_i u_i$ for $x = (x_i)_i \in \lambda_1(A)$. Since $u \in k_\infty(A)$, there is some $n_0 \in \mathbb{N}$ so that $\sup_i v_{n_0}(i)|u_i| < \infty$. Then, for $x \in \lambda_1(A)$, we have

$$|\psi(u)(x)| = \Big|\sum_{i=1}^\infty x_i u_i\Big| \leq \sum_{i=1}^\infty |x_i|\,|u_i|$$

$$= \sum_{i=1}^\infty |x_i| v_{n_0}(i) a_{n_0}(i) |u_i| \leq \Big(\sup_i |v_{n_0}|u_i|\Big) \sum_{i=1}^\infty a_{n_0}(i)|x_i| \,.$$

This shows that L_u is well defined and, since it is clearly linear, continuous. In other words, $L_u \in \lambda_1(A)'$. It is plain that $T(L_u) = u$, so that T is onto. Injectivity follows exactly as in Example 2.15, showing that T is a linear bijection and completing the proof. □

Example 2.23 Let $A = (a_n)_{n\in\mathbb{N}}$ be a Köthe matrix. We take $w = (w_i)_{i=1}^\infty \in \lambda_\infty(A)$ (recall (2.27)) with $w_i > 0$, for $i \in \mathbb{N}$, and write

$$B_w = \Big\{ x = (x_i)_{i=1}^\infty \in \mathbb{K}^{\mathbb{N}} \colon \sum_{i=1}^\infty \frac{|x_i|}{w_i} \leq 1 \Big\}\,. \tag{2.28}$$

We denote $\mu_n = \sup_i a_n(i) w_i < +\infty$ for every $n \in \mathbb{N}$. Finally, we deduce that for each $n \in \mathbb{N}$ and $x = (x_i)_i \in B_w$,

$$q_n(x) = \sum_{i=1}^\infty a_n(i)|x_i| = \sum_{i=1}^n a_n(i) w_i \frac{|x_i|}{w_i} \leq \mu_n \sum_{i=1}^n \frac{|x_i|}{w_i} \leq \mu_n \,.$$

This proves that B_w is bounded in $\lambda_1(A)$. □

In fact, starting from this example we can even characterize the bounded sets of $\lambda_1(A)$.

Proposition 2.11 *A subset $B \subseteq \lambda_1(A)$ is bounded if and only if there exists $w = (w_i)_{i=1}^\infty \in \lambda_\infty(A)$ with $w_i > 0$ for every $i \in \mathbb{N}$ such that $B \subseteq B_w$.*

Proof Take some $B \subseteq \lambda_1(A)$ bounded. For each $n \in \mathbb{N}$, there is some $M_n > 0$ so that

$$\sup_{x\in B} \sum_{i=1}^\infty a_n(i)|x_i| \leq M_n \,,$$

and we can consider

$$C_n = \max\left(M_n, n \max_{1\le i\le n}(a_n(i), 1)\right).$$

Now, by (K2), for a given $i \in \mathbb{N}$, we can find $n(i) \in \mathbb{N}$ so that $a_{n(i)}(i) > 0$ (and, then, by (K1) also for every $n \geq n(i)$). Then

$$w_i = \inf_{n\in\mathbb{N}} \frac{2^n C_n}{a_n(i)} < \infty$$

(here we take the convention $\frac{1}{0} = \infty$). Note, on the other hand, that $\frac{C_n}{a_n(i)} \geq n$ for every $n \geq \max(i, n(i))$ and, then,

$$w_i \geq \inf_{n\in\mathbb{N}} \frac{C_n}{a_n(i)} = \min\Big(\min_{1\le n\le \max(i,n(i))} \tfrac{C_n}{a_n(i)}, n(i)\Big) > 0.$$

The sequence $w = (w_i)_{i=1}^{\infty}$ clearly belongs to $\lambda_\infty(A)$. It is only left to see that B is contained in the set B_w defined in (2.28). To see this, note first that for each i we have

$$\frac{1}{w_i} = \sup_{n\in\mathbb{N}} \frac{a_n(i)}{2^n C_n} \leq \sum_{n=1}^{\infty} \frac{a_n(i)}{2^n C_n},$$

and the latter is finite because $C_n \geq a_n(i)$ for every $n \geq i$. Now, if $x \in B$, then

$$\sum_{i=1}^{\infty} \frac{|x_i|}{w_i} \leq \sum_{i=1}^{\infty} |x_i| \Big(\sum_{n=1}^{\infty} \frac{a_n(i)}{2^n C_n}\Big) = \sum_{n=1}^{\infty} \frac{1}{2^n}\Big(\frac{1}{C_n}\sum_{i=1}^{\infty} a_n(i)|x_i|\Big) \leq 1.$$

This proves one implication. The converse one is a straightforward consequence of Example 2.23 and Proposition 2.3. □

If the Köthe matrix is nice enough, we can even see that bounded subsets are relatively compact.

Proposition 2.12 *Let A be a Köthe matrix so that for every $n \in \mathbb{N}$ there exists $m > n$ such that $\frac{a_n}{a_m} \in c_0$. Then, every bounded subset in $\lambda_1(A)$ is relatively compact.*

Proof By Proposition 2.11, it is enough to see that the sets B_w defined as in (2.28) (with $w \in \lambda_\infty(A)$ and $w_i > 0$ for every i) are relatively compact. To see this, choose some sequence $(x^k)_k \subseteq B_w$, and let us see that it has a convergent subsequence. Note that B_w is closed in $\mathbb{K}^{\mathbb{N}}$ (with the topology of pointwise convergence) and is contained in $\prod_{i=1}^{\infty}\{x \in \mathbb{K}\colon |x| \leq w_i\}$, which by Tikhonov's theorem 2.1 is compact. Hence B_w is compact, and we can find a subsequence $(x^{k(j)})_j$ that converges (pointwise) to some $x^0 = (x_i^0)_i \in B_w$. Since $B_w \subseteq \lambda_1(A)$, we have that in fact $x^0 \in \lambda_1(A)$. Let us see that $x^{k(j)} \to x^0$ in $\lambda_1(A)$ as $j \to \infty$. To see this, fix $n \in \mathbb{N}$

and choose $m > n$ with $\frac{a_n}{a_m} \in c_0$. Since $w \in \lambda_\infty(A)$, we can find $M > 0$ so that

$$w_i a_m(i) \leq M$$

for every $i \in \mathbb{N}$. Now, given $\varepsilon > 0$, there is $i_0 \in \mathbb{N}$ such that

$$\frac{a_n(i)}{a_m(i)} < \frac{\varepsilon}{4M}$$

for every $i \geq i_0$. Finally, since $(x^{k(j)})_j$ converges pointwise, we can find $j_0 \in \mathbb{N}$ so that if $1 \leq i \leq i_0 - 1$, then

$$|x_i^{k(j)} - x_i^0| < \frac{\varepsilon}{2\sum_{i=1}^{i_0} a_n(i)}$$

for every $j \geq j_0$. Joining all these together, we finally have

$$q_n(x^{k(j)} - x^0) = \sum_{i=1}^{\infty} a_n(i)|x_i^{k(j)} - x_i^0| = \sum_{i=1}^{i_0-1} a_n(i)|x_i^{k(j)} - x_i^0| + \sum_{i=i_0}^{\infty} a_n(i)|x_i^{k(j)} - x_i^0|$$

$$< \frac{\varepsilon}{2} + \frac{\varepsilon}{4M} \sum_{i=i_0}^{\infty} a_n(i)\left(|x_i^{k(j)}| + |x_i^0|\right) \leq \frac{\varepsilon}{2} + \frac{\varepsilon}{4M} M \left(\sum_{i=i_0}^{\infty} \frac{|x_i^{k(j)}|}{w_i} + \sum_{i=i_0}^{\infty} \frac{|x_i^0|}{w_i} \right) < \varepsilon .$$

This shows that $x^{k(j)} \to x^0$ in $\lambda_1(A)$ and completes the proof. □

2.5 Normable Spaces

We have seen that, in certain cases, the topology of a locally convex space may be generated by a metric (this is what we have called 'metrizable' spaces). We now look at spaces the topology of which can be equivalently defined by a norm. If this is the case, then the norm has obviously to be continuous. We say that an l.c.s. $(E, \mathscr{P})$ *admits a continuous norm* if there exists a norm q that is continuous in the topology $\tau_{\mathscr{P}}$. This (recall 2.16) happens if and only if there are $\lambda > 0$ and some $p \in \mathscr{P}$ so that

$$q(x) \leq \lambda p(x)$$

for all $x \in E$. We say that an l.c.s. $(E, \mathscr{P})$ is *normable* if there exists a norm $q : X \to [0, \infty[$ such that the norm topology τ_q coincides with the original one.

Remark 2.15 The fact that a space is normable basically requires two conditions to be satisfied.

1. The norm q is continuous.
2. The topology $\tau_{\mathscr{P}}$ is coarser than τ_q (or τ_q is finer than $\tau_{\mathscr{P}}$).

These two conditions can be reformulated in different ways. First of all, in terms of neighbourhoods (see Remark 2.8),

1. There are $\varepsilon > 0$ and $p_0 \in \mathscr{P}$ so that $B_{p_0}(0, \varepsilon) \subseteq B_q(0, 1)$.
2. For every $p \in \mathscr{P}$ and all $\varepsilon > 0$, there exists $\delta > 0$ such that $B_q(0, \delta) \subseteq B_p(0, \varepsilon)$.

In view of Lemma 2.1, this can also be reformulated by comparing the seminorms

1. There are $\lambda > 0$ and $p_0 \in \mathscr{P}$ so that $q \leq \lambda p_0$.
2. For every $p \in \mathscr{P}$, there is $\mu > 0$ such that $p \leq \mu q$.

Proposition 2.13 *The Fréchet space ω (see Example 2.14–3) does not admit a continuous norm. In particular, it is not normable.*

Proof Suppose $q : \omega \to [0, \infty[$ is a continuous norm. Then there exist n and $\lambda > 0$ so that

$$q(x) \leq \lambda \max\{|x_1|, \ldots, |x_n|\}$$

for every $x \in \omega$. Taking $x_0 = (0, \overset{n}{\ldots}, 0, 1, 0, \ldots) \in \omega$, this implies $q(x_0) = 0$, contradicting the fact that q is a norm. □

Proposition 2.14 *The Fréchet space $(C(\Omega), \tau_{co})$ does not admit a continuous norm. In particular, it is not normable.*

Proof Suppose $q : C(\Omega) \to [0, \infty[$ is a continuous norm. Then there are some compact set $K \Subset \Omega$ and some $\lambda > 0$ so that

$$q(f) \leq \lambda \sup_{x \in K} |f(x)|$$

for all $f \in C(\Omega)$. Let $x_0 \in \Omega \setminus K$ and choose $r > 0$ such that $\overline{B(x_0, r)} \subseteq \Omega \setminus K$. Defining $f_0 : \Omega \to \mathbb{K}$ by

$$f_0(x) = \begin{cases} \frac{r - \|x - x_0\|}{r} & \text{if } \|x - x_0\| \leq r \\ 0 & \text{otherwise,} \end{cases}$$

we clearly have $f_0 \in C(\Omega)$, $f_0(x_0) = 1$ and $f_0(x) = 0$ for every $x \in K$. Then

$$0 < q(f_0) \leq \lambda \sup_{x \in K} |f_0(x)| = 0,$$

and this is a contradiction. □

We finish this section with a condition for a space to be normable.

Theorem 2.24 (Kolmogorov) *A locally convex space E is normable if and only if there is a bounded* 0*-neighbourhood in X.*

Proof To begin with, if E is normable with a norm q, then the set

$$B(0,1) = \{x \in X : q(x) < 1\}$$

is a bounded 0-neighbourhood.

Suppose now that U is a bounded 0-neighbourhood in E. We can find $\varepsilon > 0$ and $p_0 \in \mathscr{P}$ so that $B_{p_0}(0, \varepsilon) \subseteq U$ and, then, $B_{p_0}(0, \varepsilon)$ is also bounded. Therefore, for each $p \in \mathscr{P}$, there exists $\lambda_p > 0$ such that

$$\sup_{x \in B_{p_0}(0,\varepsilon)} p(x) < \lambda_p \,,$$

and this implies

$$B_{p_0}(0, \varepsilon) \subseteq B_p(0, \lambda_p) \,.$$

But now Lemma 2.1 yields

$$p(x) \leq \frac{\lambda_p}{\varepsilon} p_0(x)$$

for all $x \in E$. Since this holds for every $p \in \mathscr{P}$, it gives $p_0(x) \neq 0$ for every $x \neq 0$ and p_0 is in fact a norm, which (see Remark 2.15) defines the topology of E. □

2.6 Two Theorems on Spaces of Continuous Functions

2.6.1 Stone–Weierstraß Theorem

Weierstraß proved in 1885 his famous approximation theorem: every continuous function on a closed, bounded interval can be uniformly approximated by polynomials. Later, in 1937, Stone extended this result in the context of unital subalgebras of the space of continuous functions, a result that now is known as the *Stone–Weierstraß theorem* (see Theorem 2.27 below for a precise statement). We deduce it here from a result by Bishop. The proof is based on a result by Machado, following an approach from [26].

Let us first of all fix our setting. We take a compact Hausdorff space K, and $C(K) = C(K, \mathbb{C})$ the Banach space of continuous, complex-valued functions (recall Example 2.7, we restrict ourselves for a moment to functions taking values in $\mathbb{C}$), and A a unital subalgebra of $C(K)$. For a given $f \in C(K)$ and $F \subseteq K$ we define

$$d_f(F) = \inf_{g \in A} \sup_{x \in F} |f(x) - g(x)| . \tag{2.29}$$

Clearly, if $F_1 \subseteq F_2$, then $d_f(F_1) \leq d_f(F_2)$.
A non-empty $S \subseteq K$ is said to be A-*antisymmetric* if for every $h \in A$ with $h(x) \in \mathbb{R}$ for every $x \in S$ we have that $h|_S$ is constant.

Theorem 2.25 (Machado) *Let $f \in C(K)$, then there exists a non-empty, closed, A-antisymmetric $S \subseteq K$ such that $d_f(S) = d_f(K)$.*

Proof Let $\mathcal{F}$ denote the class of all non-empty, closed subsets $F \subseteq K$ so that $d_f(F) = d_f(K)$. We order it by inclusion and our aim is to see (using Zorn's lemma, see e.g. [19, Chapter 0]) that it has a minimal element. In order to apply Zorn's lemma, we consider C, a totally ordered subclass, and let us show that it has a lower bound. Take

$$E = \bigcup \{F : F \in C\} ,$$

which is not empty and closed. Let us see that $d_f(E) = d_f(K)$. To this purpose, pick some $g \in A$ and, for each $F \in C$, consider the set

$$K(F, g) = \{x \in F : |f(x) - g(x)| \geq d_f(K)\} .$$

Since $d_f(F) = d_f(K)$, the very definition (2.29) gives that $K(F, g) \neq \varnothing$. It is clearly closed and then, being a subset of a compact Hausdorff space, is compact. Clearly, if $F_1 \subseteq F_2$ belongs to C, then $K(F_1, g) \subseteq K(F_2, g)$. This altogether yields

$$\bigcap_{F \in C} K(F, g) \neq \varnothing ,$$

and there is some $x_0 \in E$ so that $|f(x_0) - g(x_0)| \geq d_f(K)$; hence $\sup_{x \in E} |f(x) - g(x)| \geq d_f(K)$, and (since g was arbitrary) $d_f(E) \geq d_f(K)$. Then both are equal, $E \in \mathcal{F}$, and it is a lower bound of C. By Zorn's lemma, there is a minimal element of $\mathcal{F}$, which we denote by S. It is not empty and closed and satisfies that $d_f(S) = d_f(K)$. So, to complete the proof, it is only left to see that S is also A-antisymmetric. Suppose it is not, then there is $h \in A$ such that $h(x) \in \mathbb{R}$ for every $x \in S$ and that is not constant on S. Without loss of generality (taking if necessary a convenient linear combination of h and the constant function 1), we may assume $\min_{x \in S} h(x) = 0$ and $\max_{x \in S} h(x) = 1$. Consider now the sets

$$Y = \left\{x \in S : 0 \leq h(x) \leq \tfrac{2}{3}\right\} \text{ and } Z = \left\{x \in S : \tfrac{1}{3} \leq h(x) \leq 1\right\},$$

both not empty, closed, and proper subsets of S. By the minimality of S, neither of these two belongs to $\mathcal{F}$, and then $d_f(Y) < d_f(K)$ and $d_f(Z) < d_f(K)$. With this, there are $g_Y, g_Z \in A$ so that

$$\sup_{x\in Y} |f(x) - g_Y(x)| < d_f(K) \tag{2.30}$$

and

$$\sup_{x\in Z} |f(x) - g_Z(x)| < d_f(K) . \tag{2.31}$$

Now, for each $n \in \mathbb{N}$, we define

$$h_n = (1-h^n)^{2^n} \text{ and } g_n = h_n g_Y + (1-h_n) g_Z .$$

Since A is an algebra, $h_n, g_n \in A$ for every n. If we manage to find $n_0 \in \mathbb{N}$ so that

$$\sup_{x\in S} |f(x) - g_{n_0}(x)| < d_f(K) , \tag{2.32}$$

this will give $d_f(S) < d_f(K)$ and lead to a contradiction. Before we proceed, let us note that $0 \leq h_n \leq 1$ on S for every n and write $M = \sup_{x\in K} |g_Y(x) - g_Z(x)|$. Take some $x_0 \in S$, and let us consider three possible cases.
If $x_0 \in Y \cap Z$, then, using (2.30) and (2.31), we have, for each n,

$$|f(x_0)-g_n(x_0)| \leq h_n(x_0)|f(x_0)-g_Y(x_0)|+(1-h_n(x_0))|f(x_0)-g_Z(x_0)| < d_f(K) . \tag{2.33}$$

Suppose now that $x_0 \in Y \setminus Z$ (hence $0 \leq h(x) < \frac{1}{3}$), a simple computation shows that $(1-t)^N \geq 1 - Nt$ for every $N \in \mathbb{N}$ and $0 \leq t \leq 1$ and, then,

$$h_n(x_0) \geq 1 - 2^n h(x_0) \geq 1 - \left(\frac{2}{3}\right)^n .$$

With this, we have

$$|g_n(x_0) - g_Y(x_0)| = (1 - h_n(x_0))|g_Z(x_0) - g_Y(x_0)| < \left(\frac{2}{3}\right)^n M ,$$

for each n, and

$$|f(x_0) - g_n(x_0)| \leq \sup_{x\in Y} |f(x) - g_Y(x)| + \left(\frac{2}{3}\right)^n M . \tag{2.34}$$

Finally, suppose that $x_0 \in Z\backslash Y$ (that is, $\frac{2}{3} < h(x) \leq 1$). Again, simple computations show that $1 - t \leq \frac{1}{1+t}$, and $(1+t)^N \geq Nt$ for every $0 \leq t \leq 1$ and $N \in \mathbb{N}$; then

$$h_n(x_0) \leq \Big(\frac{1}{1 + h_n(x_0)^n}\Big)^{2^n} \leq \frac{1}{2^n h_n(x_0)^n} \leq \Big(\frac{3}{4}\Big)^n .$$

Then

$$|g_n(x_0) - g_Z(x_0)| = h_n(x_0)|g_Y(x_0) - g_Z(x_0)| < \Big(\frac{3}{4}\Big)^n M ,$$

and

$$|f(x_0) - g_n(x_0)| \leq \sup_{x \in Z} |f(x) - g_Z(x)| + \Big(\frac{3}{4}\Big)^n M . \tag{2.35}$$

To complete the proof we take $n_0 \in \mathbb{N}$ such that

$$\Big(\frac{2}{3}\Big)^{n_0} < \frac{d_f(K) - \sup_{x \in Y} |f(x) - g_Y(x)|}{M}$$

and

$$\Big(\frac{3}{4}\Big)^{n_0} < \frac{d_f(K) - \sup_{x \in Z} |f(x) - g_Y(x)|}{M}$$

(which, by (2.30) and (2.31) we may). Then, using (2.33), (2.34), and (2.35) we conclude that (2.32) holds. □

As a consequence, we have the following result.

Theorem 2.26 (Bishop) *Let A be a unital subalgebra of $C(K)$. If $f \in C(K)$ is such that for every A-antisymmetric S there is $g \in A$ so that $f|_S = g|_S$, then $f \in \overline{A}$.*

Proof Let f be as in the statement. By Theorem 2.25, there is some A-antisymmetric S such that $d_f(S) = d_f(K)$. Then there is $g \in A$ so that $f|_S = g|_S$, which yields $\sup_{x \in S} |f(x) - g(x)| = 0$ and then $0 = d_f(S) = d_f(K)$. Hence, for each $\varepsilon > 0$, there is some $g \in A$ so that

$$\sup_{x \in K} |f(x) - g(x)| < \varepsilon .$$

But this is exactly the norm on $C(K)$ and, therefore, $f \in \overline{A}$. □

We can use this result to prove the Stone–Weierstraß theorem. Let us recall that a subset of $C(K)$ *separates points* if for every $x, y \in K$ with $x \neq y$ there is a function f in the set with $f(x) \neq f(y)$.

Theorem 2.27 (Stone–Weierstraß) *Let A be subalgebra of $C(K)$ that separates points, contains the constant functions, and is closed under conjugation. Then A is dense in $C(K)$.*

Proof We want to see that the only A-antisymmetric subsets of K are those of the form $\{x\}$ with $x \in K$. Each one of these subsets is clearly A-antisymmetric (actually for any A), so let us see that these are the only possible ones. Suppose that S is A-antisymmetric and that $x \neq y$ belongs to S. Then there is $f \in A$ with $f(x) \neq f(y)$. Denoting $f_1 = \operatorname{Re} f$ and $f_2 = \operatorname{Im} f$ (note that the fact that A is closed under conjugation yields that both f_1 and f_2 are in A), we have that either $f_1(x) \neq f_1(y)$ or $f_2(x) \neq f_2(x)$, giving a real-valued function in A that is not constant on S (hence contradicting the fact that S is A-antisymmetric).
Now, if $f \in C(K)$ and S is any A-antisymmetric subset of K, then $S = \{x_0\}$ for some $x_0 \in K$ and it suffices to take the constant function equal to $f(x_0)$ (which is in A) and that coincides with f on S to apply Theorem 2.26 and conclude that $f \in \overline{A}$. □

The algebra of polynomials (of a fixed number of variables) clearly satisfies the hypothesis of the previous theorem. Then, as a straightforward consequence, we have the classical Weierstraß approximation theorem (in a version for several variables).

Corollary 2.1 (Weierstraß) *Let $K \subseteq \mathbb{C}^d$ be closed and bounded and $f : K \to \mathbb{C}$ be continuous. Then, for every $\varepsilon > 0$, there is a polynomial p of d variables such that*

$$\sup_{x \in K} |f(x) - p(x)| < \varepsilon .$$

2.6.2 Ascoli Theorem

Our aim now it to describe and characterize the relatively compact subsets of the space of continuous functions. So far we have only considered the space $C(\Omega)$, where $\Omega \subseteq \mathbb{R}^N$ is open. We take now a slightly more general point of view. A topological space X is said to be *locally compact* if each point has a basis of neighbourhoods consisting of compact sets. Clearly every open $\Omega \subseteq \mathbb{R}^N$ is locally compact. Along this section, X will always denote a locally compact Hausdorff topological space. Then we consider $C(X)$, the space of continuous functions $f : X \to \mathbb{K}$ and on it, the τ_{co} topology of uniform convergence on the compact sets of X, which can be defined by the family of seminorms (p_K) given in (2.24). We also consider the space $\mathbb{K}^X$ of all functions $f : X \longrightarrow K$ (obviously $C(X) \subseteq \mathbb{K}^X$), on which we consider the topology τ_p of pointwise convergence, defined by the family of seminorms

$$q_F(f) = \max_{x \in F} |f(x)| ,$$

where $F \subseteq X$ is finite. Since every finite set is compact, we have $\tau_p \subseteq \tau_{co}$. In other words, if $f_i \overset{\tau_{co}}{\to} f$, then $f_i \overset{\tau_p}{\to} f$ for every net $(f_i)_i$. We know from Examples 1.9 and 1.8 that the converse does not hold in general.

Remark 2.16 Equicontinuity (recall (2.19)) plays a fundamental rôle in the description of the relatively compact sets. Let us note that a family $H \subseteq C(X)$ is equicontinuous at $x \in X$ if and only if for every $\varepsilon > 0$ there exists a neighbourhood V of x such that $|f(y) - f(x)| < \varepsilon$ for every $y \in V$ and $f \in H$.
Also, $H \subseteq C(X)$ is τ_{co}-bounded if and only if

$$\sup_{f \in H} \sup_{x \in K} |f(x)| < \infty$$

for every $K \subseteq X$ compact.

Proposition 2.15 *If $H \subseteq C(X)$ is equicontinuous, then the topologies τ_{co} and τ_p coincide on H.*

Proof We only have to show that $\tau_{co}|_H \subseteq \tau_p|_H$, and to do that we have to check that each element of a basis of $\tau_{co}|_H$-neighbourhood contains a $\tau_p|_H$-neighbourhood. To do that, we fix $f_0 \in H$, take $K \subseteq X$ compact, and $\varepsilon > 0$ and consider the following τ_{co}-neighbourhood of f_0 in H,

$$V_{K,\varepsilon} = \{f \in H \colon \sup_{x \in K} |f(x) - f_0(x)| < \varepsilon\} .$$

Now, since H is equicontinuous, we can find a neighbourhood V_x of x (in X) such that

$$|f(y) - f(x)| < \frac{\varepsilon}{4} \text{ for all } f \in H, \text{ and all } y \in V_x .$$

Being K compact, there are $x_1, \dots, x_n \in K$ such that $K \subseteq \bigcup_{i=1}^n V_{x_i}$. Then we consider

$$W = \big\{f \in H : \max_{i=1,\dots,n} |f(x_i) - f_0(x_i)| < \varepsilon/2\big\} ,$$

which clearly is a τ_p-neighbourhood of f_0.
Take now $f \in W$ and $x \in K$; then there is some $i = 1, \dots, n$ with $x \in V_{x_i}$ and

$$|f(x) - f_0(x)| \leq |f(x) - f(x_i)| + |f(x_i) - f_0(x_i)| + |f_0(x_i) - f_0(x)| < \varepsilon .$$

This gives $W \subseteq V_{k,\varepsilon}$ and completes the proof. □

Proposition 2.16 *If $H \subseteq C(X)$ is equicontinuous, then the closure of H in $(\mathbb{K}^X, \tau_p)$ is contained in $C(X)$ and it is also equicontinuous.*

Proof Fix $x \in X$ and $\varepsilon > 0$. Since H is equicontinuous, there is a neighbourhood V_x of x in X such that

$$|f(y) - f(x)| < \frac{\varepsilon}{3},$$

for each $f \in H$ and each $y \in V_x$. We write $\overline{H}$ for the closure of H in $(\mathbb{K}^X, \tau_p)$, and take $g \in \overline{H}$ and $y \in V_x$. Then there is $f \in H$ (which depends on x and y) such that $q_{\{x,y\}}(f - g) < \varepsilon/3$. Therefore

$$|g(y) - g(x)| \leq |g(y) - f(y)| + |f(y) - f(x)| + |f(x) - g(x)| < \varepsilon.$$

This shows that $\overline{H}$ is equicontinuous (hence g is continuous). □

Note that as a consequence we have that if $(f_n)_n$ is equicontinuous and converges pointwise, then its limit is also continuous. As we know, this is in general not true (see Example 1.5). We are now ready to give a description of the relatively compact sets of $C(X)$.

Theorem 2.28 (Ascoli) *Let X be a locally compact Hasudorff topological space. A subset $H \subseteq C(X)$ is relatively compact for τ_{co} if and only if*

1. *H is equicontinuous.*
2. *$H(x) = \{f(x) : f \in H\}$ is bounded in $\mathbb{K}$ for each $x \in X$.*

Proof We begin by assuming that the two statements hold and we want to see that $\overline{H}$ (the τ_{co}-closure) is τ_{co}-compact. First of all, by Proposition 2.15, the τ_{co}-closure and the τ_p-closure coincide. Then $\overline{H}$ is, by Proposition 2.16, equicontinuous and both topologies again coincide. So, it suffices to check that H is τ_p-compact.
We consider now, for each $x \in X$, the set $\overline{H}(x) = \{g(x) : g \in \overline{H}\} \subseteq \mathbb{K}$, that is bounded because of 2 and contained in $\overline{H(x)}^{\mathbb{K}}$. Then $\overline{H} \subseteq \prod_{x \in X} \overline{H(x)} \subseteq \mathbb{K}^X$ and this is compact (by Tikhonov's theorem 2.1). Hence, $\overline{H}$ is compact.
Let us conversely assume that H is relatively compact in $(C(X), \tau_{co})$ and see that the two statements hold. First of all, H is τ_p-bounded. This gives that $H(x)$ is bounded for every $x \in X$ since the map $(C(X), \tau_p) \to \mathbb{K}$ given by $f \rightsquigarrow f(x)$ is continuous. We see now that it is also equicontinuous. We fix $x \in X$ and $\varepsilon > 0$. We fix K, a compact neighbourhood of x and, since H is relatively compact, choose $f_1, \ldots, f_n \in H$ such that

$$H \subseteq B_{p_K}(f_1, \varepsilon/3) \cup \ldots \cup B_{p_K}(f_n, \varepsilon/3).$$

That is, for each $f \in H$, there is $j = 1, \ldots, n$ so that

$$|f(x) - f_j(x)| < \frac{\varepsilon}{3}$$

For every $x \in K$. Now, each f_i is continuous, and thus there is U_i a neighbourhood of x such that

$$|f_i(y) - f_i(x)| < \frac{\varepsilon}{3}$$

for every $y \in U_i$. We let now $V = K \cap U_1 \cap \cdots \cap U_n$ (which is a neighbourhood of x), and, for every $y \in V$, we have

$$|f(y) - f(x)| \leq |f(y) - f_j(y)| + |f_j(y) - f_j(x)| + |f_j(x) - f(x)| < \varepsilon\,.$$

Therefore, H is equicontinuous. □

2.7 A Short Introduction to Hilbert Spaces

We close this chapter with a very succinct introduction to Hilbert spaces and the L_2-theory of Fourier series. We state some of the most basic facts without proof. They are rather straightforward and may be left as an exercise to the reader.

A complex vector space H is called an *inner product space* (or *unitary space*) if there is a function $(\cdot, \cdot) : H \times H \to \mathbb{C}$ called *inner product* (or *scalar product*) such that, for every $x, y, z \in H$ and all $\alpha \in \mathbb{C}$, we have

(IP1) $(x + y, z) = (x, z) + (y, z)$.
(IP2) $(\alpha x, y) = \alpha(x, y)$.
(IP3) $(y, x) = \overline{(x, y)}$ (the complex conjugate).
(IP4) $(x, x) \geq 0$.
(IP5) $(x, x) = 0$ if and only if $x = 0$.

These conditions imply

$$(0, y) = 0 \text{ for every } y \in H$$

and

$$(x, \alpha y) = \overline{\alpha}(x, y) \text{ for every } x, y \in H \text{ and } \alpha \in \mathbb{C}\,.$$

The *norm* of $x \in H$ is defined as

$$\|x\| = (x, x)^{1/2}$$

(the non-negative square root). With this definition, H becomes a normed space. Note that $\|x\|^2 = (x, x)$, and it is straightforward to check that $(\cdot, \cdot)$ is continuous on $H \times H$.

Proposition 2.17 *Let H be an inner space and $x, y \in H$. Then the following are satisfied:*
Schwarz inequality:

$$|(x, y)| \leq \|x\| \, \|y\| \,. \tag{2.36}$$

Parallelogram identity:

$$\|x + y\|^2 + \|x - y\|^2 = 2\big(\|x\|^2 + \|y\|^2\big)\,. \tag{2.37}$$

Polarization formula:

$$(x, y) = \frac{1}{4}\Big(\|x + y\|^2 - \|x - y\|^2 + i\big(\|x + iy\|^2 - \|x - iy\|^2\big)\Big)\,.$$

Example 2.29

1. For $N \in \mathbb{N}$, we have the inner space $H = \mathbb{C}^N$ with the inner product defined by

$$(x, y) = \sum_{i=1}^{N} x_i \overline{y_i}\,.$$

   Note that the norm defined by this inner product is nothing but the $\|\cdot\|_2$ from Example 2.3.
2. On $H = \ell_2$ (recall Example 2.5), we define the inner product by

$$(x, y) = \sum_{i=1}^{\infty} x_i \overline{y_i}\,.$$

   Again, the norm defined by the inner product is the one that we already introduced in (2.8) for $p = 2$.
3. Take now $H = L_2(-\pi, \pi)$, the space of all 2π-periodic functions $f : \mathbb{R} \to \mathbb{C}$ such that $|f|^2$ is integrable on $]-\pi, \pi]$ (that is, $|f|^2 \in L_1(-\pi, \pi)$, recall Example 2.6). If $f, g \in L_2(-\pi, \pi)$, then, by Hölder's inequality (2.10), the function $f\overline{g}$ is integrable on $]-\pi, \pi]$ and so

$$(f, g) = \frac{1}{2\pi}\int_{-\pi}^{\pi} f(x)\overline{g(x)}dx$$

   defines an inner product on $L_2(-\pi, \pi)$. Note that, with this notation, the functions defined by

$$\psi_n(x) = e^{inx}$$

for $n \in \mathbb{N}$ satisfy

$$(\psi_n, \psi_m) = 0 \text{ if } n \neq m \text{ and } (\psi_n, \psi_n) = 1$$

(recall Sect. 1.3.2). Note also that $\overline{\psi_n(x)} = e^{-inx}$.

□

An inner product space such that the normed space $(H, \|\cdot\|)$ is complete is called a *Hilbert space*. Two elements x and y in a Hilbert space are said to be *orthogonal* if

$$(x, y) = 0.$$

In this case, we write $x \perp y$.

Proposition 2.18 (Pythagorean theorem) *If $x_1, \ldots, x_N$ are pairwise orthogonal vectors in a Hilbert space, then*

$$\|x_1 + \cdots + x_N\|^2 = \|x_1\|^2 + \cdots + \|x_N\|^2.$$

Moreover, if all vectors are non-zero, then $x_1, \ldots, x_N$ are linearly independent.

Given a non-empty set A of a Hilbert space H, the *orthogonal* of A is defined as

$$A^{\perp} = \{y \in H : (x, y) = 0 \text{ for every } x \in A\}.$$

Proposition 2.19 *If H is a Hilbert space and $\varnothing \neq A \subseteq H$, then $A^{\perp}$ is a closed subspace of H.*

Proof It is easy to check that it is a vector subspace. Let us see that it is also closed. Suppose that $(y_n)_n \subseteq A^{\perp}$ converges to some $y \in H$. Then, for each $x \in A$, we have

$$|(x, y)| = |(x, y - y_n) + (x, y_n)| = |(x, y - y_n)| \leq \|x\| \, \|y - y_n\|.$$

Since $y_n \to y$, this gives $(x, y) = 0$ and completes the proof. □

Theorem 2.30 *Let H be a Hilbert space and $\varnothing \neq A \subseteq H$ closed and convex. Then there is a unique $x_0 \in A$ such that $\|x_0\| \leq \|x\|$ for every $x \in A$.*

Proof Let $\delta = \inf_{x \in A} \|x\|$. For each $x, y \in A$, we have, by (2.37),

$$\frac{1}{4}\|x - y\|^2 = \frac{1}{2}\|x\|^2 + \frac{1}{2}\|y\|^2 - \left\|\frac{x + y}{2}\right\|^2.$$

Since A is convex, $\frac{x+y}{2} \in A$, and

$$\|x - y\|^2 \leq 2\|x\|^2 + 2\|y\|^2 - 4\delta^2.$$

If $\|x\| = \|y\| = \delta$, then $x = y$ and the statement (including the uniqueness) follows. Take now $(y_n)_n \subseteq A$ such that $\lim_n \|y_n\| = \delta$. Given $n, m \in \mathbb{N}$, we have

$$\|y_n - y_m\|^2 \leq 2\|y_n\|^2 + 2\|y_m\|^2 - 4\delta^2 .$$

Note that the right-hand terms tend to 0 as $n, m \to \infty$. This gives that $(y_n)_n$ is a Cauchy sequence and, then, converges to some $x_0 \in H$. Since A is closed, we have $x_0 \in A$, and $\|x_0\| = \lim_n \|y_n\| = \delta$. This completes the proof. □

Theorem 2.31 *Let H be a Hilbert space and M a closed subspace. Then there are unique maps $P : H \to M$ and $Q : H \to M^{\perp}$ so that*

$$x = Px + Qx \tag{2.38}$$

for every $x \in H$. Furthermore

1. *If $x \in M$, then $Px = x$ and $Qx = 0$. If $x \in M^{\perp}$, then $Px = 0$ and $Qx = x$.*
2. *$\|x - Px\| = \inf_{y \in M} \|x - y\|$ for every $x \in H$.*
3. *$\|x\|^2 = \|Px\|^2 + \|Qx\|^2$.*
4. *P and Q are linear mappings.*

Proof Given any $x \in H$, the set $x + M = \{x + y \colon y \in M\}$ is closed and convex. Then, by Theorem 2.30, there is a unique $Qx \in x + M$ such that $\|Qx\| \leq \|x + y\|$ for every $y \in M$. Taking $Px = x - Qx$ defines a mapping $P : H \to M$ and clearly $x = Px + Qx$. Let us see now that $Qx \in M^{\perp}$. Take $y \in M$ with $\|y\| = 1$. Then, for each $\alpha \in \mathbb{C}$, we have

$$\begin{aligned}(Qx, Qx) = \|Qx\|^2 \leq \|Qx - \alpha y\|^2 &= (Qx - \alpha y, Qx - \alpha y) \\ &= (Qx, Qx) - (\alpha y, Qx) - (Qx, \alpha y) + (\alpha y, \alpha y) ,\end{aligned}$$

hence

$$0 \leq |\alpha|^2 - \alpha(y, Qx) - \overline{\alpha}(Qx, y) .$$

Taking $\alpha = (Qx, y)$, we get $0 \leq -|(Qx, y)|^2$, which implies $(Qx, y) = 0$. Now, for an arbitrary $0 \neq y \in M$, we have

$$(Qx, y) = \|y\|\big(Qx, \tfrac{1}{\|y\|} y\big) = 0 ,$$

showing that indeed $Qx \in M^{\perp}$.
So, we have defined mappings P and Q satisfying (2.38). Let us see that they are unique. If $x = x_1 + x_2$ with $x_1 \in M$ and $x_2 \in M^{\perp}$, then $x_1 - Px = Qx - x_2$ and, since $M \cap M^{\perp} = \{0\}$, we have $x_1 = Px$ and $x_2 = Qx$, showing the uniqueness of the mappings. Let us see that the remaining statements are also satisfied. To begin with, 1 is an immediate consequence of the uniqueness of the decomposition

(2.38), 2 follows from the way we have defined P, and 3 is a straightforward consequence of Proposition 2.18. It is only left to see that both P and Q are linear. Take $x, y \in H$ and $\alpha, \beta \in \mathbb{C}$, Then, using the decomposition (2.38), we have

$$0 = (\alpha x+\beta y)-\alpha x-\beta y = P(\alpha x+\beta y)+Q(\alpha x+\beta y)-\alpha Px-\alpha Qx-\beta Py-\beta Qy,$$

and therefore

$$P(\alpha x + \beta y) - \alpha Px - \beta Py = \alpha Qx + \beta QyQ(\alpha x + \beta y) .$$

Note that the left-hand side belongs to M and the right-hand side to $M^\perp$, which implies that both are 0. □

The mapping P in the previous theorem is called the *orthogonal projection* of H onto M.

Corollary 2.2 *Let H be a Hilbert space and M a closed proper subspace; then $M^\perp \neq \{0\}$ and $(M^\perp)^\perp = M$.*

Proof The first statement is a straightforward consequence of Theorem 2.31. For the second one, let us note first that obviously $M \subseteq (M^\perp)^\perp$. To complete the proof, suppose that there is $x \in (M^\perp)^\perp \setminus M$. By (2.38), there are $Px \in M$ and $Qx \in M^\perp$ with $x = Px + Qx$, where $Qx \neq 0$, since $x \notin M$. Then, since $x \in (M^\perp)^\perp$, we get

$$0 = (x, Qx) = (Px + Qx, Qx) = (Px, Qx) + (Qx, Qx) = \|Qx\|^2 ,$$

and $Qx = 0$. This is a contradiction and completes the proof. □

Theorem 2.32 (Riesz representation theorem) *Let H be a Hilbert space and $L : H \to \mathbb{C}$ be a continuous linear functional. Then, there is a unique $y \in H$ such that $Lx = (x, y)$ for every $x \in H$.*

Proof For each fixed $y \in H$, we consider the mapping $u_y : H \to \mathbb{C}$ given by $u_y(x) = (x, y)$. This is clearly linear and, by (2.36), it is continuous.
Take now some $L : H \to \mathbb{C}$ linear and continuous. If $Lx = 0$ for every $x \in H$, we just have to take $y = 0$ to have the conclusion. Otherwise, consider

$$M = \ker L = \{x \in H : Lx = 0\} ,$$

which is a closed proper subspace of H. By Theorem 2.31 (see also Corollary 2.2), there is $z \in M^\perp$ with $\|z\| = 1$, and

$$L\big((Lx)z - (Lz)x\big) = (Lx)(Lz) - (Lz)(Lx) = 0$$

for every $x \in H$. In other words, $(Lx)z - (Lz)x \in M$ for every $x \in H$ and, then, since $z \in M^{\perp}$, we have

$$0 = \big((Lx)z - (Lz)x, z\big) = (Lx)\|z\|^2 - (Lz)(x, z)$$
$$= Lx - (Lz)(x, z) = Lx - \big(x, \overline{(Lz)}z\big).$$

Taking $y = \overline{(Lz)}z$ gives the conclusion.
To show that y is unique, suppose that $y_1 \in H$ is such that $(x, y) = (x, y_1)$ for every $x \in H$. Then $(x, y - y_1) = 0$ for every x, and $\|y - y_1\|^2 = (y - y_1, y - y_1) = 0$. This gives $y = y_1$ and completes the proof. □

Remark 2.17 Let us point out that, according to the previous result, for each $L \in H'$ (the dual space of H), there is $y \in H$ so that $L = u_y$. In other words, H' can be identified with H (as sets). From (2.36) and the fact that $(y, y) = \|y\| \times \|y\|$, we have that $\|u_y\| = \|y\|$, which shows that they can be identified as Banach spaces. But we can say more, and we may define an inner product in H' by doing $(u_y, u_x) = (x, y)$. With this, we even have $H' = H$ as Hilbert spaces.

A set of vectors $\{u_i : i \in I\}$ is said to be *orthonormal* if $(u_i, u_j) = 0$ for every $i \neq j$ and $\|u_i\| = 1$ for every $i \in I$. The following properties follow easily from the definitions.

Proposition 2.20

1. *Every orthonormal set is linearly independent.*
2. *If* $\{u_1, \ldots, u_n\}$ *is orthonormal and* $x = \sum_{i=1}^{n} c_i u_i$, *with* $c_1, \ldots, c_n \in \mathbb{C}$, *then* $c_i = (x, u_i)$ *for each* $i = 1, \ldots, n$ *and*

$$\|x\|^2 = \sum_{i=1}^{n} |(x, u_i)|^2 .$$

We can now prove the *best approximation theorem*

Theorem 2.33 *Let* H *be a Hilbert space and* $\{u_1, \ldots, u_n\} \subseteq H$ *be an orthonormal set. If* M *is the linear span of* $\{u_1, \ldots, u_n\}$, *then the orthogonal projection of* H *onto* M *is given by*

$$Px = \sum_{i=1}^{n} (x, u_i) u_i .$$

In particular,

$$\Big\| x - \sum_{i=1}^{n} (x, u_i) u_i \Big\| \leq \Big\| x - \sum_{i=1}^{n} b_i u_i \Big\| \tag{2.39}$$

for every $b_1, \ldots, b_n \in \mathbb{C}$*, and there is equality in* (2.39) *if and only if* $b_i = (x, u_i)$ *for every* i*. Moreover,*

$$\sum_{i=0}^{n} |(x, u_i)|^2 = \|x\|^2 - d(x, M)^2 , \tag{2.40}$$

for every $x \in H$*.*

Proof Given $x \in H$, set $y(x) = \sum_{i=1}^{n} (x, u_i)u_i \in M$ and note that

$$(y(x), u_j) = \sum_{i=1}^{n} (x, u_i)(u_j, u_i) = (x, u_j)$$

for each $j = 1, \ldots, n$, which implies that $x - y(x) \in M^{\perp}$. Since M is closed (see Remark 3.4), it follows that $y(x)$ is the unique element in M for which $x - y(x) \in M^{\perp}$. Therefore $y(x) = Px$ by Theorem 2.31.
Again, Theorem 2.31–2 gives (2.39), and the equality holds by the uniqueness of Px attaining the distance. Finally, if $x \in H$, let us write $\delta = d(x, M)$ and note that Theorem 2.31–3 gives

$$\|x\|^2 = \|Px\|^2 + \|x - Px\|^2 = \|Px\|^2 + \delta^2 .$$

On the other hand, by Proposition 2.20, we have

$$\|Px\|^2 = \sum_{i=0}^{n} |(x, u_i)|^2 .$$

From this, (2.40) clearly follows. □

Theorem 2.34 *Let* H *be a Hilbert space and* $\{u_k : k \in \mathbb{N}\}$ *be an orthonormal set. For each* $x \in H$*, we have*

$$\sum_{k=1}^{\infty} |(x, u_k)|^2 \leq \|x\|^2 \tag{2.41}$$

(this is known as the Bessel inequality). Furthermore,

$$\sum_{k=1}^{\infty} |(x, u_k)|^2 = \|x\|^2$$

if and only if

$$\lim_{n\to\infty}\left\|x-\sum_{k=1}^{n}(x,u_k)u_k\right\|=0\,.$$

Proof By Theorem 2.33, for each $x \in H$ and $n \in \mathbb{N}$, we have

$$0\le\left\|x-\sum_{k=1}^{n}(x,u_k)u_k\right\|^2=\|x\|^2-\sum_{k=1}^{n}|(x,u_k)|^2\,. \tag{2.42}$$

Hence

$$\sum_{k=1}^{n}|(x,u_k)|^2\le\|x\|^2\,,$$

for every n. This implies that the series converges and (2.41) holds. The remaining statement about the equality is a straightforward consequence of (2.42). □

Theorem 2.35 (Riesz–Fischer) *Let $(c_k)_k \subseteq \mathbb{C}$ be such that $\sum_{k=1}^{\infty}|c_k|^2<\infty$ and $\{u_k\colon k\in\mathbb{N}\}$ an orthonormal sequence in a Hilbert space H. Then there is $x \in H$ so that*

$$c_k=(x,u_k)\text{ for every }k\in\mathbb{N}$$

and

$$\|x\|^2=\sum_{k=1}^{\infty}|c_k|^2\,.$$

Proof For each $n \in \mathbb{N}$, define $s_n=\sum_{k=1}^{n}c_ku_k$. Let us see first that $(s_n)_n$ is a Cauchy sequence. If $n > m$, we have (recall Proposition 2.20)

$$\|s_n-s_m\|^2=\left\|\sum_{k=m+1}^{n}c_ku_k\right\|^2=\sum_{k=m+1}^{n}|c_k|^2\,,$$

and, since $\sum_{k=1}^{\infty}|c_k|^2<\infty$, this yields our claim. Being H complete, there is $x \in H$ such that $\lim_n\|x-s_n\|=0$. Then, by Theorem 2.34, it is enough to show that $(x,u_k)=c_k$ for every k. Fix, then, some k and note that if $n > k$, then $(s_k,u_k)=c_k$. Therefore

$$\left|c_k-(x,u_k)\right|=|(s_n-x,u_k)|\le\|s_n-x\|\,\|u_k\|=\|s_n-x\|\,.$$

Letting $n\to\infty$ yields $(x,u_k)=c_k$ and completes the proof. □

Theorem 2.36 *Let H be a Hilbert space and $\{u_k \colon k \in \mathbb{N}\}$ an orthonormal sequence. The following statements are equivalent:*

1. *If $x \in H$ is such that $(x, u_k) = 0$ for every k, then $x = 0$.*
2. *For each $x \in H$, we have*

$$\|x\|^2 = \sum_{k=1}^{\infty} |(x, u_k)|^2 .$$

3. *For each $x \in H$, we have*

$$x = \sum_{k=1}^{\infty} (x, u_k) u_k .$$

Proof The statements in 2 and 3 are equivalent by Theorem 2.34. Since 2 clearly implies 1, it is only left to see that this implies 3. Given $x \in H$, we know from (2.41) that $\sum_{k=1}^{\infty} |(x, u_k)|^2 \leq \|x\|^2$. Within the proof of Theorem 2.35, we have shown that there is $y \in H$ with $y = \sum_{k=1}^{\infty} (x, u_k) u_k$ and such that $(x, u_k) = (y, u_k)$ for every k. Then $(x - y, u_k) = 0$ for every k. Since we are assuming that 1 holds, this implies that $x = y$ and completes the proof. □

An orthonormal sequence in a Hilbert space satisfying any of the equivalent conditions 1–2–3 in Theorem 2.36 is called *complete*.

Corollary 2.3 *Let H be a Hilbert space and $\{u_k : k \in \mathbb{N}\}$ an orthonormal sequence. If $M = \operatorname{span}\{u_k : k \in \mathbb{N}\}$ is dense in H, then the sequence is complete.*

Proof Suppose that $x \in H$ is such that $(x, u_k) = 0$ for every k. Then $(x, y) = 0$ for every $y \in M$ and, by density $(x, z) = 0$ for every $z \in H$. In particular $\|x\|^2 = (x, x) = 0$, and, then $x = 0$. So, the sequence $(u_k)_k$ is complete. □

Proposition 2.21 *Let H be a Hilbert space and $\{u_k : k \in \mathbb{N}\}$ a complete orthonormal sequence. If $x, y \in H$, then*

$$(x, y) = \sum_{k=1}^{\infty} (x, u_k) \overline{(y, u_k)} .$$

Proof Since the sequence is complete, we have

$$x = \sum_{k=1}^{\infty} (x, u_k) u_k \text{ and } y = \sum_{j=1}^{\infty} (y, u_j) u_j .$$

Then, by the continuity of the scalar product $(\cdot,\cdot)$ and the orthonormality of the sequence $(u_k)_k$, we have

$$(x,y)=\sum_{k=1}^{\infty}\sum_{j=1}^{\infty}(x,u_k)\overline{(y,u_j)}(u_k,u_j)=\sum_{k=1}^{\infty}(x,u_k)\overline{(y,u_k)}\,.$$

□

We can look again with a new point of view at the problem of the convergence of Fourier series that we already tackled in Sect. 1.3.2. For a given function f, recall the definition of the Fourier coefficient $\widehat{f}(n)$ (with $n\in\mathbb{Z}$) given in (1.6).

Theorem 2.37 *The orthonormal sequence $\{\psi_n(x)=e^{inx} : n\in\mathbb{Z}\}$ is complete in $L_2(-\pi,\pi)$. In particular, for each $f\in L_2(-\pi,\pi)$, we have*

$$\sum_{n=-\infty}^{\infty}|\widehat{f}(n)|^2=\frac{1}{2\pi}\int_{-\pi}^{\pi}|f(x)|^2dx \tag{2.43}$$

(this is known as Parseval identity) and

$$\lim_{n\to\infty}\left\|f(x)-\sum_{k=-n}^{n}\widehat{f}(k)e^{ikx}\right\|_{L_2(-\pi,\pi)}=0\,. \tag{2.44}$$

Proof In view of Corollary 2.3, in order to show that the sequence is complete, it is enough to see that the span of $\{\psi_n : n\in\mathbb{Z}\}$ (that is, the space of trigonometric polynomials) is dense in $L_2(-\pi,\pi)$. To begin with, the space $C(-\pi,\pi)$ is dense in $L_2(-\pi,\pi)$ (see, e.g., [28, Theorem 8.14]) and moreover $\|g\|_{L_2}\le\|g\|=\sup_{x\in],\pi,\pi]}|g(x)|$ for every $g\in C(-\pi,\pi)$. Therefore, it is enough to show that the trigonometric polynomials are dense in $C(-\pi,\pi)$. But this is an immediate consequence of Féjer's theorem (see Corollary 1.2) or of the Stone–Weierstraß theorem 2.27, once we identify $C(-\pi,\pi)$ with $C(\mathbb{T})$, the space of continuous functions on the unit circle $\mathbb{T}$ (recall the paragraph right before (1.14)). Hence, $\{\psi_n : n\in\mathbb{Z}\}$ is a complete orthogonal sequence.
Once we have this, (2.43) follows from Theorem 2.36–2 and (2.44) from Theorem 2.36–3. □

Corollary 2.4 *The map $T : L_2(-\pi,\pi)\to\ell_2(\mathbb{Z})$ given by $T(f)=(\widehat{f}(n))_{n\in\mathbb{Z}}$ is an isometric isomorphism.*

Proof The map T is clearly linear and, by (2.43), satisfies $\|T(f)\|^2=\|f\|$ for every $f\in L_2(-\pi,\pi)$. This implies, in particular, that it is injective. Finally, the Riesz–Fischer theorem 2.35 shows that T is also surjective. □

2.8 Exercises

2.1 Let $(x_n)_n$ be a sequence in a Hausdorff locally convex space such that $\lim_n x_n = x$. Prove that

$$\frac{1}{n}(x_1 + x_2 + \cdots + x_n)$$

also converges to x as $n \to \infty$.

2.2 Show that a set A in a vector space is absolutely convex if and only if $\lambda x + \mu y \in B$ for every $x, y \in B$ and $\lambda, \mu \in \mathbb{K}$ with $|\lambda| + |\mu| \leq 1$. Deduce that (2.15) holds.

2.3 Let X be a metric space, $x \in X$, and $A \subseteq X$. Show that $x \in \overline{A}$ if and only if $d(x, A) = 0$.

2.4 Let $(E, \mathcal{P})$ be a metrizable locally convex space. Show that for each sequence $(A_n)_n$ of bounded subsets in E there is a sequence of positive numbers $(\lambda_n)_n$ such that $\bigcup_{n=1}^{\infty} \lambda_n A_n$ is bounded.

2.5 A sequence $(B_n)_n$ of bounded sets in a locally convex space E is called a *fundamental sequence of bounded sets* if $B_n \subseteq B_{n+1}$ for every n and for each bounded set B in E there is n such that $B \subseteq B_n$. Show that a metrizable locally convex space with a fundamental sequence of bounded sets is normable.

2.6 Show that if $1 \leq p < q \leq \infty$, then $\ell_p \subseteq \ell_q$ with continuous inclusion, but they do not coincide. Show that the following inclusions are strict:

$$\bigcup_{1 \leq q < p} \ell_q \subseteq \ell_p \subseteq \bigcap_{p < q} \ell_q .$$

2.7 Let $(E, \mathscr{P})$ be a locally convex space and $B \subseteq E$. Show that B is bounded if and only if for all $(x_n)_n \subseteq B$ and all $(\lambda_n)_n \subseteq \mathbb{K}$ with $\lambda_n \to 0$ we have $\lambda_n x_n \to 0$ in E.

2.8 Let E and F be Fréchet spaces. Show that $E \times F$ is a Fréchet space.

2.9 Let X and Y be two Banach spaces. Show that $L(X, Y)$ (with the norm defined in Remark 2.9) is complete.

2.10 Let X and Y be two Banach spaces and $K(X, Y)$ be the subspace of $L(X, Y)$ consisting of operators such that $T(B_X)$ is relatively compact in Y (called *compact* operators). Show that $K(X, Y)$ is closed in $L(X, Y)$.

2.11 Let X and Y be normed spaces with open unit balls B_X and B_Y, respectively. Prove that if $T : X \to Y$ is linear, the following statements are equivalent:

1. T is open and surjective.
2. There is $\delta > 0$ such that $\delta B_Y \subseteq T(B_X)$.

3. There is $M > 0$ such that for each $y \in Y$ there is $x \in X$ with $Tx = y$ and $\|x\| \leq M\|y\|$.

2.12 Show that a sequence converges in ω if and only if converges coordinatewise; that is, a sequence $(x^{(n)})_n$ (where $x^{(n)} = (x^{(n)})_k$ for every n) converges to $x = (x_k)_k$ in ω if and only if $x_k^{(n)} \to x_k$ as $n \to \infty$ (in $\mathbb{C}$) for every k. Deduce that the space ω is complete.

2.13 Show that $(c_0)' = \ell_1$ (recall the definitions from Example 2.5).

2.14 Let c be the Banach space of convergent sequences in $\mathbb{C}$ (with the $\|\cdot\|_\infty$ norm).

1. Prove that $u : c \to \mathbb{C}$ given by $u(x) = \lim_n x_n$ is a continuous linear form.
2. Show that $T : c \to c_0$ given by $T(x) = (u(x), x_1 - u(x), x_2 - u(x), \dots)$ is an onto isomorphism such that $\|T\| = \|T^{-1}\| = 2$.

2.15 Consider the space

$$\varphi = \{x \in \mathbb{K}^{\mathbb{N}} : \text{card}\{i : x_i \neq 0\} < \infty\}.$$

Show that $\omega' = \varphi$.

2.16 Let $A = (a_n)_n$ be a Köthe matrix, $a_n = (a_n(i))_{i=1}^\infty$ and $\alpha = (\alpha_n)_{n=1}^\infty$ a sequence of scalars.

1. Show that the diagonal operator $T_\alpha : \lambda_1(A) \to \lambda_1(A)$ given by $T_\alpha(x) = (\alpha_i x_i)_i$ is continuous if and only if for every n there is $m \geq n$ such that

$$\sup_i \frac{a_n(i)|\alpha_i|}{a_m(i)} < +\infty.$$

2. Characterise those αs for which the operator $S_\alpha a : \lambda_1(A) \to \lambda_1(A)$ given by

$$S_\alpha(x_1, \dots, x_n, \dots) = (\alpha_2 x_2, \alpha_3 x_3, \dots, \alpha_n x_n, \alpha)$$

is continuous.

2.17 Characterise the bounded subsets of $\lambda_p(A)$ for $1 < p < \infty$.

2.18 Let $A = (a_n)_n$ be a Köthe matrix, $a_n = (a_n(i))_{i=1}^\infty$.

1. Show that $\lambda_1(A)$ admits a continuous norm if and only if there exists $n_0 \in \mathbb{N}$ such that $a_{n_0}(i) > 0$ for every i.
2. Show that $\lambda_1(A)$ is normable if and only if there exists $n_0 \in \mathbb{N}$ such that for every $n \geq n_0$ there is $C_n > 0$ with $a_n(i) \leq C_n a_{n_0}(i)$ for every i.

2.19 Give an example of a Köthe matrix A for which $\lambda_1(A)$ is not normable.

2.20 Given a Köthe matrix with positive elements, define, for $1 < q < \infty$ the space $k_q(A)$, and prove that $\lambda_p(A)' = k_{p'}(A)$.

2.21 Show that the space $C(\Omega)$ of continuous functions in the open set $\Omega \subset \mathbb{R}^n$ contains bounded sets that for the compact-open topology are not compact. Give explicit examples.

2.22 With the notation from Example 2.19, prove that if A is bounded in $(C(\Omega), \tau_{co})$, then there is a continuous function $v : \Omega \to [0, +\infty[$ continuous such that $A \subseteq A_v$.
Hint: Use partitions of unity.

Further Reading
[1–12, 14–25, 27, 29–36]

References

1. Bachman, G., Narici, L.: Functional Analysis. Dover Publications, Inc., Mineola, NY (2000). Reprint of the 1966 original
2. Banach, S.: Théorie des Opérations Linéaires. Éditions Jacques Gabay, Sceaux (1993). Reprint of the 1932 original
3. Bogachev, V.I., Smolyanov, O.G.: Topological Vector Spaces and Their Applications. Springer Monographs in Mathematics. Springer, Cham (2017). https://doi.org/10.1007/978-3-319-57117-1
4. Brezis, H.: Functional analysis. In: Sobolev Spaces and Partial Differential Equations. Universitext. Springer, New York (2011)
5. Cerdà, J.: Linear functional analysis. In: Graduate Studies in Mathematics, vol. 116. American Mathematical Society, Providence, RI; Real Sociedad Matemática Española, Madrid (2010). https://doi.org/10.1090/gsm/116
6. Conway, J.B.: A course in functional analysis. In: Graduate Texts in Mathematics, vol. 96. Springer, New York (1985). https://doi.org/10.1007/978-1-4757-3828-5
7. Diestel, J.: Sequences and series in Banach spaces. In: Graduate Texts in Mathematics, vol. 92. Springer, New York (1984). https://doi.org/10.1007/978-1-4612-5200-9
8. Edwards, R.E.: Functional Analysis. Dover Publications, Inc., New York (1995). Theory and applications, Corrected reprint of the 1965 original
9. Fabian, M., Habala, P., Hájek, P., Montesinos, V., Zizler, V.: Banach Space Theory. CMS Books in Mathematics/Ouvrages de Mathématiques de la SMC. Springer, New York (2011). https://doi.org/10.1007/978-1-4419-7515-7
10. Fabian, M., Habala, P., Hájek, P., Montesinos Santalucía, V., Pelant, J., Zizler, V.: Functional analysis and infinite-dimensional geometry. In: CMS Books in Mathematics/Ouvrages de Mathématiques de la SMC, vol. 8. Springer, New York (2001). https://doi.org/10.1007/978-1-4757-3480-5
11. Floret, K., Wloka, J.: Einführung in die Theorie der lokalkonvexen Räume. Lecture Notes in Mathematics, No. 56. Springer, Berlin/New York (1968)
12. Grothendieck, A.: Topological Vector Spaces. In: Notes on Mathematics and Its Applications. Gordon and Breach Science Publishers, New York/London/Paris (1973). Translated from the French by Orlando Chaljub
13. Hardy, G.H., Littlewood, J.E., Pólya, G.: Inequalities. Cambridge Mathematical Library. Cambridge University Press, Cambridge (1988). Reprint of the 1952 edition
14. Heuser, H.G.: Functional Analysis. A Wiley-Interscience Publication. John Wiley & Sons, Ltd., Chichester (1982). Translated from the German by John Horváth
15. Horváth, J.: Topological Vector Spaces and Distributions, vol. I. Addison-Wesley Publishing Co., Reading, Mass./London-Don Mills, Ont. (1966)

16. Jameson, G.J.O.: Topology and Normed Spaces. Chapman and Hall, London; Halsted Press, New York (1974)
17. Jarchow, H.: Locally Convex Spaces. Mathematische Leitfäden. B. G. Teubner, Stuttgart (1981)
18. Kadets, V.: A Course in Functional Analysis and Measure Theory. Universitext. Springer, Cham (2018). https://doi.org/10.1007/978-3-319-92004-7. Translated from the 2006 Russian edition
19. Kelley, J.L.: General topology. In: Graduate Texts in Mathematics, No. 27. Springer, New York/Berlin (1975). Reprint of the 1955 edition
20. Köthe, G.: Topological vector spaces. I. Translated from the German by D. J. H. Garling. In: Die Grundlehren der Mathematischen Wissenschaften, Band 159. Springer, New York Inc., New York (1969)
21. Köthe, G.: Topological vector spaces. II. In: Grundlehren der Mathematischen Wissenschaften [Fundamental Principles of Mathematical Sciences], vol. 237. Springer, New York/Berlin (1979)
22. Lindenstrauss, J., Tzafriri, L.: Classical Banach spaces. I. In: Ergebnisse der Mathematik und ihrer Grenzgebiete, Band 92. Springer, Berlin/New York (1977). Sequence spaces
23. Lindenstrauss, J., Tzafriri, L.: Classical Banach spaces. II. In: Ergebnisse der Mathematik und ihrer Grenzgebiete [Results in Mathematics and Related Areas], vol. 97. Springer, Berlin/New York (1979). Function spaces
24. Meise, R., Vogt, D.: Introduction to functional analysis. In: Oxford Graduate Texts in Mathematics, vol. 2. The Clarendon Press, Oxford University Press, New York (1997). Translated from the German by M. S. Ramanujan and revised by the authors
25. Pérez Carreras, P., Bonet, J.: Barrelled locally convex spaces. In: North-Holland Mathematics Studies, vol. 131. North-Holland Publishing Co., Amsterdam (1987). Notas de Matemática, 113
26. Ransford, T.J.: A short elementary proof of the Bishop-Stone-Weierstrass theorem. Math. Proc. Camb. Philos. Soc. **96**(2), 309–311 (1984). https://doi.org/10.1017/S0305004100062204
27. Robertson, A.P., Robertson, W.: Topological vector spaces. In: Cambridge Tracts in Mathematics, vol. 53, 2nd edn. Cambridge University Press, Cambridge/New York (1980)
28. Rudin, W.: Real and Complex Analysis, 3rd edn. McGraw-Hill Book Co., New York (1987)
29. Rudin, W.: Functional Analysis, 2nd edn. In: International Series in Pure and Applied Mathematics. McGraw-Hill, Inc., New York (1991)
30. Schaefer, H.H., Wolff, M.P.: Topological vector spaces. In: Graduate Texts in Mathematics, vol. 3, 2nd edn. Springer, New York (1999). https://doi.org/10.1007/978-1-4612-1468-7
31. Taylor, A.E., Lay, D.C.: Introduction to Functional Analysis, 2nd edn. Robert E. Krieger Publishing Co., Inc., Melbourne, FL (1986)
32. Trèves, F.: Topological Vector Spaces, Distributions and Kernels. Dover Publications, Inc., Mineola, NY (2006). Unabridged republication of the 1967 original
33. Valdivia, M.: Topics in locally convex spaces. In: Notas de Matemática [Mathematical Notes], vol. 85. North-Holland Publishing Co., Amsterdam/New York (1982)
34. Wilansky, A.: Modern Methods in Topological Vector Spaces. McGraw-Hill International Book Co., New York (1978)
35. Wilansky, A.: Topology for Analysis. Robert E. Krieger Publishing Co., Inc., Melbourne, FL (1983). Reprint of the 1970 edition
36. Yosida, K.: Functional analysis. In: Classics in Mathematics. Springer, Berlin (1995). https://doi.org/10.1007/978-3-642-61859-8. Reprint of the sixth (1980) edition

Chapter 3
Duality and Linear Operators

3.1 Hyperplanes

We defined the dual of a locally convex space as $L(E, \mathbb{K})$, the space of continuous, linear scalar-valued mappings (called *functionals*). We already pointed out in Remark 2.14 that this is different from E^*, the space of linear scalar-valued mappings on E (the algebraic dual of E).

Given a linear mapping $u : E \to \mathbb{K}$, we consider its *kernel* $\ker(u) = \{x \in E : u(x) = 0\}$. A subspace H of E is called a *hyperplane* if it is proper and maximal (i.e., if F is a subspace so that $H \subseteq F \subseteq E$, then either $H = F$ or $F = E$).

Lemma 3.1 *Let E be a vector space and H a subspace. Then H is a hyperplane if and only if there is $0 \neq u \in X^*$ so that* $\ker(u) = H$.

Proof Suppose first that H is a hyperplane and choose $x_0 \notin H$. Then

$$H \oplus [x_0] = \{x + \lambda x_0 : x \in H, \lambda \in \mathbb{K}\} = E,$$

and we can define $u : E \to \mathbb{K}$ by doing $u(x + \lambda x_0) = \lambda$. This clearly satisfies $u \in E^*$ and $\ker(u) = H$.

On the other hand, if there is $0 \neq u \in E^*$ so that $\ker(u) = H$, choose $x_0 \in E$ with $u(x_0) \neq 0$. Given $x \in E$, we denote $y = x - \frac{u(x)}{u(x_0)} x_0 \in \ker(u)$, and $x = y + \frac{u(x)}{u(x_0)} x_0$, showing that $E = H \oplus [x_0]$ and that H is a hyperplane. □

Lemma 3.2 *Let $u, u_1, u_2, \dots, u_n \in E^*$. The following are equivalent:*

1. $\displaystyle\bigcap_{i=1}^{n} \ker(u_i) \subseteq \ker(u)$.
2. *There exist $\lambda_1, \lambda_2, \dots, \lambda_n \in \mathbb{K}$ so that $u = \lambda_1 u_1 + \lambda_2 u_2 + \cdots + \lambda_n u_n$.*

Proof Let us suppose that 1 holds, and let us see by induction on n that so also does 2. We begin with the case $n = 1$ and suppose that $\ker(u_1) \subseteq \ker(u)$. If $u_1 = 0$,

J. Bonet et al., *Function Spaces and Operators between them*,
RSME Springer Series 11, https://doi.org/10.1007/978-3-031-41602-6_3

then $\ker(u) = \ker(u_1) = X$ and $u = 0$. Suppose then $u_1 \neq 0$ and choose $x_0 \in E$ such that $u_1(x_0) \neq 0$. Then, by Lemma 3.1, $\ker u_1$ is a hyperplane, and for every $x \in E$, there exist $\lambda \in \mathbb{K}$ and $y \in \ker(u_1) \subseteq \ker(u)$ so that $x = \lambda x_0 + y$. Therefore

$$u(x) = \lambda u(x_0) = \lambda \frac{u(x_0)}{u_1(x_0)} u_1(x_0) = \frac{u(x_0)}{u_1(x_0)} u_1(\lambda x_0 + y) = \lambda_1 u_1(x) ,$$

where $\lambda_1 = \frac{u(x)}{u_1(x)}$. Assume now that the statement holds for $n-1$, and let us see that it holds for n. Suppose, then, that $\bigcap_{i=1}^{n} \ker(u_i) \subseteq \ker(u)$ and $u_n \neq 0$. Denote $H = \ker(u_n)$, then $\bigcap_{i=1}^{n-1} \ker(u_i|_H) \subseteq \ker(u|_H)$, and, by the induction hypothesis, we can find $\lambda_1, \dots, \lambda_{n-1} \in \mathbb{K}$ so that

$$u|_H = \sum_{i=1}^{n-1} \lambda_i u_i|_H .$$

For each $x \in H$, we have $u(x) - \sum_{i=1}^{n} \lambda_i u_i(x) = 0$ and

$$\ker\left(u - \sum_{i=1}^{n-1} \lambda_i u_i\right) \supseteq H = \ker(u_n) .$$

But the hypothesis holds for $n = 1$ and this gives some $\lambda_n \in \mathbb{K}$ so that $u - \sum_{i=1}^{n-1} \lambda_i u_i = \lambda_n u_n$. This shows that 2 holds. Since the converse implication trivially holds, this completes the proof. □

Proposition 3.1 *Let $(E, \mathscr{P})$ be a locally convex space and $u \in E^*$. Then $u \in E'$ if and only if $\ker(u)$ is closed in E.*

Proof If $u \in E'$, then $\ker(u) = u^{-1}(\{0\})$ is closed. To prove the converse, we assume $u \neq 0$ and that $\ker(u) = \overline{\ker(u)} \subsetneq E$. Then, there is $x \in E \setminus \overline{\ker(u)}$. So, there exist $\varepsilon > 0$ and $p \in \mathscr{P}$ such that

$$\big(x + B_p(0, \varepsilon)\big) \cap \overline{\ker(u)} = \varnothing . \tag{3.1}$$

Our next step is to see that $u(B_p(0, \varepsilon))$ is balanced. Take some $\lambda \in \mathbb{K}$ with $|\lambda| \leq 1$ and $t \in u(B_p(0, \varepsilon))$. We can find $y \in B_p(0, \varepsilon)$ so that $t = u(y)$ and then (recall that $B_p(0, \varepsilon)$ is balanced, see (2.14))

$$\lambda t = \lambda u(y) = u(\lambda y) \in u(B_p(0, \varepsilon)) .$$

Suppose now that $u(B_p(0, \varepsilon))$ is not bounded. Since, as we have just seen, it is balanced, $u(B_p(0, \varepsilon)) = \mathbb{K}$, and there is some $y \in B_p(0, \varepsilon)$ such that $u(y) = -u(x)$. But this implies $u(x + y) = 0$ and then $x + y \in \ker(u)$ with $y \in B_p(0, \varepsilon)$. This contradicts (3.1), so $u(B_p(0, \varepsilon))$ is bounded and we can find $M > 0$ such that $|u(y)| \leq M$ for every $y \in B_p(0, \varepsilon)$, which proves that $u \in E'$. □

3.2 The Hahn–Banach Theorem

3.2.1 Analytic Version

To formulate the theorem (and later to apply it), we need a weaker concept than that of seminorm. A mapping $q : E \to \mathbb{R}$ defined on a real vector space is called *subadditive* if $q(x + y) \leq q(x) + q(y)$ for every $x, y \in E$ and *positively homogeneous* if $q(\lambda x) = \lambda q(x)$ for every $x \in E$ and $\lambda \geq 0$. Obviously every seminorm is subadditive and positively homogeneous.

Theorem 3.1 (Hahn–Banach) *Let E be a real vector space, F a proper subspace, $q : E \to \mathbb{R}$ subadditive and positively homogeneous, and $v : F \to \mathbb{R}$ linear such that*

$$v(y) \leq q(y)$$

for every $y \in F$. Then there exists a linear $u : E \to \mathbb{R}$ so that $u|_F = v$ and

$$u(x) \leq q(x)$$

for every $x \in E$.

Proof We begin by checking that we can extend v to the subspace

$$F \oplus [x_0] = \{y + \lambda x_0 \colon y \in F,\ \lambda \in \mathbb{R}\},$$

where $x_0 \notin F$. If we take any $s \in \mathbb{R}$ and define $u : F \oplus [x_0] \to \mathbb{R}$ by $u(y + \lambda x_0) = v(y) + \lambda s$, we clearly have that u is linear and extends v. We only have to choose s in such a way that u is also dominated by q. To begin with, note that if $y \in F$, then $v(y + y) \leq q(y + x_0) + q(y - x_0)$ and we can choose

$$\sup_{z \in F} \big(v(z) - q(z - x_0)\big) \leq s \leq \inf_{z \in F} \big(q(z + x_0) - v(z)\big).$$

With this choice, it is straightforward to check that $u(z + x_0) \leq q(z + x_0)$ and $u(z - x_0) \leq q(z - x_0)$ for every $z \in F$. Now, if $\lambda > 0$, then

$$u(y + \lambda x_0) = v(y) + \lambda s = \lambda\big(v\big(\tfrac{1}{\lambda} y\big) + x_0\big) \leq \lambda q\big(\tfrac{1}{\lambda} y + x_0\big) = q(y + \lambda x_0),$$

and if $\lambda < 0$, then

$$u(y + \lambda x_0) = v(y) + \lambda s = |\lambda|\big(v\big(\tfrac{1}{|\lambda|} y\big) - x_0\big) \leq |\lambda| q\big(\tfrac{1}{|\lambda|} y - x_0\big) = q(y + \lambda x_0).$$

This altogether shows that u is dominated by q on $F \oplus [x_0]$.

Consider now the family Φ of all subspaces H that contain F and all extensions u of v to H that are dominated by q on H. We establish the following order $(H_1, u_1) \leq (H_2, u_2)$ if $H_1 \subseteq H_2$ and $u_2|_{H_1} = u_1$. Now, if $(H_\alpha, u_\alpha)_\alpha$ is a totally ordered set, we define $H = \bigcup_\alpha H_\alpha$ and $u : H \to \mathbb{R}$ by $u(y) = u_\alpha(y)$ if $y \in H_\alpha$ (note that if $y \in H_\beta$, then the fact that the set is totally ordered gives $u_\alpha(y) = u_\beta(y)$). The same argument shows that H is a vector subspace of E. Then (H, u) is an upper bound for $(H_\alpha, u_\alpha)_\alpha$. By Zorn's lemma, there is a maximal element $(\tilde{H}, \tilde{u})$ in Φ. If $\tilde{H} \neq E$, we can find $x_0 \in E \setminus H$ and then extend $\tilde{u}$ to $H \oplus [x_0]$ as we did before. This contradicts the maximality of $(\tilde{H}, \tilde{u})$ and shows that, in fact, $\tilde{H} = E$ (and $\tilde{u}$ is the extension we were looking for). □

We can now pass to complex vector spaces, but let us first make a short consideration on the linear functionals. If E is a complex vector space and $u : E \to \mathbb{C}$ is linear, then we can write $u(x) = u_1(x) + iu_2(x)$ for every $x \in E$. Note that $u_1, u_2 : E \to \mathbb{R}$ are linear and $\mathbb{R}$-linear (in the sense that, for $j = 1, 2$, $u_j(\lambda x) = \lambda u_j(x)$ for every $x \in E$ and $\lambda \in \mathbb{R}$). We have

$$iu_1(x) - u_2(x) = iu(x) = u(ix) = u_1(ix) + u_2(ix),$$

and, then, $u_1(ix) = -u_2(x)$ for every $x \in E$. This implies that u is of the form

$$u(x) = u_1(x) - iu_1(ix). \tag{3.2}$$

On the other hand, if $v : E \to \mathbb{R}$ is $\mathbb{R}$-linear (recall that E is a complex vector space) and we define

$$u(x) = v(x) - iv(ix), \tag{3.3}$$

it satisfies

$$u(ix) = v(ix) - iv(-x) = v(ix) + iv(x) = i(-iv(ix) + v(x)) = iu(x)$$

and $u(rx) = ru(x)$ for every $r \in \mathbb{R}$. This gives that u is $\mathbb{C}$-linear.

Theorem 3.2 (Hahn–Banach) *Let E be a real or complex vector space and F a proper subspace. Let $p : X \to [0, \infty[$ be a seminorm and $v : F \to \mathbb{K}$ linear such that*

$$|v(x)| \leq p(x)$$

for every $x \in F$. Then there exists $u : E \to \mathbb{K}$ linear such that $u|_F = v$ and

$$|u(x)| \leq p(x)$$

for every $x \in E$.

Proof If E is real, the statement follows immediately from Theorem 3.1, which gives an extension u satisfying $u(x) \leq p(x)$ and $-u(x) = u(-x) \leq p(-x) = p(x)$ and, then, $|u(x)| \leq p(x)$ for every $x \in E$.

If E is a complex vector space, we have seen in (3.2) that $v(x) = v_1(x) - iv_1(ix)$, where v_1 is an $\mathbb{R}$-linear form. Then $|v_1(x)| \leq |v(x)| \leq p(x)$ for every $x \in F$ and, by the previous case, it can be extended to some $\mathbb{R}$-linear $u_1 : E \to \mathbb{R}$. With this, $u : E \to \mathbb{C}$ given by $u(x) = u_1(x) - iu_1(ix)$ is a $\mathbb{C}$-linear form that extends u. Now, for each $x \in E$, we can find $\theta_x \in \mathbb{R}$ such that $|u(x)| = e^{\theta_x i} u(x)$. Then

$$|u(x)| = u(e^{\theta_x i} x) = u_1(e^{\theta_x i} x) \leq p(e^{\theta_x i} x) = |e^{\theta_x i}| p(x) = p(x) .$$

□

As an immediate consequence, we can extend any continuous linear functional to every superspace.

Corollary 3.1 *Let E be a locally convex space and F a subspace. For every $v \in F'$, there exists $u \in E'$ that extends v.*

Proof Let $\mathscr{P}$ be the family of seminorms defining the topology on E. Then the topology on F is defined by the restrictions of the seminorms in $\mathscr{P}$ to F. Take $v \in F'$. There is some $C > 0$ such that $|v(x)| \leq Cp(x)$ for every $x \in F$. But $q = Cp$ is a seminorm on E, and we can find $u \in E^*$ extending v and such that $|u(x)| \leq Cp(x)$ for every $x \in E$. Therefore $u \in E'$. □

Corollary 3.2 *Let E be a locally convex space. For every $0 \neq x \in E$, there exists $u \in E'$ such that $u(x) = 1$.*

Proof Since $x \neq 0$, there is $p \in \mathscr{P}$, with $p(x) > 0$. We consider $F = \{\lambda x : \lambda \in \mathbb{K}\}$ and define $v : F \to \mathbb{K}$ by $v(\lambda x) = \lambda$. Then, $|v(\lambda x)| = |\lambda| = \frac{|\lambda|}{p(x)} p(x) \leq \frac{1}{p(x)} p(\lambda x)$, and by the Hahn–Banach Theorem 3.2, there exists $u \in E^*$ such that $u|_H = v$ and $|u(y)| \leq \frac{1}{p(x)} p(y)$ for every $y \in E$. This implies that u is continuous and, moreover, $u(x) = v(x) = 1$. □

Remark 3.1 Let us point out that if E in Corollaries 3.1 and 3.2 is a normed space, then the family of seminorms $\mathcal{P}$ reduces to the norm $\|\cdot\|$, and the proofs show that, in fact, u satisfies $\|u\| = \|v\|$ (in Corollary 3.1) and that $u(x) = 1$ and $\|u\| = 1$ (in Corollary 3.2).

3.2.2 Separation Theorems

With help of the Hahn–Banach theorem, we can use functionals to separate sets. If the sets that we want to separate are 'small' (a point and a finite dimensional space, as in the following result), we do not need more than what we already have.

Proposition 3.2 *Let E be a locally convex space, M a finite dimensional subspace, and $x_0 \in E \setminus M$. Then there is $u \in E'$ such that $u|_M = 0$ and $u(x_0) = 1$.*

Proof Choose some basis $\{x_1, \dots, x_n\}$ of M, consider $F = \operatorname{span}\{x_0, x_1, \dots, x_n\}$, and define $v(\lambda x_0 + \mu_1 x_1 + \cdots + \mu_n x_n) = \lambda$. The argument now finishes as that in Corollary 3.2. □

With essentially the same idea, we can separate closed subspaces and points.

Proposition 3.3 *Let E be a locally convex space, F a closed proper subspace, and $x_0 \in E \setminus F$. Then there is $u \in E'$ such that $u|_F = 0$ and $u(x_0) = 1$.*

Proof Consider the space $G = F + \operatorname{span}\{x_0\}$ and endow it with the topology induced by E. For each $x \in F$, there are $y \in F$ and $\lambda \in \mathbb{C}$ so that $x = y + \lambda x_0$. By doing $v(x) = \lambda$, we define a linear mapping $v : G \to \mathbb{C}$ that, since $\ker v = F$ is closed, is continuous (recall Proposition 3.1). Corollary 3.1 gives a functional $u \in E'$ that extends v and satisfies the required conditions. □

As a straightforward consequence, we have the following.

Corollary 3.3 *Let E be a locally convex space and F a subspace. The following are equivalent:*

1. *If $u \in E'$ is such that $u(x) = 0$ for every $x \in F$, then $u = 0$.*
2. *F is dense in E.*

For 'bigger' sets, we need some more work. To begin with, for each convex, absorbing set A of some vector space E, we define the *Minkowski gauge* $p_A : E \to [0, \infty[$ as

$$p_A(x) = \inf\{t > 0 \colon x \in tA\} . \tag{3.4}$$

Clearly

$$\{x \in E \colon p_A(x) < 1\} \subseteq A \subseteq \{x \in E \colon p_A(x) \leq 1\} .$$

Lemma 3.3 *If A is convex and absorbing, then p_A is subadditive and positively homogeneous. If A is absolutely convex, then p_A is a seminorm.*

Proof Note first that $p_A(0) = 0$, then for $\lambda = 0$ we have $p(\lambda 0) = \lambda p_A(x)$. On the other hand, if $\lambda > 0$, then

$$p_A(\lambda x) = \inf\{t > 0 \colon \lambda x \in tA\} = \inf\{t > 0 \colon x \in \tfrac{t}{\lambda} A\} = \lambda p_A(x) .$$

This shows that it is positively homogeneous. To see that it is subadditive, take $\frac{x}{t}, \frac{y}{s} \in A$, and then, since the set is convex, we have

$$\frac{x+y}{t+s} = \frac{t}{t+s}\frac{x}{t} + \frac{s}{t+s}\frac{y}{s} \in A ,$$

and this implies $p_A(x+y) \leq t+s$. Taking the infimum first with respect to t and then to s, we get $p_A(x+y) \leq p_A(x) + p_A(y)$.
If A is balanced, then so also is tA for every $t > 0$ and $\lambda x \in tA$ if and only if $|\lambda| \in tA$ which gives (by the positive homogeneity) $p_A(\lambda x) p_A(|\lambda| x) = |\lambda| p_A(x)$.

□

Lemma 3.4 *Let E be a real locally convex space, $u : E \to \mathbb{R}$, non-zero and linear, and A a subset of E. If A is convex, balanced, or open, then so also is $u(A)$.*

Proof The statements about convexity and balancedness are straightforward consequences of the linearity of u. The fact that if A is open, then $u(A)$ is open requires slightly more work. It is enough to show that for every $p \in \mathscr{P}$ and all $\varepsilon > 0$ there is some $\delta > 0$ so that $]-\delta, \delta[\subseteq u(B_p(0,\varepsilon))$.
Choose x_0 with $u(x_0) > 0$, take $q \in \mathscr{P}$ so that $q(x_0) > 0$, and consider $r \in \mathscr{P}$ such that $\max\{p, q\} \leq r$. Define $\delta = \frac{u(x_0)}{r(x_0)} \varepsilon$. If $|t| < \delta$, then

$$t = t\frac{u(x_0)}{u(x_0)} = u\big(\tfrac{tx_0}{u(x_0)}\big).$$

But

$$u\big(\tfrac{tx_0}{u(x_0)}\big) = |t| \frac{r(x_0)}{u(x_0)} < \varepsilon .$$

Hence $]-\delta, \delta[\subseteq u(B_r(0,\varepsilon)) \subseteq u(B_p(0,\varepsilon))$, and this completes the proof. □

Theorem 3.3 *Let E be a real locally convex space and A and B two non-empty disjoint convex sets in E, being A open. Then there exist $u \in E'$ and $\gamma \in \mathbb{R}$ so that $u(a) < \gamma \leq u(b)$ for every $a \in A$ and $b \in B$.*

Proof Choose $a_0 \in A$ and $b_0 \in B$. Both $A - a_0$ and $B - b_0$ are convex and then so also is $K = A - B + b_0 - a_0$. On the other hand, for each $y \in B$, the set $A - y + b_0 - a_0$ is open and, then, since $K = \bigcup_{y \in B} A - y + b_0 - a_0$, it is also open. Note also that, since the sets are disjoint, $x_0 = b_0 - a_0 \notin K$.
Clearly $0 \in K$, and therefore K is an open neighbourhood of 0 and, then absorbing. We consider its Minkowski gauge that, by Lemma 3.3, is subadditive and positively homogeneous. Note that, since $x_0 \notin K$, we have $p_K(x_0) \geq 1$ and, since K is open, $p_K(x) < 1$ for every $x \in K$. Consider now $F = \{\lambda x_0 \colon \lambda \in \mathbb{R}\}$ and define $v(\lambda x_0) = \lambda$. For $\lambda \geq 0$, we have $\lambda \leq \lambda p_K(x_0) = p_K(\lambda x_0)$. Since the inequality is obvious for $\lambda < 0$, we get that v is dominated by p_K on F, and, by Theorem 3.1, it can be extended to $u \in E^*$ satisfying $u(x) \leq p_K(x)$ for every $x \in E$.
If $x \in K$, then $u(-x) \leq p_K(-x) = p_K(x) < 1$. We define $U = K \cap (-K)$, which is an open neighbourhood of 0 and satisfies $|u(x)| < 1$ for every $x \in U$. This shows that u is also continuous, that is, $u \in E'$.
Now, if $a \in A$ and $b \in B$, we have $a - b + x_0 \in K$ and, then, $u(a - b + x_0) < 1 = u(x_0)$. This implies $u(a) < u(b)$. Now, Lemma 3.4 gives that both $u(A)$ and

$u(B)$ are convex (that is, intervals in $\mathbb{R}$) and $u(A)$ is open. Also, $u(A) \cap u(B) = \varnothing$. Taking $\gamma = \inf_{b \in B} u(b)$ yields the conclusion. □

Remark 3.2 If E is a complex l.c.s. and A and B are as in Theorem 3.3, we can find a continuous, $\mathbb{R}$-linear $u_1 : E \to \mathbb{R}$ separating A and B. Then, defining u as in (3.3) (which is obviously continuous), we get $u \in E'$ and $\gamma \in \mathbb{R}$ such that $\mathrm{Re}(u(a)) < \gamma \leq \mathrm{Re}(u(b))$ for all $a \in A$ and $b \in B$.

Theorem 3.4 *Let E be a real locally convex space and A and B two non-empty disjoint convex sets in E, being A closed and B compact. Then there exists $u \in E'$ such that*

$$\sup_{a \in A} u(a) < \inf_{b \in B} u(b) .$$

Before we proceed with the proof, let us note that this result does not hold if we only ask A and B to be convex and closed. To see this, just take $A = \{(x, 0) : x \in \mathbb{R}\}$ and $B = \{(x, y) : x, y \leq 0,\ xy \geq 1\}$ as subsets of $\mathbb{R}^2$.

Proof Since A is closed and B is compact, for each $b \in B$, we can find $p_b \in \mathscr{P}$ and $\varepsilon_b > 0$ so that $B_{p_b}(b, 2\varepsilon_b) \cap A \neq \varnothing$. Also, being B compact, there are $b_1, \ldots, b_N \in B$ so that (we denote p_n and ε_n for p_{b_n} and ε_{b_n})

$$B \subseteq \bigcup_{n=1}^{N} B_{p_n}(b_n, \varepsilon_n) .$$

Define $\varepsilon = \min_{n=1,\ldots,N} \varepsilon_n$ and consider $p \in \mathscr{P}$ such that $p \geq \max_{n=1,\ldots,N} p_n$. Consider now $U = \bigcup_{b \in B} B_p(b, \varepsilon)$. Then

$$\begin{aligned} U = B + B_p(0, \varepsilon) &\subseteq \bigcup_{n=1}^{N} B_{p_n}(b_n, \varepsilon_n) + B_p(0, \varepsilon) \\ &\subseteq \bigcup_{n=1}^{N} b_n + B_{p_n}(0, \varepsilon_n) + B_{p_n}(0, \varepsilon_n) \subseteq \bigcup_{n=1}^{N} B_{p_n}(b_n, 2\varepsilon_n) , \end{aligned}$$

and thus U and A are disjoint. Since U is convex and open, we can use Theorem 3.3 to find some $u \in E'$ so that $u(U)$ and $u(A)$ are two disjoint intervals in $\mathbb{R}$. On the other hand, $u(B)$ is a compact interval contained in the open interval $u(U)$. This gives the conclusion. □

Corollary 3.4 *Let E be a real or complex locally convex space and A and B two non-empty disjoint sets in E, being A closed and absolutely convex and B compact and convex. Then there exists $u \in E'$ such that*

$$\sup_{a \in A} |u(a)| < \inf_{b \in B} |u(b)| .$$

Proof From Theorem 3.4, we can find some $u : E \to \mathbb{R}$ continuous and $\mathbb{R}$-linear so that

$$\sup_{a \in A} u(a) = \mu < \gamma = \inf_{b \in B} u(b)\,.$$

Suppose first that E is real. Then for $b \in B$, we have $\gamma \leq u(b) \leq |u(b)|$. On the other hand, for each $a \in A$, there is $\lambda_a \in \mathbb{R}$ with $|\lambda_a| = 1$ so that $\lambda_a u(a) = |u(a)|$. Since A is absolutely convex $\lambda_a a \in A$ and, then, $|u(a)| = u(\lambda_a a) \leq \mu$. This gives the conclusion.

Suppose now that E is complex and define $v \in E'$ by $v(x) = u(x) - iu(ix)$. Again, $\gamma \leq u(b) \leq |v(b)|$ for every $b \in B$. The same argument as before, finding for each $a \in A$ some $\lambda_a \in \mathbb{C}$ with $|\lambda_a| = 1$ and $|v(a)| = \lambda_a v(a)$ yields

$$|v(a)| = v(\lambda_a a) = \operatorname{Re}\big(v(\lambda_a a)\big) = u(\lambda_a a) \leq \mu\,.$$

This completes the argument. □

Corollary 3.5 *Let E be a real or complex locally convex space and $\varnothing \neq A \subseteq E$ be closed and absolutely convex. For each $x \notin A$, there exists $u \in E'$ such that $|u(a)| \leq 1$ for every $a \in A$ and $|u(x)| > 1$.*

Proof Using Corollary 3.4, we can find $v \in E'$ and $\gamma \in \mathbb{R}$ so that

$$\sup_{a \in A} |v(a)| < \gamma < |v(x)|\,.$$

Defining $u = \frac{1}{\gamma} v \in E'$ yields the claim. □

Remark 3.3 Let us note that if in Proposition 3.3 or Theorem 3.3 E is a normed space, then the operator v defined in the proofs clearly satisfies $\|v\| = 1$, and therefore, by Remark 3.1, u can be taken with $\|u\| = 1$.

3.2.3 Finite Dimensional Locally Convex Spaces

The Hahn–Banach theorem allows us to characterize those locally convex spaces with finite dimension.

Remark 3.4 First of all note that if $M = \operatorname{span}\{x_1, \ldots, x_n\}$ is a finite dimensional subspace of an l.c.s. E, then we can choose some $x_0 \notin M$, and, by Proposition 3.2, we can find $u \in E'$ such that $u(a) = 1$ and $u(x) = 0$ for all $x \in M$. Since u is continuous, this gives $a \notin \overline{M}$ and shows that M is closed.

A subset A of a locally convex space $(E, \mathscr{P})$ is said to be *precompact* if for each $\varepsilon > 0$ and each $p \in \mathscr{P}$, there exist a finite number of points $x_1, \dots, x_N$ in A (or in E) so that

$$A \subseteq \bigcup_{n=1}^{N} B_p(x_n, \varepsilon) .$$

The fact that every open covering of a compact set has a finite subcovering immediately gives that every relatively compact set is precompact.
On the other hand, if A is precompact in an l.c.s. E and p is some seminorm in $\mathscr{P}$, then there are $x_1, \dots, x_N$ in A so that

$$A \subseteq \bigcup_{n=1}^{N} B_p(x_n, 1) .$$

If $x \in A$, then $x \in B_p(x_{n_0}, 1)$ for some n_0 and $p(x) \leq p(x - a_{n_0}) + p(a_{n_0}) = 1 + p(a_{n_0})$. This gives

$$\sup_{x \in A} p(x) \leq 1 + \max_{n=1,\dots,N} p(a_n)$$

and shows that A is bounded.

Theorem 3.5 (Riesz) *Let $(E, \mathscr{P})$ be an l.c.s. with a precompact* 0*-neighbourhood. Then,* $\dim(E) < \infty$.

Proof Let U be a precompact 0-neighbourhood. Since precompact sets are bounded, Kolmogorov's Theorem 2.24 implies that E is normable. Since its open unit ball (which we denote by B_E) is bounded, there is $\lambda > 0$ such that $B_E \subseteq \lambda U$, and this implies that B_E itself is precompact. Then we can find $x_1, \dots, x_N \in B_E$ such that

$$B_E \subseteq \bigcup_{n=1}^{N} \big(x_n + \frac{1}{2} B_E\big) .$$

Define $M = \operatorname{span}\{x_1, \dots, x_N\}$, then

$$B_E \subseteq M + \frac{1}{2} B_E \subseteq M + \frac{1}{2}\big(M + \frac{1}{2} B_E\big) \subseteq M + \frac{1}{2^2} B_E .$$

Proceeding by induction, we get $B_E \subseteq M + \frac{1}{2^n} B_E$ for every $n \in \mathbb{N}$, and this gives

$$B_E \subseteq \bigcap_{n=1}^{\infty} \big(M + \frac{1}{2^n} B_E\big) = \overline{M} .$$

Since by Lemma 3.4 M is closed, $\overline{M} = M$ and $E = \operatorname{span}(B_E) \subseteq M$. This implies $\dim(E) \leq N$. □

With this, we can finally show that for each n there is essentially only one n-dimensional locally convex space.

Theorem 3.6 *A locally convex space $(E, \mathscr{P})$ has dimension $n \in \mathbb{N}$ if and only if it is isomorphic to $\mathbb{K}^n$ endowed with the sup-norm.*

Proof Let us assume first that $\dim(E) = n$ and choose a basis $x_1, \ldots, x_n$. We define $\phi : \mathbb{K}^n \to E$ by $\phi(\alpha) = \sum_{i=1}^n \alpha_i x_i$ for $\alpha = (\alpha_1, \ldots, \alpha_n) \in \mathbb{K}^n$. Clearly, ϕ is linear and bijective. Also, for each $p \in \mathscr{P}$, we have

$$p(\phi(\alpha)) = p\big(\textstyle\sum_{i=1}^n \alpha_i x_i\big) \leq \max_{i=1,\ldots,n} |\alpha_i| \displaystyle\sum_{i=1}^n p(x_i) \leq \Big(\sum_{i=1}^n p(x_i)\Big) \|\alpha\|_\infty ,$$

and ϕ is continuous. It remains to show that $\phi^{-1} : (E, \tau_{\mathscr{P}}) \to \mathbb{K}^n$ is also continuous. First of all, since $x_1, \ldots, x_n \in E$ are linearly independent, by Proposition 3.2 there are $u_1, \ldots, u_n \in E'$ such that $u_i(x_j) = \delta_{ij}$ for $i, j = 1, \ldots, n$. Now, for each $x \in E$, we can write $x = \sum_{i=1}^n u_i(x) x_i$, and the inverse of ϕ is given by

$$\phi^{-1}(x) = (u_1(x), \ldots, u_n(x))$$

for each $x \in E$. To see that this is continuous, let us note that, since $u_i \in E'$ for $i = 1, \ldots, n$, for each $p \in \mathscr{P}$, there is $C > 0$ such that $|u_i(x)| \leq C p(x)$ for all $x \in E$ and $i = 1, \ldots, n$. Hence,

$$\|\phi^{-1}(x)\|_\infty = \max_{i=1,\ldots,n} |u_i(x)| \leq C p(x)$$

for each $x \in E$. This shows that ϕ^{-1} is continuous and, since every space that is isomorphic to $(\mathbb{K}^n, \|\cdot\|_\infty)$ is obviously an n-dimensional normed (hence locally convex) space, concludes the proof. □

Note that Theorem 3.6 in particular tells us that if E is a finite dimensional space, then there is only one possible Haussdorf locally convex topology on E.

3.2.4 Banach Limits

Our aim now is to extend the notion of 'limit' to bounded (not necessarily convergent) sequences. Let us be more precise, and denote by c the space of all convergent sequences, endowed with the topology defined by the norm $\|\cdot\|_\infty$ (recall Example 2.5). Then $L(x) = \lim_n x_n$ defines a linear mapping $L : c \to \mathbb{K}$ that, since $|L(x)| \leq \sup_n |x_n|$ for every $x = (x_n)_n \in c$, is continuous (and $\|L\| \leq 1$). Moreover, denoting $\mathbb{1} = (1, 1, 1, \ldots)$ (the constant sequence), we have $L(\mathbb{1}) = 1$, so that $\|L\| = 1$. Additionally,

$$\lim_n x_n = \lim_n x_{n+1} \tag{3.5}$$

for every $(x_n)_n \in c$, and if $x_n \geq 0$, then

$$\lim_n x_n \geq 0$$

(that is, the operator L is positive). So, our aim is to define a continuous operator $L : \ell_\infty \to \mathbb{K}$ that when restricted to c gives the limit and satisfies these two last properties. Let us have a look at (3.5): what we want here is that, when evaluated at $(x_n)_n$ and at $(x_{n+1})_n$, the operator L gives the same result. To handle this, we consider the *backward shift operator* $B : \mathbb{K}^\mathbb{N} \to \mathbb{K}^\mathbb{N}$, defined as $B\big((x_n)_{n=1}^\infty\big) = (x_{n+1})_{n=1}^\infty$, that is,

$$B(x_1, x_2, x_3, \ldots) = (x_2, x_3, \ldots) \tag{3.6}$$

(this operator will be studied in more detail in Sect. 5.3.1). Then (3.5) can be reformulated as $\lim x = \lim B(x)$ for every $x \in c$, and what we want is that $L = LB$. We show the existence of such an operator (called *Banach limit*) in the following result.

Theorem 3.7 *There exists an operator $L : \ell_\infty \to \mathbb{K}$ such that*

1. $\|L\| = 1$.
2. $L(x) = \lim_n x_n$ *for every* $x = (x_n)_n \in c$.
3. *If* $x \in \ell_\infty$ *is such that* $x_n \geq 0$ *for every* n*, then* $L(x) \geq 0$.
4. $L = LB$ *on* ℓ_∞.

Proof Let us handle first the real case, that is, if $\mathbb{K} = \mathbb{R}$. Consider the subspace $M = \operatorname{Im}(I - B) = \{x - Bx : x \in \ell_\infty\}$. We proceed now in several steps. Let us see first that

$$d(\mathbb{1}, M) = 1\,. \tag{3.7}$$

To see this, note first that, since $0 \in M$, we have $d(\mathbb{1}, M) \leq \|0 - \mathbb{1}\| = 1$. Let us see now that $\|\mathbb{1} - (x - Bx)\| \geq 1$ for every $x \in \ell_\infty$. Suppose first that x is a bounded, decreasing sequence (i.e., $x_{n+1} \leq x_n$ for every n). Then it is convergent and $\lim_n (x_n - x_{n+1}) = 0$; hence

$$\|\mathbb{1} - (x - Bx)\| = \sup_n |1 - (x_n - x_{n+1})| = 1\,.$$

Now, if x is not decreasing, then there is some $m \in \mathbb{N}$ such that $x_{m+1} \geq x_m$, i.e., $x_m - x_{m+1} \leq 0$, and

$$\|\mathbb{1} - (x - Bx)\| \geq |1 - (x_m - x_{m+1})| \geq 1 .$$

This altogether shows (3.7), which clearly implies that $\mathbb{1} \notin \overline{M}$. By Proposition 3.3 (see also Remark 3.3), there is a continuous linear operator $L : \ell_\infty \to \mathbb{R}$ with $\|L\| = 1$ such that $L(\mathbb{1}) = 1$ and $L(y) = 0$ for every $y \in \overline{M}$. Note that the latter implies that $L(x-Bx) = 0$ for every $x \in \ell_\infty$. We have, then, obtained L satisfying 1 and 4. To see that the other two properties hold, we need first to prove the following

$$c_0 \subseteq \ker L$$

(recall from Example 2.5 that c_0 is the space of null sequences). Pick, then, some $x \in c_0$ and set $x^{(0)} = x$ and $x^{(n+1)} = Bx^{(n)}$ for $n \in \mathbb{N}$. Then,

$$x - x^{(n)} = \sum_{j=0}^{n-1} x^{(j)} - x^{(j+1)} = \sum_{j=0}^{n-1} x^{(j)} - Bx^{(j)} \in M$$

for every n, and therefore $L(x) = L(x^{(n)})$. Now, given $\varepsilon > 0$, there exists $n_0 \in \mathbb{N}$ such that $|x_m| < \varepsilon$ for every $m \geq n_0$. Then, if $n \geq n_0$, we have

$$|L(x)| = |L(x^{(n)})| \leq \|x^{(n)}\| = \sup_{m \geq n} |x_m| < \varepsilon .$$

This gives $L(x) = 0$ and shows our claim. With this, we can now prove that the remaining properties are satisfied. Let us see first that 2 holds. If $x = (x_n)_n$ is convergent and we write $\alpha = \lim_n x_n$, then $x - \mathbb{1}\alpha \in c_0$, and then $L(x - \alpha\mathbb{1}) = 0$. Hence

$$L(x) = L(\alpha\mathbb{1}) = \alpha L(\mathbb{1}) = \lim_n x_n .$$

To finish the proof in the case $\mathbb{K} = \mathbb{R}$, we take $x \in \ell_\infty$ such that $x_n \geq 0$ and suppose that $L(x) < 0$. Taking $z = \frac{x}{\|x\|}$, we have $0 \leq z_n \leq 1$. Then $\|\mathbb{1} - z\| \leq 1$ and

$$L(\mathbb{1} - z) = L(\mathbb{1}) - L(z) = 1 - \frac{L(x)}{\|x\|} > 1 .$$

This contradicts the fact that $\|L\| = 1$ and shows that 3 holds.

We address now the case $\mathbb{K} = \mathbb{C}$. We write $L_{\mathbb{R}}$ for the operator defined on the real space of bounded real sequences. If $x = (x_n)_n \in \ell_\infty$ (now the complex space of complex sequences), we write $\operatorname{Re} x = (\operatorname{Re} x_n)_n$ and $\operatorname{Im} x = (\operatorname{Im} x_n)_n$ and define

$$L(x) = L_{\mathbb{R}}(\operatorname{Re} x) + i L_{\mathbb{R}}(\operatorname{Im} x) .$$

This gives $L : \ell_\infty \to \mathbb{C}$ that is clearly $\mathbb{C}$-linear, continuous (with $1 \leq \|L\| \leq 2$ and $L(\mathbb{1}) = 1$) and satisfies 2–4. It remains only to see that $\|L\| \leq 1$. Choose $x \in \ell_\infty$ with $\|x\| \leq 1$, and let us see that $|L(x)| \leq 1$. Let us suppose first that x has only a finite number of different values (we denote by $\ell_{\infty,0}$ the space of such sequences), then there are $E_1, \dots, E_n \subseteq \mathbb{N}$, pairwise disjoint such that $\mathbb{N} = E_1 \cup \cdots \cup E_n$, and $\alpha_1, \dots, \alpha_m \in \mathbb{C}$ with $|\alpha_j| \leq 1$ for $j = 1, \dots, m$ so that

$$x = \sum_{j=1}^{m} \alpha_j \mathbb{1}_{E_j} ,$$

where $\mathbb{1}_{E_j}$ is the sequence whose nth coordinate is 1 if $n \in E_j$ and 0 otherwise. Note that $L(\mathbb{1}_{E_j}) = L_{\mathbb{R}}(\mathbb{1}_{E_j}) \geq 0$ for every $j = 1, \dots, m$ and that $\sum_{j=1}^{m} \mathbb{1}_{E_j} = \mathbb{1}$. Then

$$|L(x)| = \Big| \sum_{j=1}^{m} \alpha_j L(\mathbb{1}_{E_j}) \Big| \leq \sum_{j=1}^{m} |\alpha_j| \, |L(\mathbb{1}_{E_j})| \leq \sum_{j=1}^{m} L(\mathbb{1}_{E_j}) = L(\mathbb{1}) = 1 .$$

To complete the proof, we need to see that the set $\ell_{\infty,0}$ is dense in ℓ_∞. Choose, then, $x \in \ell_\infty$ and $\varepsilon > 0$. We can find $y_1, \dots, y_m \in \overline{\mathbb{D}(0, \|x\|)}$ so that $\overline{\mathbb{D}(0, \|x\|)} \subseteq \bigcup_{j=1}^{m} \mathbb{D}(y_j, \varepsilon)$. We define now a sequence $z = (z_n)_n$ in the following way: given $n \in \mathbb{N}$, there is some $1 \leq j_n \leq m$ so that $|y_{j_n} - x_n| < \varepsilon$; then we set $z_n = y_{j_n}$. Clearly $\|z - x\| < \varepsilon$ and, since there are only a finite number of ys, $z \in \ell_{\infty,0}$.
If $x \in \ell_\infty$ with $\|x\| \leq 1$ is arbitrary, we may take a sequence $(x^j)_j$ in $\ell_{\infty,0}$ converging in ℓ_∞ to x. Note that by our argument above we may assume $\|x^j\| \leq \|x\| \leq 1$. Then $|L(x^j)| \leq 1$ for every j, and this gives $|L(x)| \leq 1$, completing the proof. □

3.3 Weak Topologies

Our aim now is to see how the duality relationship can be used to define a topology. We are going to work in a fairly general setting, but since our main interest will be locally convex spaces E with their duals E', we start by looking at this situation. First of all, if $(E, \mathscr{P})$ is a locally convex space and E' is its dual, then we have

- For every $u \in E'$, $u \neq 0$, there exists $x \in E$ such that $u(x) \neq 0$.
- For every $x \in E$, $x \neq 0$, there exists $u \in E'$ such that $u(x) \neq 0$.

The first property is trivially satisfied, whereas the second one follows from the Hahn–Banach theorem. On the other hand, we can define a bilinear mapping

$$\begin{aligned} \langle \cdot, \cdot \rangle : E \times E' &\longrightarrow \mathbb{K} \\ (x, u) &\longmapsto \langle x, u \rangle = u(x). \end{aligned} \tag{3.8}$$

These are the properties that relate the two vector spaces and that we isolate. A *dual pair* consists of a pair of vector spaces (E, F) together with a bilinear form $\langle \cdot, \cdot \rangle : E \times F \longrightarrow \mathbb{K}$ satisfying

(DP1) For every $y \in F$, $y \neq 0$, there exists $x \in E$ such that $\langle x, y \rangle \neq 0$.
(DP2) For every $x \in E$, $x \neq 0$, there exists $y \in F$ such that $\langle x, y \rangle \neq 0$.

Example 3.8 If E is a locally convex space and E' is its dual, then (E, E') is a dual pair with the bilinear mapping described in (3.8). In particular (recall Example 2.16), (c_0, ℓ_1), (ℓ_1, ℓ_∞), and $(\ell_p, \ell_{p'})$ are dual pairs, where the bilinear mapping is in each case defined by

$$\langle (x_i), (y_i) \rangle = \sum_{i=1}^{\infty} x_i y_i \,.$$

□

Example 3.9 Let us consider the Fréchet space ω (recall Example 2.14) and

$$\varphi = \{ y \in \omega \colon \text{ there is } i(x) \in \mathbb{N} \text{ so that } y_i = 0 \text{ for all } i \geq i(x) \} \,.$$

We define $\langle \ , \ \rangle : \omega \times \varphi \to \mathbb{K}$ by

$$\langle (x_i), (y_i) \rangle = \sum_{i=1}^{\infty} x_i y_i \,.$$

Note that this is in fact a finite sum and, then, $\langle \ , \ \rangle$ is well defined and bilinear. Given $0 \neq x \in \omega$, there is i so that $x_i \neq 0$. Then $\langle x, e_i \rangle = x_i \neq 0$ and (DP2) holds. The same argument shows that (DP1) holds, and thus (ω, φ) is a dual pair. □

Example 3.10 If $L_1([0, 1])$ is the space of measurable functions $f : [0, 1] \to \mathbb{K}$ such that $\int_0^1 |f(x)| dx < \infty$, then $(L_1([0, 1]), C([0, 1]))$ is a dual pair with the bilinear mapping defined by

$$\langle f, g \rangle = \int_0^1 f(x) g(x) dx \,.$$

□

Example 3.11 If $F \subseteq E^*$, then a bilinear form can be defined as in (3.8). Condition (DP1) is obviously satisfied, and condition (DP2) is satisfied whenever F separates points. So, if this is the case, then (E, F) is a dual pair. □

Remark 3.5 In fact, all dual pairs can be seen as a particular case of Example 3.11. Take some dual pair (E, F), fix $x \in E$, and define $u_x : F \to \mathbb{K}$ by $u_x(y) = \langle x, y \rangle$.

This is clearly linear and then belongs to F^*. On the other hand, given $\lambda, \mu \in \mathbb{K}$, and $x_1, x_2 \in E$, by the bilinearity, we have

$$u_{\lambda x_1 + \mu x_2}(y) = \langle \lambda x_1 + \mu x_2, y \rangle = \lambda \langle x_1, y \rangle + \mu \langle x_2, y \rangle = \lambda u_{x_1}(y) + \mu u_{x_2}(y)$$

for every $y \in F$. On the other hand, if $u_x = 0$, then $0 = u_x(y) = \langle x, y \rangle$ for every $y \in F$, and by (DP1), $x = 0$. This altogether shows that the mapping defined as $x \rightsquigarrow u_x$ is a linear, bijective mapping, and then we can identify E with a subset of F^* that separates points. With the same argument, F can be identified with a subset of E^* that separates points. We will from now on always bear these identifications in mind without further notice.

What we want to do now is to see how the structure of a dual pair defines a locally convex topology on each of the two spaces. To do that we begin by defining seminorms. If $J \subseteq F$ is finite, then we define, for $x \in E$,

$$p_J(x) = \max\{|\langle x, u \rangle| : u \in J\}.$$

This is a seminorm on E. With the same idea, given $I \subseteq E$ finite, we define, for $u \in F$,

$$p_I(u) = \max\{|\langle x, u \rangle| : x \in I\},$$

which is a seminorm on F.

Proposition 3.4

1. *The family of seminorms $\{p_J : J \subseteq F$ finite$\}$ defines a (Hausdorff) locally convex topology, denoted by $\sigma(E, F)$, on E.*
2. *The family of seminorms $\{p_I : I \subseteq E$ finite$\}$ defines a (Hausdorff) locally convex topology, denoted by $\sigma(F, E)$, on F.*

Proof First of all, if J_1 and J_2 are finite subsets of F, then $K = J_1 \cup J_2$ is also finite and obviously $\max\big(p_{J_1}(x), p_{J_2}(x)\big) \leq p_K(x)$ for all $x \in E$.
On the other hand, given $x \neq 0$, there is $u \in F$ such that $\langle x, u \rangle \neq 0$, and then $p_{\{u\}}(x) = |\langle x, u \rangle| > 0$, and this finally shows that $\sigma(E, F)$ is a locally convex topology on E. The proof for $\sigma(F, E)$ is basically the same. □

Once we have the topology, we describe the neighbourhoods in each of the two topologies. First, for $y_1, \ldots, y_n \in F$ and $\varepsilon > 0$, then

$$U_{y_1,\ldots,y_n,\varepsilon} = \{x \in E : |\langle x, y_i \rangle| < \varepsilon\, ,\ i = 1, \ldots n\}$$

is a neighbourhood of 0, and the family

$$\mathscr{U} = \{V_{y_1,\ldots,y_n,\varepsilon} : n \in \mathbb{N},\ y_1, \ldots, y_n \in F,\ \varepsilon > 0\}$$

defines a basis of neighbourhoods of 0 for $\sigma(E, F)$.

In the same way, if $x_1, \dots, x_n \in E$ and $\varepsilon > 0$, then

$$V_{x_1,\dots,x_n,\varepsilon} = \{y \in F : |\langle x_i, y\rangle| < \varepsilon \,,\ i = 1, \dots n\}$$

is a neighbourhood of 0, and the family

$$\mathscr{V} = \{U_{x_1,\dots,x_n,\varepsilon} : n \in \mathbb{N},\ x_1, \dots, x_n \in E,\ \varepsilon > 0\}$$

defines a basis of neighbourhoods of 0 for $\sigma(F, E)$.

Let us note that we have defined a topology, both on E and on F, that comes only from the linear structure, plus the duality assumption. There is no requirement whatsoever of a previous topological structure on any of the two spaces. Note also that, in general, the family of seminorms (or of neighbourhoods of 0) that define the topology is far from being countable; hence we cannot expect that it is always metrizable.

Proposition 3.5

1. *Let $(x_i)_i$ be a net in E. Then $x_i \overset{\sigma(E,F)}{\longrightarrow} x$ if and only if $\langle x_i, u\rangle \to \langle x, u\rangle$ for all $u \in F$.*
2. *Let $(u_i)_i$ be a net in F. Then $u_i \overset{\sigma(F,E)}{\longrightarrow} u$ if and only if $\langle x, u_i\rangle \to \langle x, u\rangle$ for all $x \in E$.*

Proof For the first statement, if $x_i \overset{\sigma(E,F)}{\longrightarrow} x$, then, given $u \in F$, we have $p_{\{u\}}(x_i - x) \to 0$, and this immediately gives the claim. Suppose now that $\langle x_i, u\rangle \to \langle x, u\rangle$ for every $u \in F$ and fix some 0-neighbourhood $V_{u_1,\dots,u_n,\varepsilon}$ in $(E, \sigma(E, F))$. Given $\varepsilon > 0$, for each $j = 1, \dots, n$, there is $i_j \in I$ so that

$$|\langle x_i - x, u_j\rangle| = |\langle x_i, u_j\rangle - \langle -x, u_k\rangle| < \varepsilon$$

for every $i > i_j$. Choose now $i_0 \in I$ so that $i_0 > i_j$ for every $j = 1, \dots, n$ and

$$\max_{j=1,\dots,n} |\langle x_i - x, u_j\rangle| < \varepsilon$$

for every $i > i_0$. Hence $x_i - x \in V_{u_1,\dots,u_n,\varepsilon}$ and this gives the conclusion. The second statement follows in the same way. □

Proposition 3.6 $(E, \sigma(E, F))' = F$ *and* $(F, \sigma(F, E))' = E$.

Proof We prove only the first statement, being the proof of the second one analogous. We start by taking some $u \in (E, \sigma(E, F))'$ (that is, a linear $u : E \to \mathbb{K}$ that is continuous with respect to $\sigma(E, F)$). Then there are $u_1, \dots, u_n \in F$ and $c > 0$ such that

$$|u(x)| \leq c \max_{i=1,\dots n} |u_i(x)| \,.$$

Hence $\bigcap_{i=1}^{n} \ker(u_i) \subseteq \ker(u)$, and, by Lemma 3.2, there exist $\lambda_1, \ldots, \lambda_n \in \mathbb{K}$ such that $u = \sum_{i=1}^{n} \lambda_i u_i \in F$.
Conversely, we fix $u \in F$ and, as we saw in Remark 3.5, $u(x) = \langle x, u \rangle$ defines a linear form. To see that it is continuous, it is enough to take the seminorm $p_{\{u\}}$. □

We see now that if E is a locally convex space, then the original topology is stronger than $\sigma(E, E')$. For this reason, this topology is often known as the *weak topology*. Also, the topology $\sigma(E', E)$ on E' is called the *weak* topology*.

Proposition 3.7 *Let $(E, \mathscr{P})$ be a locally convex space. Then $\sigma(E, E') \subseteq \tau_{\mathscr{P}}$. In particular, for every sequence $(x_n)_n$ in E, if $x_n \overset{\tau_{\mathscr{P}}}{\longrightarrow} x$, then $x_n \overset{\sigma(E,E')}{\longrightarrow} x$.*

Proof Take $J = \{u_1, \ldots, u_n\} \subseteq E'$, and then for each k, there are $c_k > 0$ and $p_k \in \mathscr{P}$ such that $|u_k(x)| \leq c_k p_k(x)$ for every $x \in E$. Taking $c = \max(c_1, \ldots, c_n)$ and $p \in \mathscr{P}$ with $\max(p_1, \ldots, p_n) \leq p$, we have

$$p_J(x) \leq cp(x) \text{ for all } x \in E\,.$$

This, by Remark 2.8, gives the claim. The statement about the convergence of sequences follows immediately from this inequality. It also follows from Proposition 3.5 and the fact that all u are continuous. □

Proposition 3.8 *Let E be a vector space and $F \subseteq E^*$ that separates points. If τ is a locally convex topology on E so that every $v \in F$ is τ-continuous, then $\sigma(E, F) \subseteq \tau$.*

Proof Fix some net $(x_i)_i \subseteq E$ converging in τ to some $x \in E$, and let us see that it also converges to x in $\sigma(E, F)$. To do that, take any $v \in F$. If $\mathcal{P}$ is a family of seminorms defining τ, by Theorem 2.11, there is $C > 0$ so that $|v(y)| \leq Cp(y)$ for every $y \in E$. For each $\varepsilon > 0$, since the net converges, we can find i_0 such that $p(x_i - x) < \frac{\varepsilon}{C}$ for all $i \geq i_0$. Then, for such i, we have

$$|v(x_i) - v(x)| = |v(x_i - x)| \leq Cp(x_i - x) < \varepsilon\,.$$

This shows that $v(x_i) \to v(x)$ and, by Proposition 3.5, completes the proof. □

Proposition 3.9 *Let (E, F) be a dual pair and G a vector subspace of E. Then (G, F) is a dual pair if and only if G is $\sigma(E, F)$-dense in E.*

Proof Let us suppose that there exists $x_0 \in E \setminus \overline{G}^{\sigma(E,F)}$. Since (E, F) is a dual pair, by Proposition 3.6, we have that $\big(E, \sigma(E, F)\big)' = F$, and then by Proposition 3.3 (note that $\overline{G}^{\sigma(E,F)}$ is a proper closed subspace), there is $u \in F$ such that $u(x_0) \neq 0$ (hence $u \neq 0$) and $u(x) = 0$ for every $x \in G$. Then (DP1) is not satisfied and (G, F) is not a dual pair.
Assume, conversely, that G is $\sigma(E, F)$-dense in E, and let us see that (G, F) is a dual pair. Take some $0 \neq u \in F$ and suppose that $u(x) = 0$ for every $x \in G$. Given any $x \in E$, we can find a net $(x_i)_i \subseteq G$ that $\sigma(E, F)$ converges to x. This implies

(see Proposition 3.5) that $\lim_i u(x_j) = u(x)$ and, since $u(x_j) = 0$ for every j, we have that $u(x) = 0$ and then $u = 0$. This shows that condition (DP1) is satisfied. Since (DP2) is trivially satisfied (because $G \subseteq E$), this completes the proof. □

Proposition 3.10 *Let $(E, \mathscr{P})$ be an l.c.s. and $A \subseteq E$ be convex. Then, $\overline{A} = \overline{A}^{\sigma(E,E')}$.*

Proof First of all, since the original topology $\tau_{\mathscr{P}}$ is finer than $\sigma(E, E')$, we have that $\overline{A}^{\sigma(E,E')}$ is closed in E and, then, $\overline{A} \subseteq \overline{A}^{\sigma(E,E')}$. To see the converse inclusion, take some $x_0 \notin \overline{A}$ and use Theorem 3.4 to find $u \in E'$ and $\gamma \in \mathbb{R}$ so that

$$\operatorname{Re} u(x_0) < \gamma < \inf_{x \in \overline{A}} \operatorname{Re} u(x) .$$

By Proposition 3.6, $u : (E, \sigma(E, E')) \to \mathbb{K}$ is continuous and, therefore, so also is $\operatorname{Re} u$. Then the set $\{x \in E \colon \operatorname{Re} u(x) < \gamma\}$ is a $\sigma(E, E')$-neighbourhood of x_0 that does not intersect A. This shows that $x_0 \notin \overline{A}^{\sigma(E,E')}$ and completes the proof. □

Let us suppose now that E is a Banach space. Then E' is again a Banach space and (E, E') is a dual pair. In this case, the weak topology $\sigma(E, E')$ on E is often denoted by w, while the weak* topology $\sigma(E', E)$ on E' is denoted by w^*.

Example 3.12 The weak topology on c_0 is $w = \sigma(c_0, \ell_1)$ (recall Example 2.16). Take $e_n = (\delta_{in})_{i=1}^{\infty} \in c_0$ for $n \in \mathbb{N}$, and let us see that $e_n \overset{w}{\to} 0$. Indeed, if $x \in \ell_1$, then

$$\langle e_n, x \rangle = \sum_{i=1}^{\infty} x_i \delta_{in} = x_n \to 0 = \langle 0, x \rangle \text{ as } n \to \infty .$$

But obviously e_n does not converge to 0 in $\| \cdot \|_\infty$ since $\|e_n\|_\infty = 1$ for all n. □

We can go one step further. Note that the dual space E' is itself a Banach space, and we can consider its dual E'' and the dual pair (E', E''). This defines a weak topology $\sigma(E', E'')$. Then, on E', we can consider two weak topologies: the $\sigma(E', E) = w^*$ topology that comes from looking at it as dual of E and the weak topology $\sigma(E', E'') = w$ that comes from considering it as predual of E''. Since E can be in a natural way looked at as a subspace of E'' by means of the identification $J : E \to E''$ given by $J(x) : X' \to \mathbb{K}$ with $J(x)(u) = u(x)$, then $\sigma(E', E) \subseteq \sigma(E', E'')$. We see now that they may be different.

Example 3.13 The w^* topology on ℓ_1 is the one defined by the dual pair (c_0, ℓ_1). In the same way as in Example 3.12, we see that the sequence $(e_n)_n$ in ℓ_1 weak* converges to 0. On the other hand, $\ell_1' = \ell_\infty$, and the weak topology is $w = \sigma(\ell_1, \ell_\infty)$. Then $e_n \overset{w}{\to} 0$ if and only if $\langle e_n, u \rangle \to \langle 0, u \rangle$ for every $u \in \ell_\infty$. But if we take $u = (1, 1, 1, 1, 1, \dots) \in \ell_\infty$, we have $\langle e_n, u \rangle = 1$ for all n, and this shows that $(e_n)_n$ cannot w-converge to 0. □

3.4 The Bipolar Theorem

Given a dual pair (E, F), each subset A of E defines a set in F (known as *polar set* of A) in a natural way

$$A^\circ = \{u \in F : |\langle x, u\rangle| \le 1 \text{ for every } x \in A\} \subseteq F .$$

Proposition 3.11 *Let (E, F) be a dual pair and A a subset of E.*

1. *A° is absolutely convex and $\sigma(F, E)$-closed.*
2. *If $B \subseteq A$, then $A^\circ \subseteq B^\circ$.*
3. *if $\lambda \in \mathbb{K}$, $\lambda \neq 0$, then $(\lambda A)^\circ = \lambda^{-1} A^\circ$.*
4. *$A \subseteq A^{\circ\circ}$.*
5. *$A^\circ = A^{\circ\circ\circ}$.*
6. *$\left(\overline{A}\right)^\circ = A^\circ$.*
7. *$\left(\bigcup_{i\in I} A_i\right)^\circ = \bigcap_{i\in I} A_i^\circ$.*
8. *If M is a vector subspace of E, then $M^\circ = \{u \in F\colon \langle x, u\rangle = 0$ for all $x \in M\}$.*

Proof For a fixed $x \in E$, the set $\{u \in F : |\langle x, u\rangle| \le 1\}$ is absolutely convex and, since x is $\sigma(E, E')$-continuous (see Proposition 3.6), closed. Then 1 follows once we have observed that

$$A^\circ = \bigcap_{x\in A} \{u \in F : |\langle x, u\rangle| \le 1\} .$$

Statements 2–4 follow immediately from the definition. As a straightforward consequence, we have

$$A^\circ \subseteq \left(A^\circ\right)^{\circ\circ} = \left(A^{\circ\circ}\right)^\circ \subseteq A^\circ .$$

For 6, take $u \in A^\circ$ and $x \in \overline{A}$. We can find a net $(x_j) \subseteq A$ converging to x. Since $|\langle x_j, u\rangle| \le 1$ for every j and $\langle x_j, u\rangle \to \langle x, u\rangle$, this yields $u \in \left(\overline{A}\right)^\circ$. The converse inclusion follows from 2.
In order to see 7, note that $u \in \left(\bigcup_{i\in I} A_i\right)^\circ$ if and only if $|\langle x, u\rangle| \le 1$ for every $x \in A_i$ and all $i \in I$. This gives the conclusion.
Suppose now that M is a vector subspace, choose some $u \in M^\circ$ and fix $x \in M$. Since $\lambda x \in M$ for every $\lambda > 0$, we have $|\langle \lambda x, u\rangle| \le 1$ and, then, $|\langle x, u\rangle| \le 1/\lambda$ for every $\lambda > 0$. Therefore $\langle x, u\rangle = 0$. This shows 8 since the converse inclusion is obvious. □

Proposition 3.12 *Let E be an l.c.s and consider the dual pair (E, E'). Then $H \subseteq E'$ is equicontinuous if and only if there is a 0-neighbourhood $U \subseteq E$ so that $H \subseteq U^\circ$.*

Proof By Proposition 2.6, H is equicontinuous if and only if we can find $p \in \mathscr{P}$ and $C > 0$ so that $|\langle x, u\rangle| \leq Cp(x)$ for every $x \in X$ and $u \in H$. Just taking $U = B_p(0, 1/C)$ gives the conclusion. □

Proposition 3.13 *Let E be an l.c.s. and consider the dual pair (E, E^*). If $\mathscr{U}$ is any basis of 0-neighbourhoods, then $E' = \bigcup_{U \in \mathscr{U}} U^\circ$.*

Proof Let us note first of all that in this case

$$U^\circ = \{u \in E^* : |\langle x, u\rangle| \leq 1 \text{ for every } x \in U\}$$

for each $U \in \mathscr{U}$. To begin with, if $u \in U^\circ$, then u is bounded in U and, hence, is continuous. This shows $\bigcup_{U \in \mathscr{U}} U^\circ \subseteq E'$. Conversely, if $u \in E'$, then there is $U \in \mathscr{U}$ such that $|\langle x, u\rangle \leq 1$ for every $x \in U$. This completes the proof. □

This procedure of taking polars can be iterated. If (E, E') is a dual pair and F is a vector space with $E \subseteq F \subseteq (E')^*$, then (E', F) is a dual pair. With this, given $A \subseteq E$, we take first the polar with respect to the dual pair (E, E') (we denote it $A^{\circ_{EE'}}$ for a moment). This is a subset of E', and then we can consider its polar with the dual pair (E', F), that is,

$$\big(A^{\circ_{EE'}}\big)^{\circ_{E'F}} \subseteq F\,.$$

This set is called the *bipolar* of A. A simple calculation from the definition shows that

$$A^{\circ_{EE'}\circ_{E'F}} = \Big\{z \in F \colon |\langle z, u\rangle| \leq \sup_{x \in A} |\langle x, u\rangle| \text{ for every } u \in E'\Big\}\,. \tag{3.9}$$

Theorem 3.14 (Bipolar Theorem) *Let (E, E') be a dual pair and F a vector space with $E \subseteq F \subseteq (E')^*$. Then*

$$A^{\circ_{EE'}\circ_{E'F}} = \overline{\Gamma(A)}^{\sigma(F,E')}$$

for every $A \subseteq E$.

Proof In order to keep the notation as clean as possible, we write just $A^{\circ\circ}$ for the bipolar. By Proposition 3.11, $A^{\circ\circ}$ is absolutely convex and $\sigma(F, E')$-closed. This implies $\overline{\Gamma(A)}^{\sigma(F,E')} \subseteq A^{\circ\circ}$. To see the converse inclusion, choose $x_0 \in F \setminus \overline{\Gamma(A)}^{\sigma(F,E')}$. We can use Corollary 3.5 (and Proposition 3.6) to find $u \in E'$ so that $|\langle x_0, u\rangle| > 1$ and $|\langle x, u\rangle| \leq 1$ for every $x \in \overline{\Gamma(A)}^{\sigma(F,E')}$ (in particular, for every $x \in A$). This implies $u \in A^{\circ_{EE'}}$ and, then, $x_0 \notin A^{\circ_{EE'}\circ_{E'F}}$. □

An interesting particular case is when we take $F = E$. In this case, since $\Gamma(A)$ is convex, Proposition 3.10 gives

$$A^{\circ\circ} = \overline{\Gamma(A)}^{\sigma(E,E')} = \overline{\Gamma(A)}^{\tau}$$

for every $A \subseteq E$. Another interesting case is that of Banach spaces. The polar of the open unit ball of a Banach space X is easily found

$$B_X^{\circ} = \{u \in X' : |u(x)| \leq 1, \ \forall x \in B_X\} = \{u \in X' : \|u\| \leq 1\} = \overline{B}_{X'} \tag{3.10}$$

and, analogously, $(B_{X'})^{\circ} = \overline{B}_{X''}$. Then Theorem 3.14 yields

$$B_{X''} = (B_X)^{\circ\circ} = \overline{\Gamma(B_X)}^{\sigma(X'',X')} = \overline{B_X}^{\sigma(X'',X')}.$$

As a matter of fact, this is a classical fact in Banach space theory, known as *Goldstine theorem*: for each $x \in X''$ with $\|x\| \leq 1$, there is a net $(x_i) \subseteq X$ with $\|x_i\| \leq 1$ that $\sigma(X'', X')$-converges to x.

Theorem 3.15 (Alaoglu-Bourbaki) *If U is a 0-neighbourhood of an l.c.s. E, then U° is $\sigma(E', E)$-compact.*

Proof Let us denote $\mathbb{K}_x = \mathbb{K}$ for each $x \in E$ and consider the product space $\mathbb{P} = \prod_{x \in E} \mathbb{K}_x$. Define now the mapping $\phi : (E', \sigma(E', E)) \longrightarrow \mathbb{P}$ by $\phi(u) = (u(x))_{x \in E}$. It is easy to check that ϕ is a linear isomorphism onto the image $\phi(E')$ when the latter is endowed with the induced topology. So, to get the conclusion, it is enough to see that $\phi(U^{\circ})$ is compact in $\mathbb{P}$.

First of all, note that, since U is a 0-neighbourhood, it is absorbing, and for each $x \in E$, we can find $\lambda_x > 0$ so that $\lambda_x^{-1} x \in U$. Hence $|u(\lambda_x^{-1} x)| \leq 1$ and, therefore, $|u(x)| \leq \lambda_x$ for every $u \in U^{\circ}$. Thus

$$\phi(U^{\circ}) \subseteq \prod_{x \in E} \overline{B_{\mathbb{K}}(0, \lambda_x)},$$

which, by Tikhonov's theorem 2.1, is compact. This shows that $\phi(U^{\circ})$ is relatively compact; it remains to prove that it is closed. To see this, take some net $(u_j) \subseteq U^{\circ}$ so that $\big(\phi(u_i)\big)_i$ is convergent in $\mathbb{P}$ to some $y = (y_x)_{x \in E}$, and let us see that $y \in \phi(U^{\circ})$. We define $u : E \to \mathbb{K}$ by $u(x) = y_x$. By the definition of ϕ, we have that $u_j(x) \to y_x$ (in $\mathbb{K}$) for every $x \in E$ and

$$\begin{aligned} u(\alpha_1 x_1 + \alpha_2 x_2) = y_{\alpha_1 x_1 + \alpha_2 x_2} &= \lim_j u_j(\alpha_1 x_1 + \alpha_2 x_2) \\ &= \lim_j \alpha_1 u_j(x_1) + \lim_j \alpha_2 u_j(x_2) = \alpha_1 u(x_1) + \alpha_2 u(x_2). \end{aligned}$$

This shows that u is linear. Since $u_j \in U^\circ$ for every j, we have $|u_j(x)| \leq 1$ for every $x \in U$. Taking limits yields $|u(x)| \leq 1$ for every $x \in U$. Then u is continuous (hence $u \in E'$) and $u \in U^\circ$. Since obviously $\phi(u) = y$, the set $\phi(U^\circ)$ is closed and this completes the proof. □

A straightforward consequence of this result and (3.10) is the following.

Corollary 3.6 *Let X be a normed (or Banach) space. Then $\overline{B}_{X'}$ is $\sigma(X', X)$-compact. In particular, every bounded subset of X' is $\sigma(X', X)$-relatively compact.*

3.5 The Mackey–Arens Theorem

Given a dual pair (E, F), we say that a locally convex topology τ on E is *of the dual pair* if $(E, \tau)' = F$. Clearly $\sigma(E, F)$ is a topology of the dual pair (recall Proposition 3.6) and, by Proposition 3.8, if τ is a topology of the dual pair (E, F), then $\sigma(E, F) \subseteq \tau$. If $(E, \mathscr{P})$ is a locally convex space, then $\tau_{\mathscr{P}}$ is an obvious example of a topology of the dual pair (E, E'). Our aim in this section is to characterize the topologies of a dual pair, showing that these correspond to topologies of uniform convergence on certain families of compact sets (see Theorem 3.16 for a precise statement).

Let $(E, \mathscr{P})$ be locally convex space and fix some equicontinuous $S \subseteq E'$. Then there is some $U \in \mathscr{U}_0(E)$ so that $S \subseteq U^\circ$, and, for each $x \in E$, we can find $\lambda > 0$ so that $x \in \lambda U$ (that is, $x/\lambda \in U$). Then

$$\sup_{u \in S} \left|u\left(\tfrac{x}{\lambda}\right)\right| \leq \sup_{u \in U^\circ} \left|u\left(\tfrac{x}{\lambda}\right)\right| \leq \frac{1}{\lambda}.$$

As a consequence,

$$q_S(x) = \sup_{u \in S} |u(x)| \tag{3.11}$$

is finite for every $x \in E$ and, therefore, defines a seminorm on E. We consider the family

$$\mathscr{Q} = \{q_S \colon S \subseteq E' \text{ equicontinuous}\}$$

and the topology defined by it.

Proposition 3.14 *For every locally convex space, we have $\tau_{\mathscr{P}} = \tau_{\mathscr{Q}}$.*

Proof We begin by showing that every $p \in \mathscr{P}$ belongs to $\mathscr{Q}$. Given some p, consider $U_p = \{x \in E \colon p(x) \leq 1\}$ and its polar $U_p^\circ \subseteq E'$ that (see Proposition 3.12) is equicontinuous. We take the associated seminorm and observe that $q_{U_p^\circ}(x) \leq 1$ if and only if $|u(x)| \leq 1$ for every $u \in U_p^\circ$, and this happens if and only if

$x \in U_p^{\circ\circ}$. Note that U_p is absolutely convex and closed and, then, by the Bipolar Theorem 3.14, $U_p = U_p^{\circ\circ}$. Summarizing, $q_{U_p^\circ}(x) \leq 1$ if and only if $p(x) \leq 1$. Hence, by Lemma 2.1, $p = q_{U_p}$ and (recall Remark 2.8) $\tau_{\mathscr{P}} \subseteq \tau_{\mathscr{Q}}$.
Fix now some equicontinuous $S \subseteq E'$ and (using Proposition 3.12) take U, an absolutely convex neighbourhood of 0 such that $S \subseteq U^\circ$. Choose $\varepsilon > 0$ and $p \in \mathscr{P}$ so that $\{x \in E : p(x) \leq \varepsilon\} \subseteq U$. Now, if $x \in E$ is such that $p(x) \leq \varepsilon$, then $|u(x)| \leq 1$ for every $u \in S$ and, $q_S(x) \leq 1$. Hence $\{x \in E : p(x) \leq \varepsilon\} \subseteq \{x \in E : q_S(x) \leq 1\}$. Using again Lemma 2.1, this gives $q_S \leq \frac{1}{\varepsilon} p$ and, by Remark 2.8, $\tau_{\mathscr{Q}} \subseteq \tau_{\mathscr{P}}$ and completes the proof. □

Let us now observe that if τ is a locally convex topology on E of the dual pair (E, F) (that is, so that $F = (E, \tau)'$) and $U, V \subseteq E$ are τ-closed, absolutely convex neighbourhoods, then, by the Bipolar Theorem 3.14,

$$\begin{aligned} \overline{\Gamma(U^\circ \cap V^\circ)}^{\sigma(F,E)} &= \big(\Gamma(U^\circ \cup V^\circ)\big)^{\circ\circ} = \Big[\big(\Gamma(U^\circ \cup V^\circ)^\circ\Big]^\circ \\ &= \big(U^{\circ\circ} \cap V^{\circ\circ}\big)^\circ = \big(U \cap V\big)^\circ . \end{aligned} \tag{3.12}$$

Theorem 3.16 (Mackey–Arens) *Let (E, F) be a dual pair and τ a Hausdorff locally convex topology on E. The following statements are equivalent:*

1. *τ is a topology for the dual pair.*
2. *There is a family $\mathscr{A}$ of $\sigma(F, E)$-relatively compact subsets of F satisfying*

$$\bigcup_{S \in \mathscr{A}} S = F \tag{3.13}$$

$$\overline{\Gamma(S \cup T)}^{\sigma(F,E)} \in \mathscr{A} \text{ for every } S, T \in \mathscr{A} \tag{3.14}$$

and such that

$$\tau = \tau_{\mathscr{Q}_{\mathscr{A}}}, \tag{3.15}$$

where $\mathscr{Q}_{\mathscr{A}} = \{q_S : S \in \mathscr{A}\}$ and the seminorms are defined in (3.11).

Proof We begin by assuming that τ is a topology for the dual pair (E, F). We consider the family

$$\mathscr{A} = \{S \subseteq F : (E, \tau)\text{-equicontinuous in } F\}.$$

Let us note first of all that, by the Alaoglu-Bourbaki Theorem 3.15, every (E, τ)-equicontinuous set is (E, τ)-relatively compact. Since every point in F is (E, τ)-equicontinuous, we immediately have (3.13). On the other hand, if $S, T \in \mathscr{A}$, we can find $U, V \in \mathscr{U}_0(E)$ so that $S \subseteq U^\circ$ and $T \subseteq V^\circ$ (recall Proposition 3.12). Using (3.12), we have

$$S \cup T \subseteq (U \cap V)^\circ = \overline{\Gamma(U^\circ \cap V^\circ)}^{\sigma(F,E)},$$

and this implies $\overline{\Gamma(S \cup T)}^{\sigma(F,E)} \in \mathscr{A}$. Finally, (3.15) follows immediately from Proposition 3.14.

Assume now that 2 holds and that there is a family $\mathscr{A}$ satisfying (3.13) and (3.14). Our aim is to see that $\tau_{\mathscr{Q}_{\mathscr{A}}}$ is a topology for the dual pair, that is, $(E, \tau_{\mathscr{Q}_{\mathscr{A}}})' = F$. Observe first of all that (3.13) implies $\sigma(E, F) \subseteq \tau_{\mathscr{Q}_{\mathscr{A}}}$ and then

$$F = \big(E, \sigma(E, F)\big)' \subseteq (E, \tau_{\mathscr{Q}_{\mathscr{A}}})' .$$

This gives one of the inclusions. To see the converse, one takes some $v \in (E, \tau_{\mathscr{Q}_{\mathscr{A}}})' \subseteq E^*$ and finds some $S \in \mathscr{A}$ and $\lambda > 0$ so that $|v(x)| \leq \lambda q_S(x)$ for every $x \in E$. This gives

$$\left| \frac{1}{\lambda} v(x) \right| \leq \sup_{u \in S} |u(x)|$$

for every $x \in E$, and (see (3.9) and the Bipolar Theorem 3.14)

$$\frac{1}{\lambda} v \in S^{\circ FE \circ EE^*} = \overline{\Gamma(S)}^{\sigma(E^*,E)} .$$

On the other hand, $\overline{\Gamma(S)}^{\sigma(F,E)}$ is by (3.14) $\sigma(F, E)$-compact and $\sigma(E^*, E)|_F = \sigma(F, E)$. Hence

$$S^{\circ FE \circ EE^*} = \overline{\Gamma(S)}^{\sigma(F,E)} ,$$

from which we deduce $\frac{1}{\lambda} v \in F$ and $v \in F$. This shows that $(E, \tau_{\mathscr{Q}_{\mathscr{A}}})' \subseteq F$ and completes the proof. □

Notice that, with this point of view, the family

$$\mathscr{A} = \{S \subseteq F : \text{ subset of the absolutely convex hull of finite subsets of } F\} ,$$

which clearly consists of $\sigma(F, E)$-compact sets, defines the topology $\sigma(E, F)$. We consider now the family

$$\mathscr{M} = \{S \subseteq F : \text{ subset of an absolutely convex } \sigma(F, E)\text{-compact set}\} . \tag{3.16}$$

This satisfies both (3.13) and (3.14), and then $\tau_{\mathscr{Q}_{\mathscr{A}}}$ is a topology of the dual pair. We call it the *Mackey topology* and denote it by $\mu(E, F)$.

Corollary 3.7 *Given a dual pair (E, F), the Mackey topology is the finest locally convex topology of the dual pair. That is, a locally convex topology τ is of the dual pair if and only if $\sigma(E, F) \subseteq \tau \subseteq \mu(E, F)$.*

Proof Let τ be a topology of the dual pair (E, F). Clearly τ is finer than the topology $\sigma(E, F)$ (recall Proposition 3.8). On the other hand, by Theorem 3.16, there is a family $\mathscr{A}$ consisting of $\sigma(F, E)$-relatively compact subsets of F satisfying (3.13) and (3.14) such that the topology τ is the topology $\tau_{\mathscr{Q}_{\mathscr{A}}}$. Since $\mathscr{A}$ is contained in the family $\mathscr{M}$ from (3.16), it readily follows that τ is coarser than $\mu(E, F)$. □

3.6 The Banach–Steinhaus Theorem

Theorem 3.17 (Banach–Steinhaus) *Let E be a Fréchet space and F a locally convex space. If $H \subseteq L(E, F)$ is such that*

$$H(x) = \{Tx : T \in H\}$$

is bounded in F for every $x \in E$, then H is equicontinuous.

Proof Let $(p_n)_n$ be an increasing countable family of seminorms defining the topology of E. We proceed by contradiction, and suppose that H is not equicontinuous. Then, there exists a continuous seminorm q on F such that for every n and every $c > 0$ we may find $T_{n,c} \in H$ and $x_{n,c} \in E$ satisfying

$$q(T_{n,c}x_{n,c}) > c \text{ and } p_n(x_{n,c}) < \frac{1}{2^n}.$$

Take, then, $x_1 \in E$ and $T_1 \in H$ such that

$$q(T_1x_1) > 1 \text{ and } p_1(x_1) < \frac{1}{2}.$$

Now, since T_1 is continuous, we may find $n(2) > 1 = n(1)$ so that $q(T_1x) \leq p_{n(2)}(x)$ for every $x \in E$. Then, denoting $M_1 = \sup_{T \in H} q(Tx_1) < \infty$, we can take $x_2 \in E$ and $T_2 \in H$ in such a way that

$$q(T_2x_2) > 2 + M_1 + 1 \text{ and } p_{n(2)}(x_2) < \frac{1}{2^2}.$$

Repeating this, we can by induction define sequences $(x_n)_n \subseteq E$, $(T_n)_n \subseteq H$ and $(n(k))_k \subseteq \mathbb{N}$ strictly increasing so that

$$p_{n(k)}(x_k) < \frac{1}{2^k} \tag{3.17}$$

for every k,

$$q(T_k x_k) > k + 1 + \sum_{j=1}^{k-1} M_j \tag{3.18}$$

(where $M_k = \sup_{T\in H} q(Tx_k) < \infty$) for every k, and

$$q(T_j x) \le p_{n(k)}(x) \tag{3.19}$$

for every $x \in E$, every $k \in \mathbb{N}$, and all $j \le k$. Our aim now is to see that $\sum_{k=1}^{\infty} x_k$ converges in E. Fix $m \in \mathbb{N}$ and find j so that $m < n(j)$. For each $i > j$ and $l \in \mathbb{N}$, we get, using (3.19) and (3.17),

$$p_m\Big(\sum_{k=i}^{i+l} x_k\Big) \le \sum_{k=i}^{i+l} p_{n(j)}(x_k) \le \sum_{k=i}^{i+l} p_{n(k)}(x_k) < \sum_{k=i}^{i+l} \frac{1}{2^j} < \frac{1}{2^{i-1}}\,.$$

This shows that the sequence $\big(\sum_{k=1}^{n} x_k\big)_n$ is Cauchy and, hence, convergent. Denote $x = \sum_{k=1}^{\infty} x_k$. Given $j \in \mathbb{N}$, we have (using now (3.18), (3.19) and (3.17))

$$\begin{aligned} q(T_j x) &\ge q(T_j x_j) - \sum_{k=1}^{j-1} q(T_j x_k) - \sum_{k=j+1}^{\infty} q(T_j x_k) \\ &> j + 1 + \sum_{i=1}^{j-1} M_i - \sum_{k=1}^{j-1} q(T_j x_k) - \sum_{k=j+1}^{\infty} p_{n(k)}(x_k) \\ &= j + 1 + \sum_{i=1}^{j-1} \big(M_i - q(T_j x_k)\big) - \sum_{k=j+1}^{\infty} p_{n(k)}(x_k) \\ &> j + 1 - \sum_{k=j+1}^{\infty} p_{n(k)}(x_k) \ge j + 1 - \sum_{k=1}^{\infty} \frac{1}{2^k} = j\,. \end{aligned}$$

This implies that $H(x)$ is not bounded and completes the proof. □

In particular, we have the following version for Banach and normed spaces.

Theorem 3.18 *Let X be a Banach space and Y a normed space. If $H \subseteq L(X,Y)$ is such that $H(x)$ is bounded in Y for every $x \in X$, then*

$$\sup_{T\in H} \|T\| < \infty\,.$$

This means that if a set is pointwise bounded, then it is also uniformly bounded. For this reason this theorem is also known as the *uniform boundedness principle*.

As a consequence of Theorem 3.17, we show that the bounded sets on every topology of a dual pair are the same (see Theorem 3.19). We need some preliminary work. Given an absolutely convex, bounded set B of a locally convex space E, we define

$$E_B = \bigcup_{n\in\mathbb{N}} nB = \operatorname{span}(B).$$

The set B is clearly absorbing in E_B, and we may consider on it the Minkowski gauge p_B (recall the definition in (3.4)).

Proposition 3.15 *Let $(E, \mathscr{P})$ be a locally convex space and $B \subseteq E$ bounded and absolutely convex. Then (E_B, p_B) is a normed space so that $(E_B, p_B) \hookrightarrow (E, \mathscr{P})$ is continuous.*

Proof By Lemma 3.3, p_B is a seminorm. To see that it is a norm, we pick $x \in E_B$ so that $p_B(x) = 0$ (note that this implies that $x \in \varepsilon B$ for every $\varepsilon > 0$) and we want to see that $x = 0$. Take $U \in \mathscr{U}_0(E)$ and find $\lambda > 0$ so that $B \subseteq \lambda U$. Then $x \in (1/\lambda)B \subseteq U$, and this implies that $x \in \cap\{U : U \in \mathscr{U}_0(E)\} = \{0\}$ (because E is Hausdorff). Therefore $x = 0$ and p_B is a norm. If $U \in \mathscr{U}_0(E)$, it is easy to see (since B is bounded) that there is $\varepsilon > 0$ such that $p_B(x) < \varepsilon$ implies $x \in U$. This gives the continuity of the inclusion. □

Proposition 3.16 *Let E be a locally convex space and $B \subseteq E$ bounded, absolutely convex and closed. If either E is sequentially complete or B is compact, then E_B is a Banach space.*

Proof Let $(x_n)_n$ be a Cauchy sequence in E_B. This means that for every $\varepsilon > 0$ there is n_0 so that

$$x_n - x_m \in \varepsilon B \text{ for every } n, m \geq n_0. \tag{3.20}$$

Suppose in first place that E is sequentially complete. Since $E_B \hookrightarrow E$ is continuous, the sequence $(x_n)_n$ is also Cauchy in E and, then, $\tau_{\mathscr{P}}$-converges to some $x_0 \in E$. Now, the sequence, being Cauchy, is bounded in E_B, and we can find some $\lambda > 0$ such that $x_n \in \lambda B$ for every n. Then, since B is $\tau_{\mathscr{P}}$-closed, $x_0 \in \lambda B \subseteq E_B$. Taking limits (with respect to $\tau_{\mathscr{P}}$) as $m \to \infty$ in (3.20), we have that for every $\varepsilon > 0$ there is n_0 so that $x_n - x_0 \in \varepsilon B$ for every $n \geq n_0$, and $x_n \to x_0$ in E_B.
Suppose now that B is $\tau_{\mathscr{P}}$-compact. As before, there is some $\lambda > 0$ such that $x_n \in \lambda B$ for every n. Then (by compactness) $(x_n)_n$ has a $\tau_{\mathscr{P}}$-cluster point $x_0 \in E_B$. Since $(x_n)_n$ is Cauchy in E, this implies that the sequence converges (in $\tau_{\mathscr{P}}$) to x_0. The proof goes on as in the previous case. □

Theorem 3.19 (Banach–Mackey) *Let (E, F) be a dual pair and τ a topology (on E) of the dual pair. For each $A \subseteq E$, the following statements are equivalent:*

1. *A is $\mu(E, F)$-bounded.*
2. *A is τ-bounded.*
3. *A is $\sigma(E, F)$-bounded.*

Proof As a straightforward consequence of Corollary 3.7, we have 1 $\Rightarrow$ 2 $\Rightarrow$ 3. To complete the proof, assume that A is $\sigma(E, F)$-bounded, and let us see that it is $\mu(E, F)$-bounded. To do this, it suffices to show that for every $\sigma(F, E)$-compact absolutely convex set $B \subseteq F$ (recall (3.16)) and $M = \overline{\Gamma(A)}^{\sigma(E,F)}$, we have

$$\sup_{x\in M}\sup_{u\in B}|u(x)| < \infty\,. \tag{3.21}$$

So, we fix any such B and consider F_B that by Proposition 3.16 (applied to $(F, \sigma(F, E))$ and B) is a Banach space. Since each $x \in M \subseteq E$ is $\sigma(F, E)$-continuous (Proposition 3.6) and $F_B \hookrightarrow \big(F, \sigma(F, E)\big)$ is continuous (Proposition 3.15), the restriction $x|_{F_B}$ is again continuous and

$$H = \{x|_{F_B} : x \in M\} \subseteq (F_B)'\,.$$

The set M is $\sigma(E, F)$-bounded, and this easily implies that $H(u) = \{x(u) : x \in M\}$ is bounded in $\mathbb{K}$ for every $u \in F$. The Banach–Steinhaus theorem (in its version of Theorem 3.18) yields (note that B is the unit ball of F_B)

$$\sup_{x\in M}\sup_{u\in B}|x(u)| < \infty\,.$$

This gives (3.21) and completes the proof. □

As a consequence of the Banach–Steinhaus theorem, we can conclude (as we already announced in Sect. 1.3.2) that there are continuous functions whose Fourier series does not converge at 0. We define $C(\mathbb{T})$ as the space of 2π-periodic continuous functions $f : \mathbb{R} \to \mathbb{C}$ which, with the norm

$$\|f\|_\infty = \sup_{t\in[-\pi,\pi]} |f(t)|\,,$$

is a Banach space. Note that $\|f\|_\infty = \sup_{t\in\mathbb{R}} |f(t)|$ for every $f \in C(\mathbb{T})$. For each $n \in \mathbb{N}$, we define $u_n : C(\mathbb{T}) \to \mathbb{C}$ as (recall (1.7) for the definition of s_n)

$$u_n(f) = s_n(f, 0) = \sum_{|j|\le n} \widehat{f}(j)\,.$$

This is clearly well defined and linear. Taking the Dirichlet kernel (recall (1.10)) and using Proposition 1.5, we can reformulate this as

$$u_n(f) = \frac{1}{2\pi} \int_{-\pi}^{\pi} f(-t) D_n(t) dt \,. \tag{3.22}$$

From this, we have

$$|u_n(f)| \leq \frac{1}{2\pi} \int_{-\pi}^{\pi} |f(-t)| \, |D_n(t)| dt \leq \|f\|_\infty \frac{1}{2\pi} \int_{-\pi}^{\pi} |D_n(t)| dt \tag{3.23}$$

for every $f \in C(\mathbb{T})$. Hence u_n is continuous and $u_n \in C(\mathbb{T})'$ for every $n \in \mathbb{N}$.

Lemma 3.5 *For each $n \in \mathbb{N}$, we have*

$$\|u_n\| = \frac{1}{2\pi} \int_{-\pi}^{\pi} |D_n(t)| dt \,.$$

Proof We fix n and define $f : [-\pi, \pi] \to \mathbb{C}$ as

$$f(t) = \begin{cases} 1 & \text{if } D_n(t) \geq 0 \\ -1 & \text{if } D_n(t) < 0. \end{cases}$$

Since this is real-valued and discontinuous only on a finite number of points, there is a sequence $(f_j)_j \subseteq C(\mathbb{T})$ so that $-1 \leq f_j(t) \leq 1$ for every j and t and $\lim_j f_j(t) = f(t)$ for every $t \in [-\pi, \pi]$. From (3.22) and the theorem of the dominated convergence, we get

$$\begin{aligned} \lim_j u_n(f_j) &= \lim_j \frac{1}{2\pi} \int_{-\pi}^{\pi} f_j(-t) D_n(t) dt \\ &= \frac{1}{2\pi} \int_{-\pi}^{\pi} f(-t) D_n(t) dt = \frac{1}{2\pi} \int_{-\pi}^{\pi} |D_n(t)| dt \,. \end{aligned}$$

Hence

$$\|u_n\| \geq \sup_{j \in \mathbb{N}} |u_n(f_j)| \geq \frac{1}{2\pi} \int_{-\pi}^{\pi} |D_n(t)| dt \,.$$

The converse inequality follows immediately from (3.23), and this completes the proof. □

Our next step is to estimate the behaviour of these norms as n grows.

Lemma 3.6 *There exist $c_1, c_2 > 0$ so that*

$$c_1 \sum_{k=0}^{2n} \frac{1}{k+1} \le \frac{1}{2\pi} \int_{-\pi}^{\pi} |D_n(t)| dt \le c_2 \sum_{k=0}^{2n} \frac{1}{k+1},$$

for every $n \ge 2$.

Proof We begin by noting that from the fact that D_n is even and (1.11), we have

$$\frac{1}{2\pi} \int_{-\pi}^{\pi} |D_n(t)| dt = \frac{1}{\pi} \int_{0}^{\pi} \left| \frac{\sin((n+1)/2)t)}{\sin(t/2)} \right| dt = \frac{2}{\pi} \int_{0}^{\pi/2} \left| \frac{\sin((2n+1)u)}{\sin u} \right| du. \tag{3.24}$$

Since

$$\frac{2}{\pi} u \le \sin u \le u \tag{3.25}$$

for every $0 \le u \le \pi/2$, then

$$\begin{aligned}
\frac{1}{2\pi} \int_{-\pi}^{\pi} |D_n(t)| dt &\ge \frac{2}{\pi} \int_{0}^{\pi/2} \frac{|\sin((2n+1)u)|}{u} du \\
&= \frac{2}{\pi} \int_{0}^{(2n+1)\pi/2} \frac{|\sin w|}{w} dw = \frac{2}{\pi} \sum_{k=0}^{2n} \int_{k\pi/2}^{(k+1)\pi/2} \frac{|\sin w|}{w} dw \\
&\ge \frac{4}{\pi^2} \sum_{k=0}^{2n} \frac{1}{k+1} \int_{k\pi/2}^{(k+1)\pi/2} |\sin w| dw = \frac{4}{\pi^2} \sum_{k=0}^{2n} \frac{1}{k+1}.
\end{aligned}$$

On the other hand, using again (3.25) after (3.24), we have

$$\begin{aligned}
\frac{1}{2\pi} \int_{-\pi}^{\pi} |D_n(t)| dt &\le \int_{0}^{\pi/2} \frac{|\sin((2n+1)u)|}{u} du = \int_{0}^{(2n+1)\pi/2} \frac{|\sin w|}{w} dw \\
&= \sum_{k=0}^{2n} \int_{k\pi/2}^{(k+1)\pi/2} \frac{|\sin w|}{w} dw \le \frac{\pi}{2} + \sum_{k=1}^{2n} \int_{k\pi/2}^{(k+1)\pi/2} \frac{|\sin w|}{w} dw \\
&\le \frac{\pi}{2} + \sum_{k=1}^{2n} \frac{1}{k} \le c_2 \sum_{k=0}^{2n} \frac{1}{k+1},
\end{aligned}$$

for some $c_2 \ge 1$. □

We have now everything at hand to state and prove our claim.

Theorem 3.20 *There exists a function in $C(\mathbb{T})$ whose Fourier series does not converge at* 0.

Proof Let us suppose that this is not the case, and $\lim_n u_n(f) = \lim_n s_n(f, 0) = f(0)$ for every $f \in C(\mathbb{T})$. This implies that the set $\{u_n(f) : n \in \mathbb{N}\}$ is bounded and, by Theorem 3.18,

$$\sup_{n\in\mathbb{N}} \|u_n\| < \infty ,$$

and this contradicts Lemmas 3.5 and 3.6. □

3.7 The Banach-Schauder Theorem

If E and F are locally convex spaces, then the *graph* of a mapping $T : E \to F$ is defined as the set

$$\mathsf{G}(T) = \{(x, y) \in E \times F : Tx = y\} .$$

Suppose that T is continuous. If a net $(x_i)_i$ converges to some $x \in E$, then the net $(T_i x)_i$ converges to some $y \in F$ and $y = Tx$. Let us note that this then implies that if (x_i, Tx_i) converges in $E \times F$ to (x, y), and then $y = Tx$ and $\mathsf{G}(T)$ is closed. Our aim in this section is to show that in some cases (if E and F are Fréchet spaces and T is linear) this implication can be reversed, and the fact that the graph is closed gives the continuity of T. In this way, we obtain another characterization of the continuity of linear operators (recall that we already have one in Theorem 2.11) between Fréchet spaces. This result, known as the *closed graph theorem* (see Theorem 3.23), follows from another important result, known as the *Banach-Schauder* or *open mapping theorem* (see Theorem 3.22), which tells that every continuous linear mapping is open. The proof of this relies on the following result, known as the *Baire category principle*.

Theorem 3.21 (Baire) *Let M be a complete metric space. If $\{G_n\}_{n\in\mathbb{N}}$ is a sequence of dense open subsets of M, then $\bigcap_{n=1}^{\infty} G_n$ is dense in M.*

Proof If we write $A = \bigcap_{n=1}^{\infty} G_n$, what we have to show is that $A \cap G \neq \varnothing$ for every open, non-empty G. First of all, since G_1 is dense, we have $G_1 \cap G \neq \varnothing$ and, being the intersection open, we can find $x_1 \in M$ and $0 < r_1 < 1$ so that

$$\overline{B}(x_1, r_1) = \{x \in M : d(x, x_1) \leq r_1\} \subseteq G_1 \cap G .$$

With the same idea, using the denseness of G_2, we may find $x_2 \in M$ and $0 < r_2 < 1/2$ so that $\overline{B}(x_2, r_2) \subseteq G_2 \cap B(x_1, r_1)$. Proceeding in this way, by induction, we can produce two sequences $(x_n)_n \subseteq M$ and $(r_n)_n \subseteq \mathbb{R}$ so that

$$\overline{B}(x_n, r_n) \subseteq G_n \cap B(x_{n-1}, r_{n-1}) \text{ and } r_n < \frac{1}{n}.$$

Given n_0, we have that $x_n, x_m \in \overline{B}(x_{n_0}, r_{n_0}) \subseteq \overline{B}(x_{n_0}, 1/n_0)$ for every $n, m \geq n_0$. Hence the sequence $(x_n)_n$ is Cauchy and, since M is complete, converges to some $x \in M$.
If we fix $n_0 \in \mathbb{N}$, then $x_n \in \overline{B}(x_{n_0}, r_{n_0})$ for every $n \geq n_0$ and, being the ball closed, $x \in \overline{B}(x_{n_0}, r_{n_0}) \subseteq G_{n_0}$. Thus $x \in A$. On the other hand, $\overline{B}(x_{n_0}, r_{n_0}) \subseteq \overline{B}(x_1, r_1) \subseteq G$ and completes the proof. □

Remark 3.6 A straightforward argument shows that an open set G of a topological space X is dense if and only if $X \setminus G$ (that is, closed) has empty interior. Then as a consequence of Theorem 3.21, if $\{C_n\}_n$ is a sequence of closed sets with empty interior of a complete metric space, then $\bigcup_{n=1}^{\infty} C_n$ has empty interior.

A mapping between topological spaces $T : X \to Y$ is said to be *open* if it sends open sets to open sets. That is, if $T(G)$ is open in Y for every open, $G \subseteq X$. We start by giving a characterization of open linear mappings on locally convex spaces, in terms of absolutely convex neighbourhoods, in the same spirit of Theorem 2.11.

Lemma 3.7 *Let E and F be locally convex spaces and $T : E \to F$ linear. Then the following are equivalent:*

1. *T is open.*
2. *If U is a neighbourhood of 0 in E, then $T(U)$ is a neighbourhood of 0 in F.*
3. *For every U, absolutely convex neighbourhood of 0 in E, there exists V, an absolutely convex neighbourhood of 0 in F such that $V \subseteq T(U)$.*

Proof Let $U \subseteq E$ be a neighbourhood of 0. Then there is some open G so that $0 \in G \subseteq U$. Then $0 \in T(G) \subseteq T(U)$. If T is open, $T(G)$ is an open set, and $T(U)$ is a neighbourhood of 0.
Now, if U is absolutely convex, the linearity of T implies that so also is $T(U)$. Then 3 is a straightforward consequence of 2.
Finally, assume that 3 holds, and let us see that T is open. If $G \subseteq E$ is open and we take $y \in T(G)$, we can find $x \in G$ so that $Tx = y$ and U, an absolutely convex neighbourhood of 0, with $x + U \subseteq G$. Then, by assumption, we can find V, an absolutely convex neighbourhood of 0 in F with $V \subseteq T(U)$. Then

$$y + V \subseteq Tx + T(U) \subseteq T(x + U) \subseteq T(G).$$

This shows that $T(G)$ is open. □

Lemma 3.8 *Let E and F be complete metrisable locally convex spaces and $T : E \to F$ be linear and surjective. For each $r > 0$, consider the sets*

$$V_r = \{x \in E : d_E(x, 0) < r\} \text{ and } W_r = \{y \in F : d_F(y, 0) < r\},$$

where d_E and d_F are the metric defining the topologies of E and F, respectively. Then, for every $r > 0$, there is $\rho > 0$ such that $W_\rho \subseteq \overline{T(V_r)}$. In particular, $\overline{T(V_r)}$ is a neighbourhood of 0 for each $r > 0$.

Proof Fix $r > 0$ and take $x \in E$. The sequence $\left(\frac{1}{n}\right)_n$ tends to 0 as $n \to \infty$, and $\frac{1}{n}x \in V_{r/2}$ for some n. This shows that $E = \bigcup_{n\in\mathbb{N}} nV_{r/2}$ and, since T is surjective,

$$F = T(E) = \bigcup_{n\in\mathbb{N}} nT\left(V_{r/2}\right) = \bigcup_{n\in\mathbb{N}} n\overline{T\left(V_{r/2}\right)}\,.$$

As a consequence of Baire's Theorem 3.21 (see Remark 3.6), $n_0\overline{T\left(V_{r/2}\right)}$ has non-empty interior for some n_0, which gives that $\overline{T\left(V_{r/2}\right)}$ has non-empty interior. Then there are $y_0 \in F$ and $\rho > 0$ so that $y_0 + W_\rho \subseteq \overline{T\left(V_{r/2}\right)}$, and if $y \in W_\rho$, there is $w \in \overline{T\left(V_{r/2}\right)}$ with $y = w - y_0$. Choose sequences $(x_n)_n$ and $(x_n')_n$ in $V_{r/2}$ such that $Tx_n \to w$ and $Tx_n' \to y_0$, and define $y_n = x_n - x_n'$ for each n. Simple computations show that $(y_n)_n \in V_r$ and $T(y_n) \to y$. Hence $y \in \overline{T(V_r)}$, and this completes the proof. □

Theorem 3.22 (Banach-Schauder/open mapping) *Let E, F be Fréchet spaces and $T : E \to F$ be linear, continuous and such that $T(E) = F$. Then T is open. In particular, if T is bijective, linear and continuous, then T^{-1} is continuous as well.*

Proof Let us denote by d_E and d_F the metrics defining the topology on E and F, respectively. Let $U \subseteq E$ be an absolutely convex neighbourhood of 0. Since T is linear, $T(U)$ is again absolutely convex. So, in view of Lemma 3.7, if we show that $T(U)$ is a neighbourhood of 0 in F, we are done.
To begin with, there is $r > 0$ so that

$$V = \{x \in E : d_E(x, 0) < r\} \subseteq V\,.$$

Now, for each $n \in \mathbb{N}$, consider the set

$$V_n = \left\{x \in E : d_E(x, 0) < \tfrac{r}{2^n}\right\}.$$

Our aim is to see that $\overline{T(V_1)} \subseteq T(U)$. Since $\overline{T(V_1)}$ is a neighbourhood of 0 in F (see Lemma 3.8), this implies that $T(U)$ is again a neighbourhood of 0.
Using Lemma 3.8, we can find a decreasing sequence of positive numbers, $(\delta_n)_n$ tending to 0 such that, denoting

$$W_n = \{y \in F : d_F(y, 0) < \delta_n\}\,,$$

we have $W_{n+1} \subseteq W_n$ and $W_n \subseteq \overline{T(V_{n+1})}$ for every n. Take now $y_1 \in \overline{T(V_1)}$. Then $(y_1 - W_1) \cap T(V_1) \neq \varnothing$, and we can find $x_1 \in V_1$ such that $Tx_1 \in y_1 - W_1$, that is, $d(Tx_1 - y_1, 0) < \delta_1$. We take now $y_2 = y_1 - Tx_1 \in W_1 \subseteq \overline{T(V_2)}$ to find $x_2 \in V_2$ such that $d(Tx_2 - y_2, 0) < \delta_2$ and define $y_3 = y_2 - Tx_2$. Proceeding in this way, we define inductively two sequences $(x_n)_n$ and $(y_n)_n$ satisfying $x_n \in V_n$, $y_n \in W_n$

and $Tx_n = y_n - y_{n+1}$ for every $n \geq 1$. If $m > n$, we have, using (2.21),

$$d_E\Big(\sum_{i=1}^{n} x_i, \sum_{i=1}^{m} x_i\Big) = d_E\Big(\sum_{i=n+1}^{m} x_i, 0\Big) \leq \sum_{i=n+1}^{m} d_E(x_i, 0) < \sum_{i=n+1}^{m} \frac{r}{2^i}.$$

This shows that the sequence $\Big(\sum_{i=1}^{n} x_i\Big)_n$ is Cauchy and, then (since E is complete), converges to some $x \in E$. Moreover

$$d_E(x, 0) = d_E\Big(\sum_{i=1}^{\infty} x_i, 0\Big) \leq \sum_{i=1}^{\infty} d_E(x_i, 0) < \sum_{i=1}^{\infty} \frac{r}{2^i} = r,$$

and $x \in V \subseteq U$. On the other hand, for each $n \in \mathbb{N}$, we have

$$T\Big(\sum_{i=1}^{n} x_i\Big) = \sum_{i=1}^{n} Tx_i = \sum_{i=1}^{n} (y_i - y_{i+1}) = y_1 - y_{n+1}.$$

The fact that $d_F(y_n, 0) < \delta_n$ implies $\lim_n y_{n+1} = 0$. Hence, the continuity of T yields

$$Tx = T\Big(\lim_n \sum_{i=1}^{n} x_i\Big) = y_1 - \lim_n y_{n+1} = y_1,$$

and $y_1 = Tx \in T(U)$. This gives our claim and completes the proof. □

Theorem 3.23 (Closed Graph Theorem) *Let E and F be Fréchet spaces. A linear map $T : E \to F$ is continuous if and only if $\mathsf{G}(T)$ is closed in $E \times F$.*

Proof As we already pointed out, if T is continuous, then $\mathsf{G}(T)$ is closed. So, it remains only to prove the converse implication. We know (see 2.8) that $E \times F$ is a Fréchet space. Now, if $\mathsf{G}(T)$ is closed in $E \times F$, it is a Fréchet space itself (with the restriction of the product topology), and the restrictions of the projections $\pi_1 : \mathsf{G}(T) \to E$ and $\pi_2 : \mathsf{G}(T) \to F$ given by $\pi_1(x, Tx) = x$ and $\pi_2(x, Tx) = Tx$ are continuous linear maps. Note that π_1 is bijective, and, by Theorem 3.22 $\pi_1^{-1} : E \to \mathsf{G}(T)$ given by $x \rightsquigarrow (x, Tx)$ is continuous. Thus $T = \pi_2 \circ \pi_1^{-1}$, being the composition of two continuous maps, is continuous. □

3.8 Topologies on the Space of Continuous Linear Mappings

If E and F are locally convex spaces, then the set $L(E, F)$ of continuous linear mappings (recall (2.18)) is clearly a vector space. We can consider different topologies in this space.

In the first place, we consider the topology of *uniform convergence on bounded sets* is given by the family of seminorms defined for $T \in L(E, F)$ as

$$p_{B,q}(T) = \sup_{x \in B} q(Tx) ,$$

where $B \subseteq E$ is bounded, and q is a continuous seminorm on F. Note that this supremum is finite because, being T continuous and B bounded, the set $T(B)$ is bounded in F. We denote by $L_\beta(E, F)$ the space $L(E, F)$ endowed with this topology. If $\mathcal{V}$ is a basis of neighbourhoods of F, then for each $V \in \mathcal{V}$ and $B \subseteq E$ bounded, the set

$$W(B, V) = \{T \in L(E, F) \colon T(B) \subseteq V\}$$

is a neighbourhood of $0 \in L(E, F)$ for this topology, and

$$\{W(B, V) \colon V \in \mathcal{V},\ B \subseteq E \text{ bounded}\}$$

is a basis of neighbourhoods of $0 \in L_\beta(E, F)$. If $F = \mathbb{K}$, then $L(E, \mathbb{K})$ is E' (the dual of E). In this case, the topology of uniform convergence on bounded sets is called the *strong topology* of E' and is denoted by $\beta(E', E)$. Let us note that the seminorms defining the topology in this case are of the form

$$q_B(u) = \sup_{x \in B} |u(x)| , \tag{3.26}$$

where B ranges over all bounded subsets of E, and the basis of neighbourhoods is given by

$$W(B) = \{u \in E' \colon \sup_{x \in B} |u(x)| < 1\} .$$

It is easy to check that $\sigma(E', E) \subseteq \mu(E', E) \subseteq \beta(E', E)$. We write $E'_\beta = (E', \beta(E', E))$ and call it the *strong dual* of E.
If both E and F are normed spaces, then $L_\beta(E, F)$ is the space $L(E, F)$ endowed with the operator norm defined as

$$\|T\| = \sup_{\|x\|_E \leq 1} \|Tx\|_F ,$$

which we briefly introduced in Remark 2.9. In this case, if $F = \mathbb{K}$, then E'_β is the dual with the topology defined by the norm

$$\|u\| = \sup_{\|x\|_E \leq 1} |u(x)| .$$

The second topology that we are going to consider is the topology of *uniform convergence on precompact sets*, given by the family of seminorms defined by

$$p_{K,q}(T) = \sup_{x \in K} q(Tx) , \tag{3.27}$$

for $T \in L(E, F)$, where $K \subseteq E$ is precompact, and q is a continuous seminorm on F. We write $L_{co}(E, F)$ for the space $L(E, F)$ endowed with this topology.

Given a continuous seminorm q on F and $x_1, \dots, x_n \in E$, then

$$p_{\{x_1,\dots,x_n\},q}(T) = \max_{i=1,\dots,n} q(Tx_j) , \tag{3.28}$$

for $T \in L(E, F)$ defines a seminorm on $L(E, F)$. The *strong operator topology* (also known as *simple topology*) is the one defined by these seminorms as $\{x_1, \dots, x_n\}$ varies on the finite subsets of E and q on the continuous seminorms on F. We write $L_s(E, F)$ for the space $L(E, F)$ endowed with this topology. It is not difficult to check that a net $(T_i)_i$ converges to some T in the strong operator topology if and only if $T_i x \to Tx$ (in F) for every $x \in E$. If $F = \mathbb{K}$, then we recover the weak topology on E', that is, $L_s(E, \mathbb{K}) = (E', \sigma(E', E))$.

The *weak operator topology* is defined as the simple topology on $L(E, (F, \sigma(F, F')))$ and is given by the seminorms

$$p_{\{x_1,\dots,x_n\},\{u_1,\dots,u_m\}}(T) = \max\{|u_j(Tx_i)| : i = 1, \dots, n\,,\ j = 1, \dots, m\} ,$$

where $\{x_1, \dots, x_n\}$ runs over all finite subsets of E and $\{u_1, \dots, u_m\}$ over all finite subsets of F'. We write $L_w(E, F)$ for $L(E, F)$ endowed with the weak operator topology.

The following inclusions are clearly continuous:

$$L_\beta(E, F) \hookrightarrow L_{co}(E, F) \hookrightarrow L_s(E, F) \hookrightarrow L_w(E, F) ;$$

that is, each one of the topologies on the left hand side of an arrow is finer than the topology on the right-hand side of the arrow. We see now how equicontinuous sets behave with these topologies.

Proposition 3.17 *Every equicontinuous subset of $L(E, F)$ is bounded in $L_\beta(E, F)$.*

Proof Fix some equicontinuous $H \subseteq L(E, F)$ and take $B \subseteq E$ bounded and q a continuous seminorm on F. By Proposition 2.6, we can find some seminorm on E (call it p) and some $C > 0$, so that

$$q(Tx) \leq Cp(x) ,$$

for every $x \in E$ and $T \in H$. On the other hand, being B bounded, $\sup_{x\in B} p(x) = \lambda < \infty$. With this, we have

$$p_{B,q}(T) = \sup_{x\in B} q(Tx) \leq C \sup_{x\in B} p(x) = C\lambda$$

for every $T \in H$. Since this holds for every bounded B and every seminorm q, the claim follows. □

Proposition 3.18 *Let $H \subseteq L(E, F)$ be equicontinuous, and $N \subseteq E$ such that* span N *is dense in E. Then the strong operator topology on $L(E, F)$ and the topology defined by the family of seminorms*

$$\left\{p_{\{x_1,\ldots,x_n\},q} : \{x_1, \ldots, x_n\} \subseteq N \textit{ finite}, \ q \textit{ continous seminorm on } F\right\}$$

(that is, the topology of simple convergence on N) coincide on H.
In particular, a net $(T_i)_i \subseteq H$ converges to T in the strong operator topology if and only if $T_i x \to Tx$ for every $x \in N$.

Proof Since the property of being equicontinous is maintained by translation, we may assume that $0 \in H$. We fix $\{x_1, \ldots, x_s\} \subseteq E$ and a continuous seminorm q on F. Considering the seminorm $p_{\{x_1,\ldots,x_s\},q}$ defined in (3.28), we have that

$$U = \{T \in H : \max_{i=1,\ldots,s} q(Tx_i) \leq 1\} \tag{3.29}$$

is a neighbourhood of 0 in H for the strong operator topology. We want to find a neighbourhood V of 0 for the topology of simple convergence on N contained in U. Since H is equicontinuous, we can find a seminorm on E (say p) and a constant $C > 0$ so that $q(Tx) \leq Cp(x)$ for every $x \in E$ and $T \in H$. By density, for each $i = 1, \ldots, s$, we can choose $z_i \in$ span N so that

$$p(x_i - z_i) < \frac{1}{2C} .$$

Note that for each i there are $n(i) \in \mathbb{N}$ and $\alpha_{i,k} \in \mathbb{K}$, $y_{i,k} \in N$ for $k = 1, \ldots, n(i)$ so that

$$z_i = \sum_{k=1}^{n(i)} \alpha_{i,k} y_{i,k} .$$

We now consider

$$G = \{y_{i,k} : i = 1, \ldots, s, \ k = 1, \ldots, n(i)\} ,$$

set

$$\sigma > \max_{i=1,\dots,s} \sum_{k=1}^{n(i)} |\alpha_{i,k}|,$$

and define

$$V = \left\{T \in H \colon \sup_{w \in G} q(Tw) < \tfrac{1}{2\sigma}\right\},$$

which is a neighbourhood of 0 for the topology of simple convergence on N. Take now some $T \in V$, and fix $i = 1, \dots, s$; then

$$q(Tz_i) \le \sum_{k=1}^{n(i)} |\alpha_{i,k}| q(y_{i,k}) < \frac{1}{2\sigma} \sum_{k=1}^{n(i)} |\alpha_{i,k}| < \frac{1}{2}.$$

On the other hand,

$$q(T(x_i - z_i)) \le Cp(x_i - z_i) < \frac{1}{2},$$

and

$$q(Tx_i) \le q(T(x_i - z_i)) + q(Tz_i) < 1,$$

so that $T \in U$. This shows that $V \subseteq U$. Note now that the sets U defined as in (3.29), where $\{x_1, \dots, x_s\}$ varies on all the finite subsets of E and q on all continuous seminorms on F is a basis of neighbourhoods of 0 for the strong operator topology. Then the proof is complete. □

Proposition 3.19 *Let $H \subseteq L(E, F)$ be equicontinuous. The strong operator topology and the topology of uniform convergence on precompact subsets coincide on H.*

Proof Since the property of being equicontinous is maintained by translation, we may assume that $0 \in H$. Each neighbourhood of 0 in H for the topology of uniform convergence on precompact subsets contains a neighbourhood of the form

$$U = \{T \in H \colon \sup_{x \in K} q(Tx) \le 1\},$$

for some precompact $K \subseteq E$ and some q continuous seminorm on F (recall the definition of the seminorms in (3.27)). Since H is equicontinuous, we can find a seminorm on E (say p) and a constant $C > 0$ so that $q(Tx) \le Cp(x)$ for every

$x \in E$ and $T \in H$. Now, being K precompact, there are $x_1, \ldots, x_n \in K$ so that for each $x \in K$ there is some $i \in \{1, \ldots, n\}$ with $p(x - x_i) < \frac{1}{2C}$. Define now

$$V = \{T \in H\colon \max_{i=1,\ldots,n} q(Tx_i) < \tfrac{1}{2}\},$$

which is a neighbourhood of 0 in H of the strong operator topology. If $T \in V$ and $x \in K$, then

$$q(Tx) \leq q(T(x - x_i)) + q(Tz_i) \leq Cp(x - x_i) + q(Tx_i) < 1,$$

and $V \subseteq U$. This yields the claim. □

Remark 3.7 If $(T_n)_n \subseteq L(E, F)$ is equicontinuous and $(T_n x)_x$ converges (in F) for every $x \in E$, we already pointed out in Remark 2.10 that $Tx = \lim_n T_n x$ defines an operator $T \in L(E, F)$. It is straightforward to check that $\{T_n\}_n \cup \{T\}$ is again equicontinuous and, as a consequence of Proposition 3.19, $T_n \to T$ uniformly on the compact sets of E. We see now that if E is Fréchet, we can drop the assumption of $\{T_n\}_n$ being equicontinuous.

Lemma 3.9 *Let $H \subseteq L(E, F)$ be equicontinuous and $S : E \to F$ linear such that*

$$Sx \in \overline{\{Tx\colon T \in H\}}^F \tag{3.30}$$

for every $x \in E$. Then S is continuous. Furthermore, the set

$$\overline{H} = \left\{S : E \to F \text{ linear} \mid Sx \in \overline{\{Tx\colon T \in H\}} \text{ for every } x \in E\right\}$$

is equicontinuous in $L(E, F)$.

Proof Fix some continuous seminorm q on F. Since H is equicontinuous, we can find a seminorm on E (say p) and a constant $C > 0$ so that $q(Tx) \leq Cp(x)$ for every $x \in E$ and $T \in H$. Then, for each $x \in E$, by (3.30), we have

$$q(Sx) \leq \sup_{T \in H} q(Tx) \leq Cp(x), \tag{3.31}$$

which gives (recall Theorem 2.11) that S is continuous. Observe that neither p nor C in (3.31) depends on S; hence $\overline{H}$ is equicontinuous. □

Theorem 3.24 *Let E be a Fréchet space and F a locally convex space. Suppose $(T_n)_n \subseteq L(E, F)$ is such that $(T_n x)_n$ converges (in F) for every $x \in E$. Then there exists $T \in L(E, F)$ so that $T_n \to T$ uniformly on the compact sets.*

Proof We know that the mapping $T : E \to F$ given by $Tx = \lim_n T_n x$ is well defined and linear. Now, for each $x \in E$, the sequence $(T_n x)_n$ converges and, then, is bounded in F. By the Banach–Steinhaus theorem 3.17, the set $\{T_n\colon n \in \mathbb{N}\}$ is equicontinuous. Now, $Tx \in \overline{\{T_n x : n \in \mathbb{N}\}}$ (the closure taken in F) for every $x \in E$.

Lemma 3.9 gives that T is continuous (i.e., $T \in L(E, F)$) and $\{T_n : n \in \mathbb{N}\} \cup \{T\}$ is equicontinuous. Since $T_n x \to Tx$ for every x (i.e., $T_n \to T$ in the strong operator topology), Proposition 3.19 gives the claim. □

3.9 Transpose of an Operator

If E and F are two vector spaces and $T : E \to F$ is a linear mapping, its *transpose* is the mapping ${}^*T : F^* \to E^*$ defined, for $v \in F^*$, by ${}^*T(v) = v \circ T$; that is,

$$\langle {}^*T(v), x\rangle = \langle v, T(x)\rangle$$

for every $x \in E$. Then, for $v_1, v_2 \in F^*$ and $\alpha_1, \alpha_2 \in \mathbb{K}$, we have

$$\begin{aligned}\langle {}^*T(\alpha_1 v_1 + \alpha_2 v_2), x\rangle &= \langle \alpha_1 v_1 + \alpha_2 v_2, T(x)\rangle \\ &= \alpha_1\langle v_1, T(x)\rangle + \alpha_2\langle v_2, T(x)\rangle = \alpha_1\langle {}^*T(v_1), x\rangle + \alpha_2\langle {}^*T(v_2), x\rangle \\ &= \langle \alpha_1^* T(v_1) + \alpha_2^* T(v_2), x\rangle\end{aligned}$$

for every $x \in E$. Then ${}^*T$ is again linear.

Example 3.25 If $E = \mathbb{K}^n$ and $F = \mathbb{K}^m$, then for each linear mapping $T : E \to F$ there exists an $n \times m$ matrix $A = (a_{ij})$ so that $T(e_i) = \sum_{j=1}^m a_{ij} e_j$ for every $i = 1, \dots, n$. Then for $i = 1, \dots, n$ and $j = 1, \dots, m$ we have

$$\langle {}^*T(e_j), e_i\rangle = \langle e_j, T(e_i)\rangle = \langle e_j, \sum_{k=1}^m a_{ik} e_k\rangle = \sum_{k=1}^m a_{ik}\langle e_j, e_k\rangle = a_{ij}\,.$$

This shows that ${}^*T(e_j) = \sum_{i=1}^n a_{ij} e_i$, that is, the matrix associated with ${}^*T$ is the transpose of A. □

Proposition 3.20 *Let (E, E') and (F, F') be two dual pairs and $T : E \to F$ a linear mapping. Then $T : (E, \sigma(E, E')) \to (F, \sigma(F, F'))$ is continuous if and only if ${}^*T(E') \subseteq F'$.*

Proof Let us note first that ${}^*T(v) = v \circ T \in E^*$ for every $v \in F'$. Now, if T is continuous, then the composition is continuous and ${}^*T(v) \in F'$. On the other hand, if ${}^*T(v) = v \circ T \in E'$ for every $v \in F'$, then it is continuous from $(E, \sigma(E, E'))$ to $\mathbb{K}$ for every v and this easily gives that $T : (E, \sigma(E, E')) \to (F, \sigma(F, F'))$ is continuous. □

If T is $\sigma(E, E')$-$\sigma(F, F')$ continuous , we denote

$${}^tT = {}^*T|_{F'} : (F', \sigma(F', F)) \to (E', \sigma(E', E))\,.$$

Now we take $^*({}^tT) : (E', \sigma(E', E))^* \to (F', \sigma(F', F))^*$ and observe that if $x \in E = (E', \sigma(E', E))'$, then $^*({}^tT)(x) = x \circ {}^tT : F' \to \mathbb{K}$ and

$$\langle ^*({}^tT)(x), v\rangle = (x \circ {}^tT)(v) = x(v \circ T) = v \circ T(x) = v(Tx) = \langle Tx, v\rangle$$

for every $v \in F'$. Hence $^*({}^tT) = T$ on E and $^*({}^tT)(E) \subseteq F = (F', \sigma(F', F))'$. By Proposition 3.20, $^*({}^tT)$ is $\sigma(E', E)$-$\sigma(F', F)$ continuous and, of course, ${}^t({}^tT) = T$.

Proposition 3.21 *Let E and F be l.c.s. and $T : E \to F$ be linear and continuous. Then ${}^tT : F' \to E'$ is $\sigma(F', F)$-$\sigma(E', E)$ continuous*

Proof For every $v \in F'$, we have $v \circ T \in E'$ and, therefore, $^*T(F') \subseteq E'$. Proposition 3.20 gives the conclusion. □

Proposition 3.22 *Let (E, E') and (F, F') be two dual pairs and $T : (E, \sigma(E, E')) \to (F, \sigma(F, F'))$ linear and continuous.*

1. *If $A \subseteq E$, then $[T(A)]^\circ = ({}^tT)^{-1}(A^\circ)$.*
2. *If $A \subseteq E$ and $T(A) \subseteq B$, then ${}^tT(B^\circ) \subseteq A^\circ$.*
3. *if $A \subseteq E$ and $B \subseteq F$ are weakly closed and absolutely convex and ${}^tT(B^\circ) \subseteq A^\circ$, then $T(A) \subseteq B$.*

Proof The proofs are rather straightforward from the definitions. In the first place, for $A \subseteq E$, we have

$$\begin{aligned}
[T(A)]^\circ &= \{v \in F' : |\langle v, Tx\rangle| \leq 1 \text{ for all } x \in A\} \\
&= \{v \in F' : |\langle {}^tTv, x\rangle| \leq 1 \text{ for all } x \in A\} \\
&= \{v \in F' : {}^tTv \in A^\circ\} = ({}^tT)^{-1}(A^\circ)\,.
\end{aligned}$$

On the other hand, if $T(A) \subseteq B$, then $B^\circ \subseteq T(A)^\circ = ({}^tT)^{-1}(A^\circ)$ (recall Proposition 3.11) and ${}^tT(B^\circ) \subseteq A^\circ$. Note that, from this, using 1, we have $T(A^{\circ\circ}) ={}^{tt} T(A^{\circ\circ}) \subseteq B^{\circ\circ}$. If both A and B are weakly closed and absolutely convex, each one of them coincides with its bipolar and this gives the third claim. □

Proposition 3.23 *Let E and F be l.c.s. and $T : E \to F$ be linear and continuous. Then ${}^tT : F' \to E'$ is $\beta(F', F)$-$\beta(E', E)$ continuous.*

Proof This follows immediately from the definition of the β-topology (recall (3.26)) and the fact that if $B \subseteq E$ is bounded, then $T(B)$ is bounded. □

Let us recall that the kernel and the image of a linear operator $T : E \to F$ are defined as the sets

$$\ker T = \{x \in E : Tx = 0\} \text{ and } \operatorname{Im} T = \{Tx \in F : x \in E\}\,.$$

Then

$$\ker {}^tT = \{v \in F' : {}^tTv = v \circ T = 0\} = \{v \in F' : v|_{\operatorname{Im} T} = 0\} = (\operatorname{Im} T)^\circ ,$$

and

$$\begin{aligned}(\operatorname{Im} {}^tT)^\circ &= \{x \in E : \langle {}^tTv, x\rangle = 0 \text{ for all } v \in F'\} \\ &= \{x \in E : \langle v, Tx\rangle = 0 \text{ for all } v \in F'\} = \ker T ,\end{aligned}$$

where the last equality holds as a consequence of the Hahn–Banach theorem.

3.10 Exercises

3.1 Show that if E is an infinite-dimensional Banach space, then $(E, \sigma(E, E'))$ is not metrizable.

3.2 Prove Schur's theorem: every $\sigma(\ell_1, \ell_\infty)$-convergent sequence in ℓ_1 is norm convergent.

3.3 Let E be a metrizable locally convex space and $U \subseteq E$ an absolutely convex set which absorbs bounded sets. Prove that U is a neighbourhood of the origin.
Hint: Let $(U_n)_n$ be a basis of absolutely convex neighbourhoods in E and assume by contradiction that U_n is not contained in nU for each n.

3.4 Prove that the locally convex topology τ of a metrizable space (E, τ) coincides with the Mackey topology $\mu(E, E')$
Hint: If $A \subseteq E'$ is $\sigma(E', E)$-compact and absolutely convex, then A° is absolutely convex and absorbs bounded sets.

3.5 Let c_{00} be the subspace of c_0 consisting of sequences that are 0 except for a finite number of coordinates, endowed with the induced norm, $\|(x_n)_n\| = \max_n |x_n|$, whose topology we denote by τ. Prove that

1. $c_{00}' = \ell_1$.
2. $\mu(c_{00}, \ell_1) = \tau \neq \beta(c_{00}, \ell_1)$.
3. $\sigma(\ell_1, c_{00})$ is not metrizable.
4. For each bounded subset $A \subseteq c_0$, there is a bounded $B \subseteq c_{00}$ such that $A \subseteq \overline{B}^{c_0}$.
5. $\beta(\ell_1, c_0) = \beta(\ell_1, c_{00})$, and this is the norm topology on ℓ_1.
6. $A \subseteq \ell_1$ is $\sigma(\ell_1, c_{00})$-bounded if and only if for each n there is $m_n > 0$ such that

$$\sup\{|x_n| : x = (x_n)_n \in A\} \leq m_n .$$

7. Let $A = \{e_n : n \in \mathbb{N}\}$, then

   a. $\sigma(\ell_1, c_{00}) - \lim_n e_n = 0$.

b. $A \cup \{0\}$ is $\sigma(\ell_1, c_{00})$-compact.
c. A is not $\beta(\ell_1, c_{00})$-bounded.

3.6 Let (E, τ) be a metrizable locally convex space. Show that if there is a sequence $(A_n)_n$ of bounded sets such that $E = \bigcup A_n$, then E is normable.

3.7 Let φ be the (vector) space of all sequences which are 0 except for a finite number of coordinates. Consider the dual pair (φ, φ) with the bilinear form given by

$$\langle (x_n)_n, (y_n)_n \rangle = \sum_{n=1}^{\infty} x_n y_n .$$

Prove that $M \subseteq \varphi$ is

1. $\sigma(\varphi, \varphi)$-bounded if and only if $\sup_{x \in M} |x_n| < \infty$ for each $n \in \mathbb{N}$.
2. $\beta(\varphi, \varphi)$-bounded if and only if it is $\sigma(\varphi, \varphi)$-bounded and there is $n_0 \in \mathbb{N}$ such that $x_n = 0$ for every $n \geq n_0$ and all $x \in M$.

3.8 Let $(E, \mathscr{P})$ be a locally convex metrizable space and $(F, \mathscr{Q})$ a locally convex space. Show that a linear map $T : E \longrightarrow F$ is continuous if and only if for each bounded set B in E, $T(B)$ is bounded in F.

3.9 Show that every open linear map is surjective. Show that if E is a Fréchet space and $T : E \to F$ is open (thus surjective), then F is also Fréchet.

3.10 Show that every separable Banach space is isometric to a quotient of ℓ_1.
Hint: Use the open mapping theorem.

Further Reading
[1–31]

References

1. Bachman, G., Narici, L.: Functional Analysis. Dover Publications, Inc., Mineola, NY (2000). Reprint of the 1966 original
2. Banach, S.: Théorie des Opérations Linéaires. Éditions Jacques Gabay, Sceaux (1993). Reprint of the 1932 original
3. Bogachev, V.I., Smolyanov, O.G.: Topological Vector Spaces and Their Applications. Springer Monographs in Mathematics. Springer, Cham (2017). https://doi.org/10.1007/978-3-319-57117-1
4. Brezis, H.: Functional analysis. In: Sobolev Spaces and Partial Differential Equations. Universitext. Springer, New York (2011)
5. Cerdà, J.: Linear functional analysis. In: Graduate Studies in Mathematics, vol. 116. American Mathematical Society, Providence, RI; Real Sociedad Matemática Española, Madrid (2010). https://doi.org/10.1090/gsm/116
6. Conway, J.B.: A course in functional analysis. In: Graduate Texts in Mathematics, vol. 96. Springer, New York (1985). https://doi.org/10.1007/978-1-4757-3828-5

7. Dunford, N., Schwartz, J.T.: Linear Operators. Part I. Wiley Classics Library. John Wiley & Sons, Inc., New York (1988). Reprint of the 1958 original.
8. Edwards, R.E.: Functional Analysis. Dover Publications, Inc., New York (1995). Theory and applications, Corrected reprint of the 1965 original
9. Fabian, M., Habala, P., Hájek, P., Montesinos, V., Zizler, V.: Banach Space Theory. CMS Books in Mathematics/Ouvrages de Mathématiques de la SMC. Springer, New York (2011). https://doi.org/10.1007/978-1-4419-7515-7
10. Fabian, M., Habala, P., Hájek, P., Montesinos Santalucía, V., Pelant, J., Zizler, V.: Functional analysis and infinite-dimensional geometry. In: CMS Books in Mathematics/Ouvrages de Mathématiques de la SMC, vol. 8. Springer, New York (2001). https://doi.org/10.1007/978-1-4757-3480-5
11. Floret, K., Wloka, J.: Einführung in die Theorie der lokalkonvexen Räume. Lecture Notes in Mathematics, No. 56. Springer, Berlin/New York (1968)
12. Grothendieck, A.: Topological vector spaces. In: Notes on Mathematics and its Applications. Gordon and Breach Science Publishers, New York/London/Paris (1973). Translated from the French by Orlando Chaljub
13. Heuser, H.G.: Functional Analysis. A Wiley-Interscience Publication. John Wiley & Sons, Ltd., Chichester (1982). Translated from the German by John Horváth
14. Horváth, J.: Topological Vector Spaces and Distributions, vol. I. Addison-Wesley Publishing Co., Reading, Mass./London-Don Mills, Ont. (1966)
15. Jameson, G.J.O.: Topology and Normed Spaces. Chapman and Hall, London; Halsted Press, New York (1974)
16. Jarchow, H.: Locally Convex Spaces. Mathematische Leitfäden. B. G. Teubner, Stuttgart (1981)
17. Kadets, V.: A Course in Functional Analysis and Measure Theory. Universitext. Springer, Cham (2018). https://doi.org/10.1007/978-3-319-92004-7. Translated from the 2006 Russian edition
18. Kelley, J.L.: General topology. In: Graduate Texts in Mathematics, No. 27. Springer, New York/Berlin (1975). Reprint of the 1955 edition
19. Köthe, G.: Topological Vector Spaces. I. Translated from the German by D. J. H. Garling. Die Grundlehren der mathematischen Wissenschaften, Band 159. Springer, New York Inc., New York (1969)
20. Köthe, G.: Topological vector spaces. II. In: Grundlehren der Mathematischen Wissenschaften [Fundamental Principles of Mathematical Sciences], vol. 237. Springer, New York/Berlin (1979)
21. Lindenstrauss, J., Tzafriri, L.: Classical Banach spaces. I. In: Ergebnisse der Mathematik und ihrer Grenzgebiete, Band 92. Springer, Berlin/New York (1977). Sequence spaces
22. Lindenstrauss, J., Tzafriri, L.: Classical Banach spaces. II. In: Ergebnisse der Mathematik und ihrer Grenzgebiete [Results in Mathematics and Related Areas], vol. 97. Springer, Berlin/New York (1979). Function spaces
23. Meise, R., Vogt, D.: Introduction to functional analysis. In: Oxford Graduate Texts in Mathematics, vol. 2. The Clarendon Press, Oxford University Press, New York (1997). Translated from the German by M. S. Ramanujan and revised by the authors
24. Pérez Carreras, P., Bonet, J.: Barrelled locally convex spaces. In: North-Holland Mathematics Studies, vol. 131. North-Holland Publishing Co., Amsterdam (1987). Notas de Matemática, 113
25. Robertson, A.P., Robertson, W.: Topological vector spaces. In: Cambridge Tracts in Mathematics, vol. 53, 2nd edn. Cambridge University Press, Cambridge/New York (1980)
26. Rudin, W.: Functional analysis, 2nd edn. In: International Series in Pure and Applied Mathematics. McGraw-Hill, Inc., New York (1991)
27. Schaefer, H.H., Wolff, M.P.: Topological vector spaces. In: Graduate Texts in Mathematics, vol. 3, 2nd edn. Springer, New York (1999). https://doi.org/10.1007/978-1-4612-1468-7
28. Taylor, A.E., Lay, D.C.: Introduction to Functional Analysis, 2nd edn. Robert E. Krieger Publishing Co., Inc., Melbourne, FL (1986)

29. Valdivia, M.: Topics in locally convex spaces. In: Notas de Matemática [Mathematical Notes], vol. 85. North-Holland Publishing Co., Amsterdam/New York (1982)
30. Wilansky, A.: Topology for Analysis. Robert E. Krieger Publishing Co., Inc., Melbourne, FL (1983). Reprint of the 1970 edition
31. Yosida, K.: Functional analysis. In: Classics in Mathematics. Springer, Berlin (1995). https://doi.org/10.1007/978-3-642-61859-8. Reprint of the sixth (1980) edition

Chapter 4
Spaces of Holomorphic and Differentiable Functions and Operators Between Them

So far, the only examples of spaces of functions we have dealt with are those consisting of continuous functions, either on a compact set ($C(K)$) the spaces which are Banach spaces (recall Theorem 2.8) or on an open set ($C(\Omega)$) the spaces which we know that are Fréchet (Theorem 2.18) not normable (Proposition 2.14) spaces. Our aim now is to extend this, finding other classes of functions that appear naturally in the classical theory, looking at them from the point of view of the general theory that we presented in Chap. 2.

4.1 Space of Holomorphic Functions

4.1.1 Locally Convex Structure

We begin by looking at holomorphic functions of one complex variable. We choose some open and connected set in the complex plane (that, for what topology is concerned, we identify with $\mathbb{R}^2$) $\Omega \subseteq \mathbb{C}$. A function $f : \Omega \to \mathbb{C}$ is *holomorphic* if for every $z_0 \in \Omega$ the following limit exists

$$f'(z_0) = \lim_{z \to z_0} \frac{f(z) - f(z_0)}{z - z_0} .$$

We define the space

$$\mathrm{H}(\Omega) = \{f : \Omega \to \mathbb{C} \text{ holomorphic on } \Omega\} .$$

J. Bonet et al., *Function Spaces and Operators between them*,
RSME Springer Series 11, https://doi.org/10.1007/978-3-031-41602-6_4

Every holomorphic function is continuous; hence, this is a subspace of $(C(\Omega), \tau_{co})$, and on it, we consider the induced topology. That is, on $H(\Omega)$, we consider the topology of uniform convergence on compact sets, defined by the seminorms

$$p_K(f) = \sup_{z \in K} |f(z)|, \text{ for } K \Subset \Omega .$$

Obviously, in this way, $(H(\Omega), \tau_{co})$ is a metrizable l.c.s. Our aim now is to see that it is in fact a Fréchet space. We address this question by seeing that it is a closed subspace of $C(\Omega)$. We have to see that the limit point of every convergent sequence in $(H(\Omega), \tau_{co})$ also belongs to $H(\Omega)$. We recall (without proof) some classical results, the proof of which can be found, for example, in [1, 3–5, 8] (or in any other text on basic function theory of one complex variable).

Theorem 4.1 (Cauchy Theorem) *Let $\Omega \subseteq \mathbb{C}$ be open and simply connected and $f \in H(\Omega)$. Then*

$$\int_\gamma f(z)dz = 0,$$

for every closed piecewise smooth curve $\gamma \subseteq \Omega$.

Theorem 4.2 (Cauchy Integral Formula) *Let $f \in H(\Omega)$ and $\overline{\mathbb{D}(z_0, r)} \subseteq \Omega$; then, for every $z \in \mathbb{D}(z_0, r)$ and all $n \in \mathbb{N} \cup \{0\}$, we have*

$$f^{(n)}(z) = \frac{n!}{2\pi i} \int_{C(z_0,r)} \frac{f(w)}{(w-z)^{n+1}} dw ,$$

where $C(z_0, r)$ is the circle centred at z_0 of radius r with positive orientation. In particular,

$$|f^{(n)}(z_0)| \leq \frac{n!}{r^n} \sup_{z \in \mathbb{D}(z_0,r)} |f(z)| , \tag{4.1}$$

for every $n \in \mathbb{N} \cup \{0\}$.

Theorem 4.3 (Morera's Theorem) *If f, a continuous function on $\mathbb{D}(z_0, r)$ for some $z_0 \in \mathbb{C}$ and $r > 0$, is such that for each triangle $T \subseteq \mathbb{D}(z_0, r)$, we have*

$$\int_T f(z)dz = 0,$$

then f is holomorphic on $\mathbb{D}(z_0, r)$.

Theorem 4.4 *Let $(f_n)_n \subseteq H(\Omega)$ such that $f_n \overset{\tau_{co}}{\to} f$; then $f \in H(\Omega)$. That is, $(H(\Omega), \tau_{co})$ is closed in $(C(\Omega), \tau_{co})$. In particular, $(H(\Omega), \tau_{co})$ is a Fréchet space.*

Proof Let $z_0 \in \Omega$, and we choose an open disc $\mathbb{D}(z_0, r) \subset \Omega$. Obviously (since the sequence converges uniformly on the compact sets), $f \in C(\mathbb{D}(z_0, r))$. Let T be a triangle in $\mathbb{D}(z_0, r)$. Since T is compact, the sequence $(f_n)_n$ converges to f uniformly on T, and this implies that $\int_T f_n(z)dz \to \int_T f(z)dz$. On the other hand, since each f_n is holomorphic, Theorem 4.1 gives $\int_T f_n(z)dz = 0$ for every $n \in \mathbb{N}$. Hence, $\int_T f(z)dz = 0$, and by Theorem 4.3, $f \in \mathrm{H}(\mathbb{D}(z_0, r))$. In particular, f is holomorphic at z_0, and since this holds for every $z_0 \in \Omega$, the conclusion follows. □

But we have more. If the sequence converges uniformly on compact sets, then so also do all the derivatives.

Theorem 4.5 (Weierstraß Theorem) *Let $(f_n)_n \subseteq H(\Omega)$ be such that $f_n \overset{\tau_{co}}{\to} f$. Then $f_n^{(k)} \to f^{(k)}$ uniformly on compact sets for every $k \in \mathbb{N}_0$. That is, if $f_n \to f$ in $(H(\Omega), \tau_{co})$, then $f_n^{(k)} \to f^{(k)}$ in $(C(\Omega), \tau_{co})$ for every k.*

Proof Fix $k \in \mathbb{N}$. Since each compact set can be covered by a finite family of closed discs, it is enough to prove that $f_j^{(k)} \longrightarrow f^{(k)}$ uniformly on any closed disc. So, fix $z_0 \in \Omega$ and choose $r > 0$ such that $\overline{\mathbb{D}(z_0, r)} \subseteq \Omega$. We take $R > r$ such that $\overline{\mathbb{D}(z_0, R)} \subset \Omega$. By Theorem 4.2, for each $|z - z_0| \leq r$, we have

$$f_n^{(k)}(z) - f^{(k)}(z) = \frac{k!}{2\pi i} \int_{C(z_0,R)} \frac{f_n(w) - f(w)}{(w - z)^{k+1}} dw .$$

This gives

$$\begin{aligned} |f_n^{(k)}(z) - f^{(k)}(z)| &\leq \frac{k!}{2\pi} 2\pi R \frac{\max_{|w-z_0|=R} |f_n(w) - f(w)|}{(R - r)^{k+1}} \\ &= \frac{k!R}{(R - r)^{k+1}} \max_{|w-z_0|=R} |f_n(w) - f(w)| , \end{aligned}$$

and

$$\max_{z \in \overline{\mathbb{D}(z_0,r)}} |f_n^{(k)}(z) - f^{(k)}(z)| \leq \frac{k!R}{(R - r)^{k+1}} \max_{|w-z_0|=R} |f_n(w) - f(w)| .$$

Since $C(z_0, R)$ is a compact set, the last term tends to 0 as $n \to \infty$ and this completes the proof. □

Remark 4.1 Since every compact subset of $\mathbb{D}(0, R)$ is contained in $\overline{\mathbb{D}(0, r)}$ for some $0 < r < R$, Theorem 1.10 gives that the sequence of partial sums $S_N(z) = \sum_{n=0}^{N} a_n z^n$ of a power series converges uniformly over the compact sets of $\mathbb{D}(0, R)$ to the function $f(z) = \sum_{n=0}^{\infty} a_n z^n$. Each S_N is a polynomial (hence, holomorphic) and Theorem 4.4 implies that f is holomorphic on $\mathbb{D}(0, R)$. Also, Theorem 4.5 immediately gives $f^{(n)}(0) = n!a_n$ for every n.

So, the function defined by a convergent power series is always holomorphic. As a matter of fact, the converse also holds, and every holomorphic function can be represented by a power series (called the *Taylor series* of the function).

Theorem 4.6 *Suppose that f is holomorphic on $\mathbb{D}(z_0, r)$ for some $r > 0$. Then there exist $R \geq r$ such that*

$$f(z) = \sum_{n=0}^{\infty} a_n (z - z_0)^n$$

for every $z \in \mathbb{D}(z_0, R)$, where $a_n = \frac{f^{(n)}(z_0)}{n!}$ for every n. Moreover, the series converges uniformly on $\mathbb{D}(z_0, s)$ for every $0 < s < R$.

Proposition 4.1 *The space $(H(\mathbb{C}), \tau_{co})$ is separable.*

Proof Fix $0 < R$ and $\varepsilon > 0$. By Theorem 4.6, there is some N such that

$$\sup_{|z| \leq R} \left| f(z) - \sum_{n=0}^{N} a_n z^n \right| < \varepsilon .$$

Now, for each $j = 1, \ldots, N$, we can find $q_j, r_j \in \mathbb{Q}$ so that $|\operatorname{Re} a_j - q_j| < \varepsilon$ and $|\operatorname{Im} a_j - r_j| < \varepsilon$. With this, if $|z| < R$, we have

$$\left| \sum_{n=0}^{N} a_n z^n - \sum_{n=0}^{N} (q_j + i r_j) z^n \right| < \varepsilon (R^N + \cdots + R + 1) .$$

Joining these two facts, we conclude that for every compact $K \subseteq \mathbb{C}$ and $\varepsilon > 0$, we can find a polynomial P_N whose coefficients have rational real and imaginary parts so that

$$\sup_{z \in K} |f(z) - P_N(z)| < \varepsilon .$$

Hence, the set of such polynomials (which is obviously countable) is dense in $H(\mathbb{C})$. □

We have seen that, like $C(\Omega)$, the space of holomorphic functions is a Fréchet space. Our aim now is to show that it is not normable, but, unlike the case for the space of continuous functions, it does admit continuous norms. We again need a classical result from the elementary theory of holomorphic functions in one complex variable.

Theorem 4.7 (Identity Theorem) *Let Ω be an open and connected set. Let $(z_n)_n$ be a sequence with a limit point in Ω. If $f(z_n) = 0$ for all $n \in \mathbb{N}$, then $f(z) = 0$ for every $z \in \Omega$.*

Proposition 4.2 *Let Ω be an open and connected set. The space $(H(\Omega), \tau_{co})$ admits continuous norms but is not normable.*

Proof Let us see first that it admits continuous norms. We choose an infinite compact set K in Ω. We let $p_K(f) = \sup_{z\in K} |f(z)|$. Then p_K is obviously a continuous seminorm on $(\mathrm{H}(\Omega), \tau_{co})$. Since every sequence in K has a convergent subsequence, Theorem 4.7 easily gives that q is a norm.

Let us suppose now that $(\mathrm{H}(\Omega), \tau_{co})$ is normable, and let us call q the norm that induces the topology. By Remark 2.15, we can find a compact $K_0 \Subset \Omega$ and $\lambda > 0$ so that

$$q(f) \leq \lambda \sup_{z\in K_0} |f(z)|$$

for every $f \in \mathrm{H}(\Omega)$. On the other hand (also by Remark 2.15), for every compact $K \Subset \Omega$, there is $\mu > 0$ such that

$$\sup_{z\in K} |f(z)| \leq \mu q(f)$$

for all $f \in \mathrm{H}(\Omega)$. Combining these two facts, we have that for every K there is $c_K > 0$ such that

$$\sup_{z\in K} |f(z)| \leq c_K \sup_{z\in K_0} |f(z)| \,. \tag{4.2}$$

We want to see now that this is not possible. We consider two cases.

First of all, if $\Omega = \mathbb{C}$, then we choose $R > 0$ such that $K_0 \subseteq \mathbb{D}(0, R)$ and consider $K = \overline{\mathbb{D}(0, R+1)}$. By (4.2), there exists $c_R > 0$ such that

$$\sup_{|z|\leq R+1} |f(z)| \leq c_R \sup_{|z|\leq R} |f(z)|$$

for every $f \in \mathrm{H}(\mathbb{C})$. Take now $f_k(z) = z^k$ for $k \in \mathbb{N}$, and we have $(R+1)^k \leq c_R R^k$ for every $k \in \mathbb{N}$. But $\left(\frac{R+1}{R}\right)^k \xrightarrow{k} +\infty$, and this is a contradiction.

Suppose now $\Omega \neq \mathbb{C}$. We choose $\xi \in \partial\Omega$ and define $g : \Omega \to \mathbb{C}$ by $g(z) = \frac{1}{z-\xi}$. Obviously, $g \in \mathrm{H}(\Omega)$ and $\lim_{z\to\xi} |g(z)| = \infty$. We set now $M = \max_{z\in K_0} |g(z)|$. Then there exists $z_0 \in \Omega$ such that $|g(z_0)| > 2M$. If $h(z) = \frac{g(z)}{M}$, we have $h \in \mathrm{H}(\Omega)$ and $|h(z)| \leq 1$ for every $z \in K_0$, which gives $|h(z)^k| \leq 1$ for all $k \in \mathbb{N}$ and $z \in K_0$. Taking $K = \{z_0\}$, (4.2) gives some $c_K > 0$ such that $|f(z_0)| \leq c_K \sup_{z\in K_0} |f(z)|$ for every $f \in \mathrm{H}(\Omega)$. If we evaluate in $f = h^k$, we obtain

$$2^k < \left| \frac{g(z_0)}{M} \right|^k \leq c_K \sup_{z\in K_0} |h(z)|^k \leq c_K$$

for all $k \in \mathbb{N}$. This gives a contradiction and completes the proof. □

4.1.2 Representation as a Sequence Space

Our aim now is to see that the spaces of holomorphic functions on open discs can be represented as (that is, are isomorphic to) Köthe echelon spaces for a convenient choice of Köthe matrix.

Given $0 < R \leq \infty$, we consider the set $\mathbb{D}_R = \{z \in \mathbb{C} : |z| < R\}$ and the space $\mathrm{H}(\mathbb{D}_R)$ of holomorphic functions on $\mathbb{D}_R$. Note that with this notation $\mathrm{H}(\mathbb{D}_1)$ is the space of holomorphic functions on the open unit disc $\mathrm{H}(\mathbb{D})$, and $\mathrm{H}(\mathbb{D}_\infty)$ is the space of entire functions $\mathrm{H}(\mathbb{C})$. Given $(r_n)_n$, an increasing sequence of positive numbers such that $0 < r_n \nearrow R$, the sequence of seminorms

$$p_n(f) = \sup_{|z| \leq r_n} |f(z)|$$

for $n \in \mathbb{N}$, defines a fundamental sequence of seminorms for the compact-open topology on $\mathrm{H}(\mathbb{D}_R)$.

Theorem 4.8 *Let $0 < R \leq \infty$ and $0 < r_n \nearrow R$. Define $a_n(k) = r_n^k$ for each $n \in \mathbb{N}$ and $k \in \mathbb{N}_0$ and the Köthe matrix $A = (a_n)_n$. Then the spaces $H(\mathbb{D}_R)$ and $\lambda_1(A)$ are isomorphic.*

Proof We define the operator

$$T : \begin{array}{ccc} \mathrm{H}(\mathbb{D}_R) & \longrightarrow & \lambda_1(A) \\ \sum_{k=0}^\infty c_k z^k & \longmapsto & (c_k)_{k \in \mathbb{N}_0} \end{array},$$

where $c_k = \dfrac{f^{(k)}(0)}{k!}$. That is, for a function f, we define Tf as the sequence of the coefficients of the Taylor series expansion around 0. We have to see that T defines an isomorphism. We divide the proof into several steps.

First of all, we have to see that the operator is well defined. Given $f \in \mathrm{H}(\mathbb{D}_R)$ with Taylor expansion $\sum_{k=0}^\infty c_k z^k$, we have, by the Cauchy Integral Formula,

$$c_k = \frac{1}{2\pi i} \int_{C(0, r_{n+1})} \frac{f(w)}{w^{k+1}} dw$$

for each $k \geq 0$. This gives

$$|c_k| \leq \frac{1}{r_{n+1}^k} \max_{|w| = r_{n+1}} |f(w)| \leq \frac{1}{r_{n+1}^k} p_{n+1}(f).$$

Therefore,

$$q_n((c_k)_k) = \sum_{k=0}^\infty a_n(k)|c_k| \leq p_{n+1}(f) \sum_{k=0}^\infty \left(\frac{r_n}{r_{n+1}}\right)^k \leq c p_{n+1}(f). \tag{4.3}$$

This shows that $(c_k)_k \in \lambda_1(A)$, and then T is well defined. Let us note that (4.3) also shows that T is continuous.

If $Tf = 0$, then $f^{(k)}(0) = 0$ for every $k \in \mathbb{N}$ and $f = 0$ in $\mathbb{D}(0, R)$. This gives $\ker T = \{0\}$, and therefore, T is injective. We see now that it is also surjective and that T^{-1} is continuous. To do that, we take some $(c_k)_k \in \lambda_1(A)$ and define

$$f(z) = \sum_{k=0}^{\infty} c_k z^k .$$

Now, given $z \in \mathbb{D}_R$, we take n so that $|z| \leq r_n$, and we have

$$\sum_{k=0}^{\infty} |c_k z^k| \leq \sum_{k=0}^{\infty} |c_k| r_n^k = q_n((c_k)_k) .$$

Then the series converges absolutely on $\mathbb{D}_R$ and then uniformly on the compact sets. By Weierstraß theorem, $f \in \mathrm{H}(\mathbb{D}_R)$ and clearly $Tf = (c_k)_k$. Hence, T is surjective. Finally, for each n, we have

$$p_n(T^{-1}(c_k)_k) = p_n\Big(\sum_{k=0}^{\infty} c_k z^k\Big) \leq \sum_{k=0}^{\infty} |c_k| r_n^k = q_n((c_k)_k)$$

and T^{-1} is continuous: □

Corollary 4.1

1. $H(\mathbb{D}) \cong \lambda_1(B)$, *with* $B = (b_n)_n$ *defined by* $b_n(k) = (1 - \frac{1}{n})^k$ *for* $k = 0, 1, 2, \ldots$.
2. $H(\mathbb{C}) \cong \lambda_1(C)$, *with* $C = (c_n)_n$ *defined by* $c_n(k) = n^k$ *for* $k = 0, 1, 2, \ldots$.

Remark 4.2 Let us make a short comment on the seminorms defining the topology of the space $\mathrm{H}(\mathbb{C})$ of entire functions. First of all, since $\{z \in \mathbb{C} \colon |z| \leq n\}$ is a fundamental sequence of compact sets in $\mathbb{C}$, the τ_{co} topology is generated by the sequence of seminorms defined as

$$p_n(f) = \sup_{|z| \leq n} |f(z)|$$

for $f \in \mathrm{H}(\mathbb{C})$. But Corollary 4.1 provides us with an alternative family of seminorms: if $f(z) = \sum_{n=0}^{\infty} a_n z^n$, then the family

$$\|f\|_k = \sum_{n=0}^{\infty} |a_n| k^n ,$$

for $k \in \mathbb{N}$, also defines the τ_{co} topology. We can give even a third family defining the topology. Letting

$$|f|_k = \sup_{n \in \mathbb{N}_0} |a_n| n^k ,$$

for $k \in \mathbb{N}$, it is straightforward to check that $|f|_k \leq \|f\|_k \leq \frac{\pi^2}{6} |f|_{2k}$ for every $f \in \mathrm{H}(\mathbb{C})$, so that the family $(|\cdot|_k)_k$ defines the same topology.

4.1.3 Montel Theorem

In Theorem 2.28, we gave a clear picture of how the compact sets of continuous functions are. We see now what do compact sets of holomorphic functions look like.

Lemma 4.1 *Every bounded subset H of $(H(\Omega), \tau_{co})$ is equicontinuous.*

Proof For each compact $K \subseteq \Omega$, there is $\lambda_K > 0$ (see Remark 2.16) such that

$$\sup_{f \in H} \sup_{z \in K} |f(z)| \leq \lambda_K .$$

We fix now $a \in \Omega$, and we want to see that H is equicontinuous at a. We fix $r > 0$ and $\delta > 0$ so that $K = \overline{\mathbb{D}(a, r)} \subseteq L = \overline{\mathbb{D}(a, r + \delta)} \subseteq \Omega$. Note that, for each $z \in K$, the circle with centre z and radius δ is contained in L. Then for each $f \in H$, we have

$$|f'(z)| = \left| \frac{1}{2\pi i} \int_{C(z,\delta)} \frac{f(\omega)}{(\omega - z)^2} d\omega \right| \leq \frac{\lambda_L}{\delta} ,$$

and $|f(z) - f(a)| \leq |z - a| \sup_{w \in K} |f'(w)| \leq \frac{\lambda_L}{\delta} |z - a|$. Then, given $\varepsilon > 0$ and z such that $|z - a| < \frac{\varepsilon\delta}{\lambda_L}$, we have $|f(z) - f(a)| \leq \varepsilon$. This shows that H is equicontinuous at a. □

As an immediate consequence of this lemma and Theorem 2.28, we have a description of the relatively compact sets of holomorphic functions.

Theorem 4.9 (Montel) *Every bounded subset H of $(H(\Omega), \tau_{co})$ is relatively compact.*

In particular, if $(f_n)_n$ is τ_{co}-bounded in $H(\Omega)$, there are a subsequence $(f_{n_j})_j$ and $f \in H(\Omega)$ such that $f_{n_j} \longrightarrow f$ for τ_{co}.

We finish this section with an extension of this result. We need a short topological lemma.

Lemma 4.2 *Let (K, τ) be a Hausdorff compact topological space. If σ is a Hausdorff topology on K and $\sigma \subseteq \tau$, then $\tau = \sigma$.*

Proof We just have to see that every τ-closed set is σ-closed. Take then F τ-closed. Since (K, τ) is compact, F is τ-compact. Now, since every σ-open set is also τ-open, every covering by σ-open sets admits a finite subcovering and F is σ-compact. Being σ Hausdorff, this implies that F is σ-closed. □

Theorem 4.10 (Vitali) *Let $\Omega \subset \mathbb{C}$ be an open connected set. Set $(f_n)_n \subseteq H(\Omega)$, τ_{co}-bounded. If there exists $A \subset \Omega$ with an accumulation point such that for each $x \in A$ the pointwise limit $\lim_{n\to\infty} f_n(x)$ exists, then there is $f \in H(\Omega)$ such that $f_n \longrightarrow f$ in τ_{co}.*

Proof The set $H = \overline{\{f_n : n \in \mathbb{N}\}}^{\tau_{co}}$ is τ_{co}-bounded. Then, by Theorem 4.9, there are $f \in \mathrm{H}(\Omega)$ and $(f_{n_j})_j$ such that $f_{n_j} \overset{j\to\infty}{\longrightarrow} f$ in $(\mathrm{H}(\Omega), \tau_{co})$.

What we want to do now is to see that, in fact, $f_n \longrightarrow f$ in τ_{co}. To do that we consider on $\mathrm{H}(\Omega)$ the topology of pointwise convergence on A, that we denote by τ_A. This is defined by the family of seminorms $p_F(f) = \max_{z\in F} |f(z)|$ for $F \subseteq A$ finite. Let us note that a sequence $(g_n)_n$ converges to 0 in τ_A if and only if $g_n(x) \to 0$ for every $x \in A$. We clearly have that $\tau_A \subseteq \tau_{co}$. Let us see that it is also Hausdorff. If $g \in \mathrm{H}(\Omega)$ is such that $g(z) = 0$ for every $z \in A$, then, by the identity principle of holomorphic functions (Theorem 4.7), the fact that A has an accumulation point gives $g(z) = 0$ for every $z \in \Omega$. Proposition 2.2 shows then that the topology τ_A is Hausdorff. Therefore, by Lemma 4.2, $\tau_A|_H = \tau_{co}|_H$.

Now, if $z \in A$, we have on the one hand that $\big(f_n(z)\big)_n$ is convergent and on the other hand that $\lim_j f_{k_j}(x) = f(x)$. Hence, $\lim_n f_n(z) = f(z)$ and $f_n \overset{\tau_A}{\longrightarrow} f$. But, since both topologies coincide on H, this implies that f_n converges to f on τ_{co}. □

Remark 4.3 Montel's theorem is a consequence of Vitali's theorem. Let $\Omega \subset \mathbb{C}$ be open and connected, and let $H \subseteq (\mathrm{H}(\Omega), \tau_{co})$ be bounded. Fix $(f_n)_n \subseteq H$. Select a sequence $(z_j)_j \subseteq \Omega$, $z_j \neq z_k$ if $j \neq k$ and $z_j \longrightarrow z_0 \in \Omega$. Set $A = (z_j)_j$. By a diagonal procedure, we can select a subsequence $(f_{n_k})_k$ such that $(f_{n_k}(z_j))_k$ is convergent for each $j = 1, ..$ By Vitali's theorem, there is $f \in \mathrm{H}(\Omega)$ such that $f_{n_k} \longrightarrow f$ in $(\mathrm{H}(\Omega), \tau_{co})$. This implies that H is relatively compact in $(\mathrm{H}(\Omega), \tau_{co})$.

4.1.4 Dual of the Space of Entire Functions

Our aim now is to give a representation of the dual of $\mathrm{H}(\mathbb{C})$. Let us note in the first place that Corollary 4.1 and Proposition 2.10 already give a representation in terms of a space of sequences. But we are interested in a representation in terms of holomorphic functions. In fact, we give two such representations: in terms of germs of holomorphic functions and in terms of holomorphic functions of exponential type by means of the Fourier–Borel transform.

Germs of holomorphic functions

Let us consider the class of complex-valued functions f that are holomorphic at some open disc $\mathbb{D}(0, \varepsilon)$ for $\varepsilon > 0$ and, on it, define a relation $f \sim g$ if there is some $\varepsilon > 0$ so that $f(z) = g(z)$ for every $z \in \mathbb{D}(0, \varepsilon)$. This clearly defines an equivalence relation, and we define H($\{0\}$) to be the space of equivalence classes, which we call *germs* at 0. We identify each class of equivalence (germ) with any of its representatives. If $f \in$ H($\{0\}$), then it is holomorphic on some disc and, then, $f(z) = \sum_{n=0}^{\infty} \lambda_n z^n$ for every $|z| < \varepsilon$. Now, for $g \sim f$, we have that $f(z) = g(z)$ in some $\mathbb{D}(0, \delta)$ and, hence, $g(z) = \sum_{n=0}^{\infty} \lambda_n z^n$ for every $|z| < \delta$. In other words, the sequence of coefficients $(\lambda_n)_n$ is independent of the choice of representative of the germ, and we write $f = \sum_{n=0}^{\infty} \lambda_n z^n$. This power series converges on some disc $\mathbb{D}(0, \varepsilon)$, and therefore,

$$\sup_{n \in \mathbb{N}_0} |\lambda_n| \delta^n < \infty\,, \tag{4.4}$$

for $0 < \delta < \varepsilon$. Note also that if the power series $\sum_{n=0}^{\infty} \lambda_n z^n$ converges at some point, then it automatically defines a germ in H($\{0\}$).

Theorem 4.11 *If $L \in H(\mathbb{C})'$, then $\sum_{n=0}^{\infty} L(z^n) z^n$ defines a germ in $H(\{0\})$. Moreover, the mapping $T : H(\mathbb{C})' \to H(\{0\})$ defined by $T(L) = \sum_{n=0}^{\infty} L(z^n) z^n$ is a linear bijection.*

Proof Given any $L \in$ H($\mathbb{C}$)$'$, there are $k \in \mathbb{N}$ and $C > 0$ so that (recall the seminorms defined in Remark 4.2)

$$|L(f)\| \leq C\|f\|_k = C \sum_{n=0}^{\infty} |a_n| k^n\,,$$

for every $f(z) = \sum_{n=0}^{\infty} a_n z^n \in$ H($\mathbb{C}$). Taking $f = z^n$, we have

$$|L(z^n)|\Big(\frac{1}{k^n}\Big) \leq C$$

for every $n \in \mathbb{N}_0$. So, the power series $\sum_{n=0}^{\infty} L(z^n) z^n$ converges (absolutely) on $\mathbb{D}\big(0, \frac{1}{2k}\big)$ and, then, defines a germ in H($\{0\}$). This also shows that the mapping T is well defined. It is clearly linear, so it is only left to see that it is injective and surjective. For the injectivity, suppose that $T(L) = 0$ for some $L \in$ H($\mathbb{C}$)$'$. This means that $L(z^n) = 0$ for every $n \in \mathbb{N}_0$, and since the polynomials are dense in H($\mathbb{C}$), we have that $L(f) = 0$ for every entire function f. Thus $L = 0$ and T is injective.

In order to see that T is surjective, pick any $g \in \mathrm{H}(\{0\})$, so that $g(z) = \sum_{n=0}^{\infty} \lambda_n z^n$ for every $|z| < \varepsilon$, and define $L_g : \mathrm{H}(\mathbb{C}) \to \mathbb{C}$ by

$$L_g(f) = \sum_{n=0}^{\infty} a_n \lambda_n ,$$

for $f(z) = \sum_{n=0}^{\infty} a_n z^n \in \mathrm{H}(\mathbb{C})$. Choose some $k \in \mathbb{N}$ such that $\frac{1}{k} < \varepsilon$ and denote (recall (4.4))

$$\sup_{n \in \mathbb{N}_0} |\lambda_n| \frac{1}{k^n} = S < \infty .$$

Then

$$|L_g(f)| \leq \sum_{n=0}^{\infty} |a_n \lambda_n| \leq S \sum_{n=0}^{\infty} |a_n| k^n = S \|f\|_k .$$

This shows that L_g is well defined and continuous. Since it is clearly linear, we get $L_g \in \mathrm{H}(\mathbb{C})'$. Finally, since one easily has $L_g(z^n) = \lambda_n$, we immediately get $T(L_g) = g$, which shows that T is surjective and completes the proof. □

Remark 4.4 We can also give an integral representation of the duality between $\mathrm{H}(\mathbb{C})$ and $\mathrm{H}(\{0\})$. Given a germ $g(z) = \sum_{n=0}^{\infty} \lambda_n z^n$ that is holomorphic at some $\mathbb{D}(0, \varepsilon)$, we define for $|\zeta| > \frac{1}{\varepsilon}$,

$$G(\zeta) = \frac{1}{\zeta} g\big(\tfrac{1}{\zeta}\big) = \sum_{n=0}^{\infty} \lambda_n \frac{1}{\zeta^{n+1}} ,$$

which is a holomorphic function on $\mathbb{C} \setminus \overline{\mathbb{D}(0, 1/\varepsilon)}$. Now we fix some $0 < \delta < \varepsilon$ and define (denoting by γ_δ the circle centred at 0 and with radius $1/\delta$) $L_G : \mathrm{H}(\mathbb{C}) \to \mathbb{C}$ by

$$L_G(f) = \frac{1}{2\pi i} \int_{\gamma_\delta} f(z) G(z) dz .$$

Note that

$$|L_G(f)| \leq \frac{1}{\delta} \sup_{|\zeta| = \delta} |g(\zeta)| \sup_{|z| \leq 1/\delta} |f(z)| .$$

This shows that L_G is well defined and continuous. Since it is clearly linear, we get $L_G \in H(\mathbb{C})'$. Finally, if $g(z) = \sum_{n=0}^{\infty} \lambda_n z^n$, then for each $n = 0, 1, 2, \ldots$, we have

$$L_G(z^n) = \frac{1}{2\pi i} \int_{\gamma_\delta} z^n G(z) dz = \frac{1}{2\pi i} \sum_{k=0}^{\infty} \lambda_k \int_{\gamma_\delta} \frac{z^n}{z^{k+1}} dz = \lambda_n .$$

Accordingly, L_G coincides with the functional L_g defined in the proof of Theorem 4.11.

Let us finish by noting that the representation given in Theorem 4.11 is a particular case of the Köthe–da Silva Dias theorem on the duality of $H(\Omega)$ for an open set $\Omega \subseteq \mathbb{C}$. The interested reader is referred to [6, Chapter Six].

Functions of exponential type
An entire function f is of *exponential type* if there exist $k \in \mathbb{N}$ and $C > 0$ so that

$$|f(z)| \leq Ce^{k|z|}$$

for every $z \in \mathbb{C}$. We write $\mathrm{Exp}(\mathbb{C})$ for the space of all $f \in H(\mathbb{C})$ of exponential type. Our aim now is to show that $H(\mathbb{C})'$ can be identified with $\mathrm{Exp}(\mathbb{C})$. Functionals in $H(\mathbb{C})'$ are now denoted by μ and, if necessary, we will use a subscript to emphasize the variable in the function on which μ is acting. To be more precise, if f is a function that depends on several variables, we will write $\mu_w(f) = \mu([w \rightsquigarrow f(w)])$. We define the *Fourier–Borel* transform of $\mu \in H(\mathbb{C})'$ as the function $\widehat{\mu} : \mathbb{C} \to \mathbb{C}$ given by

$$\widehat{\mu}(z) = \mu_w(e^{zw}) .$$

Theorem 4.12 *The Fourier–Borel transform* $\mathcal{F} : H(\mathbb{C})' \to \mathrm{Exp}(\mathbb{C})$ *given by* $\mathcal{F}(\mu) = \widehat{\mu}$ *is a linear bijection.*

Proof Let us check in first place that it is well defined, and we begin by seeing that $\widehat{\mu}$ is entire for every $\mu \in H(\mathbb{C})'$. Take one such μ and note that, by (2.22) (recall also Remark 4.2), we can find $k \in \mathbb{N}$ and $C > 0$ so that

$$|\mu(f)| \leq C \sup_{|z| \leq k} |f(z)| , \tag{4.5}$$

for every $f \in H(\mathbb{C})$. Applying this to the monomial $f(w) = w^n$, we get

$$|\mu_w(w^n)| \leq Ck^n , \tag{4.6}$$

for every $n \in \mathbb{N}_0$. Note now that for each $\lambda \in \mathbb{C}$, the series $\sum_{n=0}^{\infty} \frac{\lambda^n}{n!} \zeta^n$ converges in $H(\mathbb{C})$ (that is, absolutely and uniformly on compact sets, as a function on ζ) to $e^{\lambda\zeta}$.

Using this and the fact that from (4.6), we have

$$\frac{|\mu_w(w^n)|\,|z|^n}{n!} \le C\frac{k^n}{n!}|z^n|,$$

we obtain that the series

$$\sum_{n=0}^{\infty} \frac{\mu_w(w^n)}{n!} z^n \tag{4.7}$$

converges in H(ℂ) to some entire function g. On the other hand, we have

$$\widehat{\mu}(z) = \mu_w(e^{wz}) = \mu_w\Big(\textstyle\sum_{n=0}^{\infty} \frac{z^n}{n!} w^n\Big) = \displaystyle\sum_{n=0}^{\infty} \frac{\mu_w(w^n)}{n!} z^n = g(z)\,,$$

for each $z \in \mathbb{C}$. That is, $\widehat{\mu} = g$, and it is an entire function (note also that its Taylor expansion at 0 is given by (4.7)). Let us see now that it is of exponential type. This is a straightforward consequence of (4.5), from which we have

$$|\widehat{\mu}(z)| = |\mu_w(e^{zw})| \le C \sup_{|z|\le k} |e^{zw}| \le Ce^{k|z|}\,,$$

for every $z \in \mathbb{C}$. This finally shows that $\mu \in \mathrm{Exp}(\mathbb{C})$, and therefore, $\mathcal{F}$ is well defined. It is clearly linear.

If $\widehat{\mu} = 0$, all Taylor coefficients are 0 and, from (4.7), $\mu_w(w^n) = 0$ for every n. Since the polynomials are dense in H(ℂ), this implies that $\mu(f) = 0$ for every $f \in \mathrm{H}(\mathbb{C})$ and shows that $\mathcal{F}$ is injective.

It is only left to complete the proof to see that the $\mathcal{F}$ is onto. To this end, fix some $g \in \mathrm{Exp}(\mathbb{C})$ with Taylor expansion $\sum_{n=0}^{\infty} g_n z^n$ and fix $k \in \mathbb{N}$ and $C > 0$ so that $|g(z)|e^{-k|z|} \le C$. With this and the Cauchy inequalities, we have

$$|g_n| \le \frac{1}{r^n} \sup_{|z|=r} |g(z)| \le \frac{Ce^{kr}}{r^n}\,,$$

for every $r > 0$ and $n \in \mathbb{N}$. In particular,

$$|g_n| \le \frac{Ce^{kn}}{n^n}$$

for every n. Fix now $m \in \mathbb{N}$ such that $m \ge 2e^k$, and observe that, if $f \in \mathrm{H}(\mathbb{C})$ has Taylor expansion $f(z) = \sum_{n=0}^{\infty} f_n z^n$, then as a consequence of the Cauchy inequalities, we have

$$|f_n| \le \frac{1}{m^n} \sup_{|z|=m} |f(z)|\,.$$

Thus

$$\sum_{n=0}^{\infty} |f_n| n! |g_n| \leq C \Big(\sup_{|z|=m} |f(z)| \Big) \sum_{n=0}^{\infty} \frac{e^{kn}}{m^n} \frac{n!}{n^n} \leq 2C \Big(\sum_{n=0}^{\infty} \frac{1}{2^n} \Big) \sup_{|z|\leq m} |f(z)| ,$$

and the mapping $\mu_g : \mathrm{H}(\mathbb{C}) \to \mathbb{C}$ defined by $\mu_g(f) = \sum_{n=0}^{\infty} n! f_n g_n$ is well defined and, since it is obviously linear, also continuous. In other words, $\mu_g \in \mathrm{H}(\mathbb{C})'$. Recall that the Taylor coefficients of e^{wz} (as a function of w) are $\frac{z^n}{n!}$, and this implies

$$\widehat{\mu_g}(z) = \big(\mu_g\big)_w (e^{wz}) = \sum_{n=0}^{\infty} n! \frac{z^n}{n!} g_n = \sum_{n=0}^{\infty} g_n z^n = g(z) .$$

Hence, $\mathcal{F}(\mu_g) = g$, and this completes the proof. □

4.2 Spaces of Differentiable Functions

We move on and consider (real) differentiable functions of several (real) variables. We are going to consider different spaces of such functions. We start by fixing some basic notation. Given $d \in \mathbb{N}$, a multi-index is a vector $\alpha = (\alpha_1, \ldots, \alpha_d) \in \mathbb{N}_0^d$, and its *order* is defined to be $|\alpha| = |\alpha_1| + \cdots + |\alpha_d|$. With this, we write the partial derivatives of a function f defined on some open $\Omega \subseteq \mathbb{R}^d$ as

$$\partial^\alpha f = \frac{\partial^{|\alpha|} f}{\partial^{\alpha_1} x_1 \cdots \partial^{\alpha_d} x_d} .$$

For each $k \in \mathbb{N}$, we define $C^k(\Omega)$ as the space of all functions $f : \Omega \to \mathbb{K}$ such that all partial derivatives up to order k exist and are continuous on Ω; that is,

$$C^k(\Omega) = \big\{ f : \Omega \to \mathbb{K} : \partial^\alpha f \text{ exist and are continuous functions on } \Omega \text{ for all } |\alpha| \leq k \big\} .$$

We choose now a fundamental sequence of compact sets $(K_n)_{n=1}^{\infty}$ of Ω and define the seminorms

$$p_n(f) = \sup_{|\alpha|\leq k} \sup_{x \in K_n} |\partial^\alpha f(x)| .$$

The space $C^0(\Omega)$ is just the space of continuous functions. By the definition, we clearly have

$$C^{k+1}(\Omega) \subseteq C^k(\Omega) \subseteq \cdots \subseteq C^1(\Omega) \subseteq C(\Omega)$$

for every k. We define now the space

$$C^{\infty}(\Omega) = \bigcap_{k \in \mathbb{N}} C^k(\Omega),$$

and we endow it with the locally convex topology given by the system of seminorms

$$p_n(f) = \sup_{|\alpha| \leq n} \sup_{x \in K_n} |\partial^\alpha f(x)|, \text{ for } n \in \mathbb{N}. \tag{4.8}$$

Proposition 4.3 *A sequence $(f_k)_k$ converges to f in $\tau_{(p_n)}$ if and only if $\partial^\alpha f_k \to \partial^\alpha f$ in τ_{co} for every α.*

Proof Suppose that $f_k \to f$ in $\tau_{(p_n)}$, then $p_n(f_k - f) \overset{k}{\to} 0$ for every n. Fix now some α and some compact $K \Subset \Omega$. There is some n_0 so that $|\alpha| \leq n_0$ and $K \subseteq K_{n_0}$. Hence

$$\sup_{x \in K} |\partial^\alpha f_k(x) - \partial^\alpha f(x)| \leq \sup_{|\beta| \leq n_0} \sup_{x \in K_{n_0}} |\partial^\beta f_k(x) - \partial^\beta f(x)| \leq p_{n_0}(f_k - f) \to 0.$$

This shows that $\partial^\alpha f_k \to \partial^\alpha f$ uniformly on compact sets.

Suppose now that $\partial^\alpha f_k$ converges to $\partial^\alpha f$ in τ_{co} for every α. Fix $n \in \mathbb{N}_0$ and $\varepsilon > 0$. Since K_n is compact, for each α with $|\alpha| \leq n$, we can find k_α so that

$$\sup_{x \in K_n} |\partial^\alpha f_k(x) - \partial^\alpha f(x)| < \varepsilon$$

for every $k \geq k_\alpha$. But the set $\{\alpha \in \mathbb{N}_0^d \colon |\alpha| \leq n\}$ is finite, and we may take k_0 the maximum of all k_αs. Then, for $k \geq k_0$,

$$\sup_{|\alpha| \leq n} \sup_{x \in K_n} |\partial^\alpha f_k(x) - \partial^\alpha f(x)| < \varepsilon.$$

□

Theorem 4.13 *$(C^\infty(\Omega), (p_n)_n)$ is a Fréchet space.*

Proof From Theorem 2.13, it follows immediately that the space is metrizable. In order to see that it is also complete, we choose a Cauchy sequence $(f_k)_k$. Proceeding as in Proposition 4.3, for each fixed α, the sequence $(\partial^\alpha f_k)_k$ is τ_{co}-Cauchy. Then, by Theorem 2.18, there is $f \in C(\Omega)$ so that $f_k \to f$ uniformly on compact sets, and for each α, there is $f_\alpha \in C(\Omega)$ so that $\partial^\alpha f_k \to f_\alpha$ uniformly on compact sets. What we have to see now is that $f \in C^\infty(\Omega)$ and $\partial^\alpha f = f_\alpha$ for every α.

Take $\alpha = (1, 0, \dots, 0)$, and let us see that $\partial^\alpha f$ exists and $\partial^\alpha f(a) = f_\alpha(a)$ for every $a \in \Omega$. Fix, then, $a = (a_1, \dots, a_d) \in \Omega$ and choose $r > 0$ such that

$$K = \{(a_1 + t, a_2, \dots, a_d) \colon t \in [-r, r]\} \subseteq \Omega.$$

Now we define $g : [-r, r] \to \mathbb{K}$ by $g(t) = f(a_1+t, a_2, \dots, a_d)$ and, in an obvious way (replacing f by f_k or f_α), g_k and g_α for $k \in \mathbb{N}$. Note that all these functions are continuous and that every g_k is differentiable on $]-r, r[$ (and $g_k' = \partial^\alpha f_k$). Then, by the Fundamental Theorem of Calculus,

$$g_k(t) - g_k(-r) = \int_{-r}^{t} g_k'(s)ds\,,$$

for $t \in [-r, r]$. On the other hand, the set K is compact, and the fact that $(\partial^\alpha f_k)_k$ converges to f_α uniformly on compact sets gives that $g_k' \to g_\alpha$ uniformly on $[-r, r]$. Therefore,

$$\lim_k \int_{-r}^{t} g_k'(s)ds = \int_{-r}^{t} \lim_k g_k'(s)ds = \int_{-r}^{t} g_\alpha(s)ds\,.$$

This altogether gives

$$g(t) - g(-r) = \int_{-r}^{t} g_\alpha(s)ds$$

and this implies that g is differentiable and $g' = g_\alpha$ for every $t \in [-r, r]$. In particular, g is differentiable at 0, and $g'(0) = g_\alpha(0)$. This gives that $\partial^\alpha f(a)$ exists and $\partial^\alpha f(a) = f_\alpha(a)$. Iterating this procedure a finite number of times, differentiating each time once with respect to one variable, we finally get the conclusion. □

With a similar proof, we also have the following.

Proposition 4.4 *$(C^k(\Omega), (p_n)_n)$ is a Fréchet space.*

Theorem 4.14 *Every bounded subset of $C^\infty(\Omega)$ is relatively compact.*

Proof Let $H \subseteq C^\infty(\Omega)$ be bounded. We want to see that every sequence in H has a convergent subsequence. We need to see (recall Proposition 4.3) that the subsequence converges uniformly on compact sets to some function f, and that all the partial derivatives converge (again in τ_{co}) to the partial derivatives of f. To do so, we are going to use several times Ascoli's Theorem 2.28. Our first step is to show that, for each fixed multi-index $\alpha \in \mathbb{N}_0^d$, the set

$$\{\partial^\alpha f : f \in H\}$$

is equicontinuous at every $x \in \Omega$. Note that, since H is bounded, for each compact $K \subseteq \Omega$ and such an α, there is $\mu_{\alpha,K}$ such that

$$|\partial^\alpha f(x)| \leq \mu_{\alpha,K} \tag{4.9}$$

for every $x \in K$ and $f \in H$. Fix now some $\alpha \in \mathbb{N}_0^d$ and $x_0 \in \Omega$. There is $\delta_0 > 0$ such that $K = \overline{B(x_0, \delta_0)} \subseteq \Omega$. Given $y \in K$, the segment $[x_0, y]$ is contained in K, and for each $f \in H$ the function $\partial^\alpha f$ is continuous. We may then apply the mean value theorem for several variables to have

$$|\partial^\alpha f(x_0) - \partial^\alpha f(y)| \leq \sup_{c \in [x_0, y]} \left|\nabla \partial^\alpha f(c)\right| |x_0 - y|$$

$$= \sup_{c \in [x_0, y]} \Big(\sum_{i=1}^d \left|\partial_i (\partial^\alpha f)(c)\right|^2 \Big)^{\frac{1}{2}} |x_0 - y| \leq d\big(\max_{|\beta| \leq |\alpha|+1} \mu_{\beta,K} \big) |x_0 - y| \, .$$

This gives that $\{\partial^\alpha f : f \in H\}$ is equicontinuous (recall Remark 2.16), given $\varepsilon > 0$ just take $|x_0 - y| < \delta < \min(\delta_0, \frac{\varepsilon}{d \max \mu_{\beta,K}})$.

On the other hand, (4.9) gives that the set $\{\partial^\alpha f(x) : f \in H\}$ is bounded in $\mathbb{K}$ for every fixed $x \in \Omega$. With this, Ascoli's Theorem 2.28 gives that $\{\partial^\alpha f : f \in H\}$ is τ_{co}-relatively compact in $C(\Omega)$ for every $\alpha \in \mathbb{N}_0^d$.

Take now a sequence $(f_n)_n \subseteq H$. By what we have just shown (using Ascoli's theorem), there is a subsequence $(n(k,0))_k$ and a continuous function $f^{(0)}$ so that $f_{n(k,0)} \to f^{(0)}$ in τ_{co} as $k \to \infty$. Let us now enumerate the set of multi-indices $\mathbb{N}_0^d = \{\alpha_j : j \in \mathbb{N}_0\}$ in such a way that $\alpha_0 = (0, 0, \ldots, 0)$. We take now the sequence $\big(\partial^{\alpha_1} f_{n(k,0)}\big)_k$, and using again the compactness of the set, we find a subsequence $(n(k,1))_k$ and a continuous function $f^{(1)}$ so that $f_{n(k,1)} \to f^{(1)}$ in τ_{co} as $k \to \infty$. Repeating this procedure, for each $j \in \mathbb{N}_0$, we define by induction a sequence $(n(k,j))_k$ and a continuous function $f^{(j)}$ so that each $(n(k,j))_k$ is a subsequence of $(n(k,j-1))_k$ and $\partial^{\alpha_j} f_{n(k,j)} \to f^{(j)}$ in τ_{co} as $k \to \infty$. Now, for each $k \in \mathbb{N}$, define $n_k = n(k,k)$. Then $(f_{n_k})_k$ is a subsequence of $(f_n)_n$ and

$$\partial^{\alpha_j} f \to f^{(j)}$$

in τ_{co} as $k \to \infty$ for every j. The same argument as in Theorem 4.13 shows that in fact $f = f^{(0)} \in C^\infty(\Omega)$ and $\partial^{\alpha_j} f = f^{(j)}$ for every j. Thus $(f_{n_k})_k$ converges to f in $C^\infty(\Omega)$, and this completes the proof. □

4.3 Some Operators on Spaces of Functions

We give now some examples of continuous, linear operators defined on the spaces of functions that we have considered so far. These will appear several times in the rest of this text.

Example 4.15 (Multiplication operators) Let Ω be an open set in $\mathbb{R}^d$ and fix a continuous function $g : \Omega \to \mathbb{K}$. With this, we define the associated multiplication operator

$$M_g : C(\Omega) \to C(\Omega), \text{ by } M_g(f) := f \cdot g .$$

This is clearly well defined and linear. Let us see that it is also continuous (recall that on $C(\Omega)$, we consider the compact-open topology). Given a compact set $K \subseteq \Omega$, we have (taking $C = \sup_{x \in K} |g(x)|$)

$$p_K(M_g(f)) = \sup_{x \in K} |f(x)g(x)| \leq \sup_{x \in K} |g(x)| \, p_K(f) = C \, p_K(f)$$

and, by Theorem 2.11, M_g is continuous.

Note that, with the same argument, if $\Omega \subseteq \mathbb{C}$ and g is holomorphic, then $M_g : \mathrm{H}(\Omega) \to \mathrm{H}(\Omega)$ is also continuous. □

Example 4.16 (Composition operators) We take now an open set $\Omega \subseteq \mathbb{R}^d$ and a continuous function $\varphi : \Omega \to \Omega$. We define the composition operator C_φ with symbol φ as

$$C_\varphi : C(\Omega) \to C(\Omega), \text{ by } C_\varphi(f) = f \circ \varphi ,$$

that is clearly well defined and linear. In order to see that it is continuous, choose some compact $K \subseteq \Omega$. Since φ is continuous, $K_2 = \varphi(K)$ is again compact, and we have

$$p_K(C_\varphi(f)) = \sup_{z \in K} |(f \circ \varphi)(z)| = \sup_{z \in K} |f(\varphi(z))| = \sup_{z \in K_2} |f(u)| = p_{K_2}(f) .$$

This shows that the operator is continuous.

As in the previous example, if $\Omega \subseteq \mathbb{C}$ is open and simply connected and $\varphi : \Omega \to \Omega$ is holomorphic, the composition operator $C_\varphi : \mathrm{H}(\Omega) \to \mathrm{H}(\Omega)$ is well defined and continuous. □

Example 4.17 (The differentiation operator) We consider now the open unit disc of the complex plane and define the differentiation operator

$$D : \mathrm{H}(\mathbb{D}) \to \mathrm{H}(\mathbb{D}) \text{ by } D(f) = f' .$$

This is clearly well defined and linear. We want to see now that it is continuous. To do that, let us briefly recall (see (2.24) and (2.25)) that the topology is generated by the family of seminorms $p_n = p_{K_n}$, where the fundamental sequence of compact sets is given by

$$K_n = \left\{ z \in \mathbb{D} : |z| \leq 1 - \tfrac{1}{n} \right\} ,$$

for $n \in \mathbb{N}$. If $z \in K_n$, then we have, by the Cauchy Integral Formula for the derivatives (Theorem 4.2),

$$f'(z) = \frac{1}{2\pi i} \int_{C(0,1-\frac{1}{n+1})} \frac{f(w)}{(w-z)^2} dw \,.$$

Let us note that, for each $|w| = 1 - \frac{1}{n+1}$, we have

$$|w - z| \geq |w| - |z| = r_{n+1} - |z| \geq r_{n+1} - r_n = \frac{1}{n(n+1)} \,,$$

and this gives

$$\begin{aligned} |f'(z)| &\leq \frac{1}{2\pi} \sup_{|w|=1-\frac{1}{n+1}} \frac{|f(w)|}{|w-z|^2} 2\pi r_{n+1} \\ &\leq n^3(n+1) \sup_{|w|=1-\frac{1}{n+1}} |f(\omega)| \leq n^3(n+1) p_{n+1}(f) \,. \end{aligned}$$

Hence, $p_n(D(f)) = p_n(f') \leq n^3(n+1)p_{n+1}(f)$ for every f and D is continuous.

A similar argument shows that the differentiation operator $D : \mathrm{H}(\mathbb{C}) \to \mathrm{H}(\mathbb{C})$ is continuous. □

Example 4.18 (Integration operator) We now consider $\Omega = \mathbb{C}$. For $z \in \mathbb{C}$, we take $[0, z] = \{\lambda z : 0 \leq \lambda \leq 1\}$, the closed segment from 0 to z, and define the integration operator

$$J : \mathrm{H}(\mathbb{C}) \to \mathrm{H}(\mathbb{C}), \quad \text{as} \quad (Jf)(z) := \int_0^z f(w) dw \,.$$

Once again, it is clear that it is well defined and linear, and it is only left to see that it is continuous. To that purpose, we consider the fundamental sequence of compact sets given by $K_n = \{z \in \mathbb{C} : |z| \leq n\}$ for $n \in \mathbb{N}$ and recall that the sequence of seminorms $p_n = p_{K_n}$ defines the compact-open topology. If $z \in K_n$, then clearly $[0, z] \subseteq K_n$, and we have

$$|Jf(z)| = \left| \int_0^z f(w)dw \right| \leq |z| \max_{w \in [0,z]} |f(w)| \leq n \max_{w \in K_n} |f(w)| \,.$$

This gives $p_n(J(f)) \leq np_n(f)$ for every f and J is continuous. A similar argument shows that the integration operator $J : \mathrm{H}(\mathbb{D}) \to \mathrm{H}(\mathbb{D})$ is continuous.

Moreover, it is easy to see that $DJf = f$ and $JDf(z) = f(z) - f(0)$. □

Example 4.19 For $1 \leq j \leq d$, we write $\partial_j = \frac{\partial}{\partial x_j}$ (the partial derivative with respect to the jth variable) and, for $\alpha_j \in \mathbb{N}_0$, we write $\partial_j^{\alpha_j} = \frac{\partial^{\alpha_j}}{\partial x_j^{\alpha_j}}$. Finally, given $\alpha = (\alpha_1, \dots, \alpha_d) \in \mathbb{N}_0^d$, we write $\partial^\alpha = \partial_1^{\alpha_1} \cdots \partial_d^{\alpha_d} = \frac{\partial^\alpha}{\partial x_1^{\alpha_1} \cdots \partial x_d^{\alpha_d}}$.

If f is a function of d variables, at least m times differentiable, and $a_\alpha \in \mathbb{C}$ for $\alpha \in \mathbb{N}_0^d$ with $|\alpha| \leq m$ (with at least one $a_\alpha \neq 0$ if $|\alpha| = m$, we write

$$P(\partial) f = \sum_{|\alpha| \leq m} a_\alpha \partial^\alpha f .$$

This defines an operator, called *linear partial differential operator of order m with constant coefficients*,

$$P(\partial) : C^\infty(\Omega) \to C^\infty(\Omega)$$

(here $\Omega \subseteq \mathbb{R}^d$ is some open set). This is clearly well defined and linear. If $(K_n)_n$ is a fundamental system of compact subsets of Ω and $p_n = p_{K_n}$, then

$$\begin{aligned} p_n(P(\partial)f) &= \sup_{|\beta| \leq n} \sup_{x \in K_n} \big|\partial^\beta \big(P(\partial) f(x)\big)\big| = \sup_{|\beta| \leq n} \sup_{x \in K_n} \Big|\partial^\beta \Big(\sum_{|\alpha| \leq m} a_\alpha \partial^\alpha f(x) \Big)\Big| \\ &= \sup_{\substack{x \in K_n \\ |\beta| \leq n}} \Big| \sum_{|\alpha| \leq m} a_\alpha \partial^{\alpha+\beta} f(x) \Big| \leq \sup_{\substack{x \in K_n \\ |\gamma| \leq n+m}} |\partial^\gamma f(x)| \Big(\sum_{|\alpha| \leq m} |a_\alpha| \Big) \\ &\leq \sup_{\substack{x \in K_{n+m} \\ |\gamma| \leq n+m}} |\partial^\gamma f(x)| \Big(\sum_{|\alpha| \leq m} |a_\alpha| \Big) = \Big(\sum_{|\alpha| \leq m} |a_\alpha| \Big) \cdot p_{n+m}(f) , \end{aligned}$$

and the operator is continuous. Let us note that if the operator is of order m, we may consider it as $P(\partial) : C^{k+m}(\Omega) \to C^k(\Omega)$ for each k and, with the same argument, it is continuous.

Each linear partial differential operator with constant coefficients has an associated polynomial of N variables, called the *symbol* of the operator, given by

$$P(z) = \sum_{|\alpha| \leq m} a_\alpha z^\alpha ,$$

where we denote $z^\alpha = z_1^{\alpha_1} \cdots z_d^{\alpha_d}$ for $z = (z_1, \dots, z_d) \in \mathbb{C}^d$.

The classical theory of partial differential equations is full of such operators. We recall now just a few of the most classical ones.

The *Laplace operator*

$$P(\partial) = \Delta = \frac{\partial^2}{\partial x_1^2} + \cdots + \frac{\partial^2}{\partial x_d^2} \text{ with symbol } P(z) = z_1^2 + \cdots + z_d^2 .$$

The *heat operator*

$$P(\partial)=\frac{\partial}{\partial t}-\Delta_x = \frac{\partial}{\partial t}-(\frac{\partial^2}{\partial x_1^2}+\cdots+\frac{\partial^2}{\partial x_d^2}) \text{ with symbol } P(z,t) = t-(z_1^2+\cdots+z_d^2) .$$

The *wave operator*

$$P(\partial)=\frac{\partial^2}{\partial t^2}-\Delta_x=\frac{\partial^2}{\partial t^2}-(\frac{\partial^2}{\partial x_1^2}+\cdots+\frac{\partial^2}{\partial x_d^2}) \text{ with symbol } P(z,t)=t^2-(z_1^2+\cdots+z_d^2) .$$

□

4.4 Exercises

4.1 Let $\Omega \subseteq \mathbb{C}$ be an open, connected set and $A \subseteq \mathrm{H}(\Omega)$ bounded. Show:

1. The set $A' = \{f' : f \in A\}$ is bounded in $\mathrm{H}(\Omega)$.
2. If A' is bounded in $\mathrm{H}(\Omega)$ and there is $z_0 \in \Omega$ such that $\{f(z_0) : f \in A\}$ is bounded in $\mathbb{C}$, then A is bounded in $\mathrm{H}(\Omega)$.

4.2 Let $\Omega \subseteq \mathbb{C}$ be an open and connected set. Let $(f_n)_n$ be a sequence in $\mathrm{H}(\Omega)$ of injective functions such that $f_n \to f$ in $(\mathrm{H}(\Omega), \tau_{co})$. Show that f is either injective or constant.

4.3 Let $\Omega \subseteq \mathbb{C}$ be an open and connected set. Let $(f_n)_n \subseteq \mathrm{H}(\Omega)$ be a sequence such that $f_n \to f$ in $(\mathrm{H}(\Omega), \tau_{co})$ and f is not identically zero. Suppose that $\overline{B(a,r)} \subseteq \Omega$ for some $a \in \Omega$ and $r > 0$. If $f(z) \neq 0$ for every $z \in \mathbb{C}$ such that $|z - a| = r$, then there exists $n_0 \in \mathbb{N}$ such that for any $n \geq n_0$, f and f_n have the same zeros in $B(a,r)$.
Hint: Recall Rouche's Theorem.

4.4 For each $f \in \mathrm{H}(\mathbb{D})$, denote by $a_n(f)$ (for $n = 0, 1, 2, \ldots$) the nth Taylor coefficient. Show that $A \subseteq \mathrm{H}(\mathbb{D})$ is bounded if and only if the following two conditions hold:

1. $\sup\{|a_n(f)| : f \in A\}$ is bounded in $\mathbb{C}$ for every n.
2. $\limsup_n \big(\sup\{|a_n(f)| : f \in A\}\big)^{1/n} \leq 1$.

4.5 Obtain a characterization of the bounded subsets of $\mathrm{H}(\mathbb{C})$, similar to the one in Exercise 4.4.

4.6 For each $0 < r < 1$, let $\gamma_r : [0, 1] \to \mathbb{C}$ be given by $\gamma_r(t) = re^{2\pi i t}$. Show that $(f_n)_n \subseteq \mathrm{H}(\mathbb{D})$ converges to $f \in \mathrm{H}(\mathbb{D})$ if and only if

$$\int_{\gamma_r} |f(z) - f_n(z)| dz \longrightarrow 0 \text{ as } n \to \infty$$

for every $r \in]0, 1[$.

4.7 Let $f \in \mathrm{H}(\mathbb{D})$ be non-constant and such that $\sup_{z \in \mathbb{D}} |f(z)| \leq 1$. Prove that the function $g : \mathbb{D} \to \mathbb{C}$ defined by

$$g(z) = \sum_{n=0}^{\infty} f(z)^n$$

is holomorphic on $\mathbb{D}$. Is $g(z)$ bounded?

4.8 Let $A \subseteq \mathrm{H}(\Omega)$ be a bounded set such that $\bigcup_{f \in A} f(\Omega)$ is contained in an open subset $D \subseteq \mathbb{C}$. Let $g \in \mathrm{H}(D)$ be bounded on bounded sets. Show that $\{g \circ f : f \in A\}$ is bounded in $\mathrm{H}(D)$.

4.9 Show that $C^1(\mathbb{R})$ contains bounded sets that are not relatively compact.

4.10 Let $B(\Omega)$ be the subset of $C^\infty(\Omega)$ consisting of functions f so that all the partial derivatives of f are bounded in Ω, endowed with the locally convex topology defined by the seminorms

$$p_n(f) = \sup_{|\alpha| \leq n} \sup_{x \in \Omega} \left| \partial^\alpha f(x) \right|,$$

for $n = 0, 1, 2, \ldots$. Show that $B(\Omega)$ is a Fréchet space.

Further Reading
[1–5, 7–9]

References

1. Ahlfors, L.: Complex Analysis—An Introduction to the Theory of Analytic Functions of One Complex Variable, 3rd edn. AMS Chelsea Publishing, Providence (2021). Reprint of the 1978 original
2. Apostol, T.M.: Mathematical Analysis, 2nd edn. Addison-Wesley Publishing Co., Reading/London/Don Mills (1974)
3. Conway, J.B.: Functions of One Complex Variable. Graduate Texts in Mathematics, vol. 11, 2nd edn. Springer, New York/Berlin (1978)
4. Gamelin, T.W.: Complex Analysis. Undergraduate Texts in Mathematics. Springer, New York (2001). https://doi.org/10.1007/978-0-387-21607-2
5. Jameson, G.J.O.: A First Course on Complex Functions. Chapman and Hall, Ltd., London (1970)

6. Köthe, G.: Topological Vector Spaces. I. Translated from the German by D.J.H. Garling. Die Grundlehren der mathematischen Wissenschaften, Band 159. Springer, New York (1969)
7. Rudin, W.: Principles of Mathematical Analysis. International Series in Pure and Applied Mathematics, 3rd edn. McGraw-Hill Book Co., New York/Auckland/Düsseldorf (1976)
8. Rudin, W.: Real and Complex Analysis, 3rd edn. McGraw-Hill Book Co., New York (1987)
9. Shapiro, J.H.: Composition Operators and Classical Function Theory. Universitext: Tracts in Mathematics. Springer, New York (1993). https://doi.org/10.1007/978-1-4612-0887-7

Chapter 5
Transitive and Mean Ergodic Operators

If E is a complex locally convex (Hausdorff) space and $T : E \to E$ is an operator (in order to make reading as easy as possible, all along this chapter, we assume without further notice that all operators are linear and continuous), we may compose it with itself. Our aim in this chapter is to analyse what happens when we iterate this process, that is, if we compose T with itself n times and let n tend to ∞. We begin by fixing some notation. First of all, we write

$$\begin{aligned}
T^0 &= I \text{ (the identity on } E)\\
T^1 &= T\\
T^2 &= T \circ T\\
T^3 &= T \circ T \circ T\\
&\vdots\\
T^n &= \underbrace{T \circ \cdots \circ T}_{n \text{ times}} = T^{n-1} \circ T \text{ for } n \in \mathbb{N}\,.
\end{aligned}$$

Also, given some $x \in E$, we define its *orbit* with respect to T as

$$\mathrm{Orb}(x, T) = \{x, Tx, T^2x, T^3x, \ldots\} = \{T^n x : n = 0, 1, 2, \ldots\}\,. \tag{5.1}$$

J. Bonet et al., *Function Spaces and Operators between them*,
RSME Springer Series 11, https://doi.org/10.1007/978-3-031-41602-6_5

5.1 Transitive Operators

An operator $T \in L(E)$ is called *transitive* if for every pair of non-empty open sets $U, V \subseteq E$ there is some $n \in \mathbb{N}$ so that

$$T^n(U) \cap V \neq \varnothing .$$

Let us start by seeing some immediate consequences of this definition.

Proposition 5.1 *Let $T \in L(E)$ be a transitive operator. Then the following statements hold:*

1. *$T(E)$ is dense in E.*
2. *$T^n(E)$ is dense in E for each $n \in \mathbb{N}$.*
3. *Given $U, V \subseteq E$ non-empty and open, for every $m \in \mathbb{N}$, there is some $n > m$ so that $T^n(U) \cap V \neq \varnothing$.*
4. *Let U, V be neighbourhoods of 0 in E so that $V \neq E$. Then for every $x \in E$, there are $n \in \mathbb{N}$ and $y \in E$ satisfying $x - y \in U$ and $T^n x - T^n y \notin V$ (i.e., T has sensitive dependence on the initial conditions).*

Proof Fix $U = E$ and take $\varnothing \neq V \subseteq E$ any open set. By the transitivity of T, we can find n so that $\varnothing \neq T^n(E) \cap V \subseteq T(E) \cap V$, and $T(E)$ is dense. To check that 2 holds, let us note that if $S : E \to F$ is any operator with dense range and $A \subseteq E$ is dense, then $S(E) = S(\overline{A}) \subseteq \overline{S(A)}$ and therefore $F = \overline{S(E)} \subseteq \overline{S(A)}$. With this, an argument with induction from 1 yields our second claim.

Fix now m and some open and non-empty $U, V \subseteq E$. By 2, $T^m(E)$ is dense, and then $T^m(E) \cap V \neq \varnothing$, and $\left(T^m\right)^{-1}(V) \subseteq E$ is open and non-empty. Being T transitive, we can find $k \in \mathbb{N}$ and $x \in U$ so that $T^k x \in \left(T^m\right)^{-1}(V)$. Hence, $T^{m+k}x \in V$ and 3 holds.

We finish by seeing that 4 holds. Take U and V as in the statement, pick $U' \subseteq U$, an absolutely convex, open neighbourhood of 0, and consider $G = E \setminus \overline{V}$ (which is open and non-empty). By the transitivity of T, there is some n so that $T^n(U') \cap G \neq \varnothing$. Given $x \in E$, take $u \in U'$ and define $y = x - u$. Then $T^n x - T^n y = T^n(u) \in G$. This yields the claim and completes the proof. □

Our aim now is to find conditions that ensure that a given operator is transitive.

Lemma 5.1 *Let $T \in L(E)$ be such that there exist $E_0, F_0 \subseteq E$ dense so that for every $x \in E_0$ and $y \in F_0$, and for every absolutely convex, open 0-neighbourhood U, there are $m \in \mathbb{N}$ and $u \in U$ with $T^m x \in U$ and $T^m u - y \in U$. Then T is transitive.*

Proof Let G_1, G_2 be any non-empty, open subsets of E. By density, we can find $x \in E_0$ and $y \in F_0$, and some U, absolutely convex, open 0-neighbourhood so that

$$x + U \subseteq G_1 \text{ and } y + 2U \subseteq G_2 .$$

Now, by hypothesis, there are m and $u \in U$ so that

$$T^m u \in y + U \text{ and } T^m x \in U .$$

This altogether gives $x + u \in x + U \subseteq G_1$ and $T^m(x + u) \in y + 2U \subseteq G_2$. This implies that $T^m(G_1) \cap G_2 \neq \varnothing$, and T is transitive. □

Proposition 5.2 *Let $T \in L(E)$ be such that there exist $E_0, F_0 \subseteq E$ dense so that for every $x \in E_0$ and $y \in F_0$, and for every absolutely convex, open 0-neighbourhood U, there are $m \in \mathbb{N}$ and $u \in U$ with $T^m(x + u) \in y + U$. Then T is transitive.*

Proof Let G_1, G_2 be any non-empty, open subsets of E. By density, we can find $x \in E_0$ and $y \in F_0$, and some U, absolutely convex, open 0-neighbourhood so that

$$x + U \subseteq G_1 \text{ and } y + U \subseteq G_2 .$$

By hypothesis, there are m and $u \in U$ so that $T^m(x + u) \in y + U$. This clearly gives the conclusion. □

Theorem 5.1 (Bès–Peris) *Let $T \in L(E)$. Suppose that there exist $E_0, F_0 \subseteq E$ dense, $(n_k)_k \subseteq \mathbb{N}$ strictly increasing, and operators $S_{n_k} : F_0 \to E$ so that*

$$T^{n_k} x \underset{k}{\longrightarrow} 0 \textit{ for every } x \in E_0 \tag{5.2}$$

$$S_{n_k} y \underset{k}{\longrightarrow} 0 \textit{ for every } y \in F_0 \tag{5.3}$$

$$T^{n_k} S_{n_k} y \underset{k}{\longrightarrow} y \textit{ for every } y \in F_0 . \tag{5.4}$$

Then T is transitive.

Proof Take any $x \in E_0$, $y \in F_0$, and any absolutely convex open 0-neighbourhood U. Using (5.2), (5.3), and (5.4), we can find some $k_0 \in \mathbb{N}$ such that

$$T^{n_k} x \in \frac{1}{2} U, \; S_{n_k} y \in U, \text{ and } y - T^{n_k} S_{n_k} y \in \frac{1}{2} U ,$$

for every $k \geq k_0$. Let now $m = n_{k_0}$ and $u = S_m y \in U$. We can now finish our argument in two ways. On the one hand, we have

$$T^m x \in \frac{1}{2} U \subseteq U, \text{ and } T^m u - y \in \frac{1}{2} U \subseteq U ,$$

and the conclusion follows from Lemma 5.1. On the other hand, we have

$$T^m(x+u) \in y + \frac{1}{2}U + \frac{1}{2}U \subseteq y + U\,,$$

and Proposition 5.2 yields the conclusion. □

As a straightforward consequence, we have the following reformulation.

Corollary 5.1 *Suppose that $T \in L(E)$ satisfies any of the following two conditions:*

(1) (Godefroy–Shapiro) There exist $E_0, F_0 \subseteq E$ dense, $(n_k)_k \subseteq \mathbb{N}$ strictly increasing and operators $S_{n_k} : F_0 \to E$ so that

$$T^{n_k} \circ S_{n_k} = I_{F_0} \text{ for every } k \in \mathbb{N}$$
$$T^{n_k} x \xrightarrow[k]{} 0 \text{ for every } x \in E_0$$
$$S_{n_k} y \xrightarrow[k]{} 0 \text{ for every } y \in F_0\,.$$

(2) (Kitai) There exist $E_0, F_0 \subseteq E$ dense, an operator $S : F_0 \to E$ with $T \circ S = I_{F_0}$ and $(n_k)_k \subseteq \mathbb{N}$ strictly increasing so that

$$T^{n_k} x \xrightarrow[k]{} 0 \text{ for every } x \in E_0$$
$$S^{n_k} y \xrightarrow[k]{} 0 \text{ for every } y \in F_0\,.$$

Then T is transitive.

A notion that is closely related with transitivity is that of hypercyclicity. An operator $T : E \to E$ is said to be *hypercyclic* if there is a vector $x \in E$ (called *hypercyclic vector*) such that $\mathrm{Orb}(x, T)$ (recall the definition in (5.1)) is dense in E. Observe that if E is a locally convex Hausdorff space on which there is a hypercyclic operator, then necessarily E has to be separable.

Remark 5.1 Suppose that $T \in L(E)$ is hypercyclic, with hypercyclic vector x, and fix two non-empty open sets $U, V \subseteq E$. Since the orbit of x is dense in E, we can find some m so that $T^m x \in U$. The set $\{T^{m+1}x, T^{m+2}x, \ldots\}$ is again dense in E, and we can find k such that $T^{m+k}x \in V$. Therefore, $T^k(U) \cap V \neq \varnothing$, and this shows that T is transitive.

In fact, if the space is Fréchet (and separable), both concepts are equivalent.

Theorem 5.2 (Birkhoff) *Let E be a separable Fréchet space. Then $T \in L(E)$ is hypercyclic if and only if it is transitive.*

Proof Let us suppose that T is transitive and see that it is also hypercyclic (in view of Remark 5.1, this yields the result). Since E is separable, we can choose a dense countable set $(x_j)_{j \in \mathbb{N}}$ in E and consider the family $\{B(x_j, 1/m) : j, m \in \mathbb{N}\}$ that, being countable, we can write as $\{V_n : n \in \mathbb{N}\}$. Then, for every open set $G \subseteq E$, there is some n with $V_n \subseteq G$. Now, for each $n \in \mathbb{N}$, we define the set

$$G_n = \{x \in E : T^m x \in V_n \text{ for some } m\} .$$

On the one hand, $T^m(E)$ is dense in E for every m (because T is transitive, see Proposition 5.1) and then $T^m(E) \cap V_n \neq \varnothing$ for every n. This shows that, for each n, the set G_n is not empty. Now, V_n is open, and from this, we deduce that for each $x \in G_n$, we can find m and U an open neighbourhood of x such that $T^m U \subseteq V_n$. Hence, $U \subseteq G_n$ and G_n is open.

Our next step is to see that, moreover, G_n is dense in E. Let $V \subseteq E$ be open and non-empty and choose (by the transitivity of T) some m with $T^m V \cap V_n \neq \varnothing$. Then there is some $y \in V$ with $T^m y \in V_n$ (that is, $y \in G_n$). Hence, $G_n \cap V \neq \varnothing$, and since the open set V was arbitrary, this shows that G_n is dense in E for every n. By Baire's Theorem 3.21, $G = \bigcap_n G_n$ is dense in E.

We are going to finish the proof by seeing that every vector in G is hypercyclic. Take $z \in G$ and $U \subseteq E$ open, and find k so that $V_k \subseteq U$. Then (since $z \in G_k$) there is some m such that $T^m z \in V_k \subseteq U$, and this shows that $\mathrm{Orb}(z, T) \cap U \neq \varnothing$. Hence, $\mathrm{Orb}(z, T)$ is dense, and this completes the proof. □

Let us note that in the proof we have seen that a hypercyclic operator on a separable Fréchet space not only has a hypercyclic vector, but the set of such vectors is dense and, moreover, is the intersection of a countable family of open sets.

We can give a sort of version of Corollary 5.1 for hypercyclic operators.

Proposition 5.3 (Godefroy–Shapiro) *Let E be a complex, separable locally convex Hausdorff space and $T \in L(E)$. Suppose that the spaces*

$$E_0 = \mathrm{span}\{x \in E : Tx = \lambda x \textit{ for some } \lambda \in \mathbb{C},\ |\lambda| < 1\}$$

$$F_0 = \mathrm{span}\{y \in E : Ty = \lambda y \textit{ for some } \lambda \in \mathbb{C},\ |\lambda| > 1\}$$

are both dense in E. Then T is hypercyclic.

Proof First of all, note that $T^n x \to 0$ as $n \to \infty$ for every $x \in E_0$. On the other hand, if $y \in E$ is such that there is $\lambda \in \mathbb{C}$ with $|\lambda| > 1$ so that $Ty = \lambda y$, we define $Sy = \frac{1}{\lambda} y$ and extend it by linearity to F_0. It is straightforward to check that $TSy = y$ and $S^n y \to 0$ for every $y \in F_0$. Corollary 5.1–(1) then gives that T is transitive, and being E separable, Theorem 5.2 gives that it is hypercyclic. □

So far we have given some sufficient conditions for an operator to be transitive. We change now the point of view, looking for necessary conditions. The transpose of the operator (recall Sect. 3.9) plays a fundamental rôle.

Lemma 5.2 *Let $T \in L(E)$ be transitive. Then for every $0 \neq u \in E'$ and every non-empty open set $G \subseteq E$, the set $\bigcup_{n \in \mathbb{N}} ({}^tT)^n(u)(G)$ is dense in $\mathbb{C}$.*

Proof Fix some non-empty open set $G \subseteq E$. If $\varnothing \neq A \subseteq \mathbb{C}$ is open, then $F = u^{-1}(A)$ is open (and not empty) in E, and we can find $n \in \mathbb{N}$ and $x \in G$ so that $T^n x \in F$ (because T is transitive). Then $u\big(T^n x\big) = \big[\big({}^tT\big)^n(u)\big](x) \in A$, and this implies that $\bigcup_{n\in\mathbb{N}}({}^tT)^n(u)(G) \cap A \neq \varnothing$. Since the open set A was arbitrary, this yields the claim. □

Proposition 5.4 *Let $T \in L(E)$ be transitive. Then $\{\big({}^tT\big)^n(u) : n \in \mathbb{N}\}$ is not equicontinuous in E' for every $0 \neq u \in E'$.*

Proof Let us suppose that there is some $0 \neq u \in E'$ for which $\{\big({}^tT\big)^n(u) : n \in \mathbb{N}\}$ is equicontinuous. Then there is some open 0-neighbourhood U so that $\big({}^tT\big)^n(u) \in U^\circ$ for every n. Therefore, $\big|\big({}^tT\big)^n(u)(x)\big| \leq 1$ for every n and all $x \in U$, and $\bigcup_{n\in\mathbb{N}}({}^tT)^n(u)(G)$ cannot be dense in $\mathbb{C}$. By Lemma 5.2, the operator is not transitive. □

Corollary 5.2 *Let $T \in L(E)$ be transitive. Then $\{T^n : n \in \mathbb{N}\}$ is not equicontinuous in $L(E)$.*

Proof Suppose that $\{T^n : n \in \mathbb{N}\}$ is equicontinuous. Then for each $0 \neq v \in E'$ there is some 0-neighbourhood U so that $|v(T^n x)| \leq 1$ for every $x \in U$. Then $\{({}^tT)^n(v) : n \in \mathbb{N}\}$ is equicontinuous and, by Proposition 5.4, T is not transitive.

□

Proposition 5.5 *Let $T \in L(E)$ be transitive. Then ${}^tT$ does not have eigenvalues. In particular, ${}^tT$ is injective.*

Proof Suppose that ${}^tT$ has an eigenvalue, that is, there are $0 \neq u \in E'$ and $\lambda \in \mathbb{C}$ so that ${}^tTu = \lambda u$. Our aim is to show that we can find some open non-empty $G \subseteq E$ for which

$$\bigcup_{n\in\mathbb{N}} \big({}^tT\big)^n(u)(G)$$

is not dense in $\mathbb{C}$. This, in view of Lemma 5.2, will yield the conclusion.

In first place, note that for each $n \in \mathbb{N}$ and $x \in E$, we have

$$|({}^tT)^n(u)(x)| = |u(T^n x)| = |\lambda|^n |u(x)|\,.$$

Now we distinguish two cases. If $|\lambda| \leq 1$, we consider the open set

$$G_1 = \{x \in E : |u(x)| < 1\}\,,$$

and $|({}^tT)^n(u)(x)| < 1$ for every n and $x \in G_1$. On the other hand, if $|\lambda| > 1$, we take

$$G_2 = \{x \in E : |u(x)| > 1\}\,,$$

and $|({}^tT)^n(u)(x)| > 1$ for every n and $x \in G_2$. Hence, for each possible eigenvalue λ, we can find some open, non-empty set G for which $\bigcup_{n\in\mathbb{N}} \left({}^tT\right)^n(u)(G_1)$ is not dense. This completes the proof. □

Corollary 5.3 *If E is a finite dimensional space, then no operator $T \in L(E)$ is transitive.*

Proof If E is finite dimensional, then ${}^tT \in L(E')$ has at least one eigenvalue. □

Corollary 5.4 *Let $x \in E$ be a hypercyclic vector of an operator $T \in L(E)$. Then* $\mathrm{Orb}(x, T)$ *is linearly independent.*

Proof Suppose that there is some $n \in \mathbb{N}$, and $\alpha_0, \ldots, \alpha_n \in \mathbb{C}$ so that

$$T^{n+1}x = \sum_{j=0}^{n} \alpha_j T^j x \,.$$

Then $\mathrm{Orb}(x, T) \subseteq \mathrm{span}\{x, Tx, \ldots, T^n x\}$, which is a finite dimensional subspace of E. Since $\mathrm{Orb}(x, T)$ is dense in E, we deduce that $E = \mathrm{span}\{x, Tx, \ldots, T^n x\}$. This contradicts Corollary 5.3. □

Theorem 5.3 (Bourdon) *Let E be a separable locally convex space and $T \in L(E)$ hypercyclic. Then, there exists a dense subspace F of E such that every non-zero $x \in F$ is hypercyclic for T.*

Proof Choose some $x \in E$ so that $\overline{\mathrm{Orb}(x, T)} = E$, and define $F = \mathrm{span}(\mathrm{Orb}(x, T))$, which is obviously dense and satisfies $T(F) \subseteq F$. What we have to see is that, if $0 \neq y = \sum_{j=0}^{n} \alpha_j T^j x \in F$, then $\overline{\mathrm{Orb}(y, T)} = E$. To do this, we consider $P(z) = \sum_{j=0}^{n} \alpha_j z^j$, so that $y = P(T)(x)$, and observe that $T^k(P(T))x = P(T)(T^k x)$ for every k. Then, if $P(T)$ has dense range,

$$\overline{\mathrm{Orb}(y, T)} = \overline{P(T)\,\mathrm{Orb}(x, T)} = \overline{P(T)\overline{\mathrm{Orb}(x, T)}} = \overline{P(T)(E)} = E \,,$$

and we are done.

It suffices, then, to see that $P(T)$ has dense range. By the Fundamental Theorem of Algebra, we can find $a_m \neq 0$ and $c_1, \ldots, c_m \in \mathbb{C}$ so that

$$P(T) = a_m (T - c_1 I) \cdots (T - c_m I) \,.$$

So, if we see that $T - \lambda I$ has dense range for each $\lambda \in \mathbb{C}$, then we are done. Suppose that this is not the case and there is some $\lambda \in \mathbb{C}$ with $\overline{(T - \lambda I)(E)} \neq E$. Then, by the Hahn–Banach theorem, we can find some non-zero $u \in E'$ such that $u(z) = 0$ for every $z \in (T - \lambda I)(E)$. With this, for each $w \in E$, we have

$$({}^tTu - \lambda u)(w) = u(Tw) - u(\lambda w) = u\big((T - \lambda I)(w)\big) = 0 \,,$$

from which ${}^{t}Tu = \lambda u$, and λ is an eigenvalue of ${}^{t}T$. By Proposition 5.5, T cannot be transitive and, hence, neither hypercyclic. This is a contradiction and completes the proof. □

We finish this section by introducing another notion concerning the asymptotic behaviour of an operator. A point $x \in E$ is said to be *n-periodic* for the operator $T \in L(E)$ if $T^n x = x$, and *periodic* if it is n-periodic for some n. With this, an operator in $L(E)$ is *chaotic* (in the sense of Devaney) if it is transitive and the set of periodic points is dense in E. The set of periodic points of an operator has an interesting description.

Proposition 5.6 *Let E be a complex locally convex space, and $T \in L(E)$. Then the set of periodic points of T is exactly*

$$\operatorname{span}\{x \in E\colon \text{ there exists } \lambda \in \mathbb{C} \text{ with } \lambda^n = 1 \text{ for some } n \in \mathbb{N} \text{ so that } Tx = \lambda x\}\,. \tag{5.5}$$

Proof Let us denote the set in (5.5) by G_0. If $x \in G_0$, then $x = \sum_{k=1}^{m} \alpha_k x_k$, where, for each k, there are $\lambda_k \in \mathbb{C}$ and $n_k \in \mathbb{N}$ with $\lambda_k^{n_k} = 1$ and $Tx_k = \lambda_k x_k$. Taking n to be the least common multiple of $\{n_1, \dots, n_m\}$, it is straightforward to see that each x_k is n-periodic and then so also is x. This gives one inclusion.

Let us see now that every periodic point lies in G_0. Choose, then, some $x_0 \in E$ so that $T^n x_0 = x_0$ for some $n \in \mathbb{N}$, and let $\lambda_1, \dots, \lambda_n$ be the nth roots of 1. For each $i = 1, \dots, n$, we define

$$p_i(z) = \prod_{j \neq i} (z - \lambda_j),$$

and since all λ_ks are different, the system $\{p_1(z), \dots, p_n(z)\}$ is a basis of the space of polynomials of degree strictly less than n. Then, we can find $\alpha_1, \dots, \alpha_n \in \mathbb{C}$ so that

$$1 = \sum_{i=1}^{n} \alpha_i p_i(z)\,,$$

for every $z \in \mathbb{C}$. In other words, if we write the left hand side as $\sum_{k=0}^{n-1} a_k z^k$, we have that $a_0 = 1$ and $a_k = 0$ for $k = 1, \dots, n-1$; hence,

$$I = \sum_{i=1}^{n} \alpha_i p_i(T)\,.$$

Then, writing $y_i = p_i(T)(x_0)$, we have

$$x_0 = \sum_{i=1}^{n} \alpha_i y_i\,. \tag{5.6}$$

On the other hand, note that $(z - \lambda_i)p_i(z) = z^n - 1$ for every $i = 1, \dots, n$, which yields

$$(T - \lambda_i I)(y_i) = (T - \lambda_i I)p_i(T)(x_0) = (T^n - I)(x_0) = T^n x_0 - x_0 = 0,$$

and then $Ty_i = \lambda_i y_i$. Since $\lambda_i^n = 1$, this and (5.6) show that x_0 is in G_0 and completes the proof. □

As a straightforward consequence of Propositions 5.3 and 5.6, we have the following.

Proposition 5.7 (Godefroy–Shapiro) *Let E be a complex, separable locally convex Hausdorff space and $T \in L(E)$. Suppose that the spaces*

$$E_0 = \operatorname{span}\{x \in E\colon Tx = \lambda x \text{ for some } \lambda \in \mathbb{C},\ |\lambda| < 1\}$$

$$F_0 = \operatorname{span}\{y \in E\colon Ty = \lambda y \text{ for some } \lambda \in \mathbb{C},\ |\lambda| > 1\}$$

$$G_0 = \operatorname{span}\{x \in E\colon \text{ there exists } \lambda \in \mathbb{C} \text{ with } \lambda^n = 1 \text{ for some } n \in \mathbb{N} \text{ so that } Tx = \lambda x\}$$

are dense in E. Then T is chaotic.

Given two locally convex spaces E and F, the operators $T \in L(E)$ and $S \in L(F)$ are said to be *quasi-conjugate* if there is a continuous operator $R : F \to E$ with dense range such that $R \circ S = T \circ R$.

Proposition 5.8 *Let $T \in L(E)$ and $S \in L(F)$ be quasi-conjugate operators. If S is transitive (resp., hypercyclic or chaotic), then so also is T.*

Proof We only give the details for the transitivity, the other cases being left to the reader. Take $U, V \subseteq E$ open and not empty. Since R is continuous and has dense range, $R^{-1}(U)$ and $R^{-1}(V)$ are non-empty open subsets of F. Then (because S is transitive), there are n and $x \in R^{-1}(U)$ so that $S^n x \in R^{-1}(V)$. Therefore, $Rx \in U$ and $RS^n x \in V$. One can easily check that $R \circ S^n = T^n \circ R$. This gives the conclusion. □

Corollary 5.5 *Let $T \in L(E)$ be such that there exist a dense subspace $F \subseteq E$, invariant for T (i.e., $TF \subseteq F$), and a locally convex topology τ on F, finer than the one induced by E and so that the restriction $T|_F : (F, \tau) \to (F, \tau)$ is continuous and transitive (resp., hypercyclic or chaotic). Then, so also is $T : E \to E$.*

Proof It is enough to take the inclusion $(F, \tau) \hookrightarrow E$ as R in Proposition 5.8. □

5.2 Mean Ergodic Operators

Given an operator $T \in L(E)$, we define its *Cesàro means* as

$$T_n = \frac{1}{n}\sum_{k=1}^{n} T^k$$

for $n \in \mathbb{N}$ (we set $T_0 = I$). Clearly, each $T_n \in L(E)$ (i.e., is linear and continuous). An operator $T \in L(E)$ is said to be:

- *Power bounded* if $(T^n)_n$ is equicontinuous in $L(E)$.
- *Cesàro bounded* if $(T_n)_n$ is equicontinuous in $L(E)$.
- *Mean ergodic* if there is $P \in L(E)$ such that $T_n \to P$ in $L_s(E)$ (i.e., $T_n x \to Px$ as $n \to \infty$ for every $x \in E$).
- *Uniformly mean ergodic* if there is $P \in L(E)$ such that $T_n \to P$ in $L_\beta(E)$ (i.e., uniformly on the bounded subsets of E).

Our aim is to see how these properties relate to each other. Clearly, every uniformly mean ergodic operator is mean ergodic. Also, if T is power bounded, then $(T^n)_n$ is equicontinuous, and given a continuous seminorm q, there is some continuous seminorm p so that

$$q(T^n x) \leq p(x) \text{ for every } x \in E \text{ and } n \in \mathbb{N}\,.$$

This has two consequences. On the one hand, $q\big(\frac{1}{n}T^n x\big) \leq \frac{1}{n}p(x)$, and since this holds for every q (and every n), we have

$$\frac{1}{n}T^n x \to 0 \text{ for every } x \in E\,. \tag{5.7}$$

But we can say even more. If $B \subseteq E$ is bounded, then $\sup_{x\in B} p(x) < \infty$. This immediately gives that

$$\frac{1}{n}T^n x \to 0 \text{ uniformly on the bounded subsets of } E\,. \tag{5.8}$$

On the other hand,

$$q(T_n x) \leq \frac{1}{n}\sum_{k=1}^{n} q(T^k x) \leq \frac{1}{n}\sum_{k=1}^{n} p(x) = p(x) \tag{5.9}$$

for every x and n. This shows that $(T_n)_n$ is equicontinuous in $L(E)$. So, every power bounded operator is Cesàro bounded. For this reason, from now on, we focus on the relation between Cesàro boundedness and mean ergodicity. We begin with two simple observations that follow by direct computation from the definition. For every

$T \in L(E)$ and all $n \in \mathbb{N}$, we have

$$\frac{1}{n}T^n = T_n - \frac{n-1}{n}T_{n-1} \tag{5.10}$$

and

$$(I-T)T_n = T_n(I-T) = \frac{1}{n}(T - T^{n+1})\,. \tag{5.11}$$

As a straightforward consequence of the definitions and (5.10), we have the following.

Proposition 5.9 *Let $T \in L(E)$:*

1. *If T is mean ergodic, then*

$$\frac{1}{n}T^n \to 0 \text{ in } L_s(E)$$

(that is, $\frac{1}{n}T^n x \to 0$ for every $x \in E$).

2. *If T is uniformly mean ergodic, then*

$$\frac{1}{n}T^n \to 0 \text{ in } L_\beta(E)$$

(that is, $\frac{1}{n}T^n \to 0$ uniformly on the bounded subsets of E).

Remark 5.2 Let us note that from (5.11) we have, for each $x \in E$,

$$T_n(x - Tx) = \frac{1}{n}(Tx - T^{n+1}x) = \frac{1}{n}Tx - \frac{n+1}{n}\Big(\frac{1}{n+1}T^{n+1}x\Big)\,.$$

So, if x satisfies $\frac{1}{n}T^n x \to 0$, then $T_n(x - Tx) \to 0$. The same computations show that, if $\frac{1}{n}T^n \to 0$ uniformly on the bounded sets of E, then $T_n(I - T) \to 0$ uniformly on the bounded sets.

Remark 5.3 Suppose that E is a complex locally convex Hausdorff space and $T \in L(E)$ is such that there are $0 \neq x_0 \in E$ and $\lambda \in \mathbb{C}$ with $|\lambda| > 1$ so that $Tx_0 = \lambda x_0$. Then $\frac{1}{n}T^n x_0 = \frac{1}{n}\lambda^n x_0$ does not tend to 0 and T cannot be neither power bounded nor mean ergodic (recall (5.7) and Proposition 5.9).

Theorem 5.4 (Eberlein's mean ergodic theorem) *Let $T \in L(E)$ be Cesàro bounded and $x \in E$ such that*

$$\lim_n \frac{1}{n}T^n x = 0\,. \tag{5.12}$$

Then for each $y \in E$, the following statements are equivalent:

1. *$Ty = y$ and y belongs to the closed convex hull of* $\mathrm{Orb}(x, T)$.
2. $y = \lim_n T_n x$.
3. $y = \sigma(E, E') - \lim_n T_n x$.
4. *y is a $\sigma(E', E)$-cluster point of $(T_n x)_n$ (i.e., for every $\sigma(E', E)$-neighbourhood U of 0 and every m, there is some $n > m$ such that $y - T_n x \in U$).*

Proof The implications 2 ⇒ 3 ⇒ 4 are obvious. In order to see 1 ⇒ 2, let us take some y in the closed convex hull of $\mathrm{Orb}(x, T)$ and such that $y = Ty$. We fix some seminorm q on E and $\varepsilon > 0$. Being T Cesàro bounded, $(T_n)_n$ is equicontinuous, and we can find a seminorm p so that $q(T_n z) \leq p(z)$ for every $z \in E$ and every $n \in \mathbb{N}$. By the choice of y, there are $\alpha_0, \ldots, \alpha_m \geq 0$ with $\sum_{i=0}^m \alpha_i = 1$ satisfying

$$p\Big(y - \sum_{i=0}^m \alpha_i T^i x\Big) < \frac{\varepsilon}{2}. \tag{5.13}$$

If $k \in \mathbb{N}$ is fixed, a straightforward computation shows that

$$T_n T^k x - T_n x = \frac{1}{n} \sum_{j=1}^k T^{n+j} x - \frac{1}{n} \sum_{j=1}^k T^j x .$$

We clearly have $\lim_n \frac{1}{n} T^j x = 0$ for each $j = 1, \ldots, k$, and by (5.12),

$$\lim_n \frac{1}{n} T^{n+j} x = \lim_n \Big(\frac{n+j}{n}\Big)\Big(\frac{1}{n+j} T^{n+j} x\Big) = 0 .$$

Therefore, for each fixed k, we have $T_n T^k x - T_n x \to 0$ as $n \to \infty$, and we can find $n(k) \in \mathbb{N}$ so that

$$q(T_n T^k x - T_n x) < \frac{\varepsilon}{2}$$

for every $n \geq n(k)$. We write $s_x = \sum_{i=0}^m \alpha_i T^i x$ (recall (5.13)) and note that $T_n y = y$ for every n (since $Ty = y$). Then, taking $n \geq \max(n(1), \ldots, n(m))$ yields

$$q(y - T_n x) \leq q(T_n(y - s_x)) + q(T_n s_x - T_n x) \leq p(y - s_x) + \sum_{i=0}^m \alpha_i q(T_n T^i x - T_n x) < \varepsilon .$$

Hence, $T_n x \to y$ as $n \to \infty$.

Let us assume now that 4 holds and write C for the convex hull of $\mathrm{Orb}(x, T)$. Clearly, $T_n x \in C$, and then y is in the $\sigma(E', E)$-closure of C that, since C is convex, coincides with $\overline{C}$ (the closure in the topology of E, recall Proposition 3.10). Just as before, from (5.12), we deduce that $T_n T^k x - T_n x \to 0$ as $n \to \infty$ for each fixed k, and therefore,

$$\lim_n v(T_n T^k x - T_n x) = 0\,,$$

for every $v \in E'$. Take now any functional $v \in E'$ and, for $\varepsilon > 0$, find n_0 so that

$$|v(T_n T x - T_n x)| < \frac{\varepsilon}{3}\,,$$

for every $n \geq n_0$. The set

$$U = \left\{z \in E\colon |v(y - z)| < \tfrac{\varepsilon}{3},\ |{}^tT(v)(z - y)| < \tfrac{\varepsilon}{3}\right\}$$

is a $\sigma(E', E)$-neighbourhood of y in E, and since y is a $\sigma(E', E)$-cluster point of $(T_n x)_n$, given n_0, there is some $n \geq n_0$ such that $T_n x \in U$. Then

$$|v(y - Ty)| < |v(y - T_n x)| + |v(T_n x - T T_n x)| + |v(T T_n x - Ty)| < \varepsilon\,.$$

Since $\varepsilon > 0$ was arbitrary, this gives $v(y - Ty) = 0$, and since this holds for every $v \in E'$, the Hahn–Banach theorem gives $y - Ty = 0$. □

Corollary 5.6 *Let $T \in L(E)$ be Cesàro bounded. Then T is mean ergodic if and only if the following two conditions hold:*

1. $\frac{1}{n} T^n x \to 0$ *as* $n \to \infty$ *for every* $x \in E$.
2. $(T_n x)_n$ *is* $\sigma(E', E)$*-relatively compact for every* $x \in E$.

Proof If T is mean ergodic, then from Proposition 5.9 the statement 1 holds. On the other hand, for each $x \in E$, the sequence $(T_n x)_n$ is convergent in E, and therefore, $\sigma(E', E)$-relatively compact.

Let us conversely assume that both 1 and 2 hold. Given $x \in E$, since $(T_n x)_n$ is $\sigma(E', E)$-relatively compact, it has some $\sigma(E', E)$-cluster point y that, by Eberlein's Theorem 5.4, satisfies $y = \lim_n T_n x$. The sequence $(T_n)_n$ is equicontinuous (because T is Cesàro bounded); therefore, defining $P : E \to E$ by $Px = \lim_n T_n x$, we have $P \in L(E)$ (recall Remark 2.10) and T is mean ergodic. □

An operator $T : E \to F$ between (Hausdorff) locally convex spaces is called *reflexive* if it maps bounded subsets of E into relatively $\sigma(F, F')$-compact subsets of F. A locally convex space E is called *semi-reflexive* if I is reflexive, that is, if every bounded subset of E is relatively $\sigma(E, E')$-compact.

Corollary 5.7 *Let $T \in L(E)$ be Cesàro bounded and reflexive, satisfying*

$$\frac{1}{n} T^n x \to 0 \ \text{as}\ n \to \infty \ \text{for every}\ x \in E\,. \tag{5.14}$$

Then T is mean ergodic. In particular, every power bounded operator on a semi-reflexive locally convex space is mean ergodic.

Proof Given $x \in E$, we have

$$T_n x = T_n(x - Tx) + T_n Tx\,.$$

The first term tends to 0 as $n \to \infty$ (recall Remark 5.2). On the other hand, by the equicontinuity of $(T_n)_n$, the sequence $(T_n Tx)_n$ is bounded and then (because T is reflexive) relatively $\sigma(E, E')$-compact. These two facts imply that $(T_n x)_n$ is relatively $\sigma(E, E')$-compact. By Corollary 5.6, the operator T is mean ergodic.

If E is semi-reflexive, then every continuous operator is reflexive. Since every power bounded operator is Cesàro bounded and satisfies (5.14), the conclusion follows. □

Our aim is to characterize those Cesàro bounded operators that are mean ergodic. We need first a lemma.

Lemma 5.3 *Let $T \in L(E)$ be Cesàro bounded such that*

$$\frac{1}{n}T^n x \to 0 \text{ as } n \to \infty \text{ for every } x \in E\,. \tag{5.15}$$

Then for each $z \in E$, the following statements are equivalent:

1. $\lim_n T_n z = 0$.
2. $v(z) = 0$ *for every* $v \in \ker(I - {}^tT)$.
3. $z \in \overline{\mathrm{Im}(I - T)}$.

Proof Let us suppose that $\lim_n T_n z = 0$ and take $v \in E'$ satisfying $v = {}^tTv$. Then $v = {}^tT_n v$ for every n, and

$$v(z) = \big({}^tT_n v\big)(z) = v(T_n z)\,.$$

This tends to 0 as $n \to \infty$, showing that 1 implies 2. Assume now that 2 holds, and suppose that $z \notin \overline{\mathrm{Im}(I - T)}$. By the Hahn–Banach theorem, we can find $v \in E'$ so that $v(z) = 1$ and $v(y) = 0$ for every $y \in \mathrm{Im}(I - T)$. This implies $v(x - Tx) = 0$ for every $x \in E$ and, then, $v = {}^tTv$. So, we have $v \in \ker(I - {}^tT)$ but $v(z) = 1$. This contradicts 2. Finally, if $y \in \mathrm{Im}(I - T)$, we can find some $x \in E$ so that $y = x - Tx$. Then $T_n y = T_n(x - Tx)$, which, by (5.15) and Remark 5.2, tends to 0 as $n \to \infty$. Since this holds for every $y \in \mathrm{Im}(I - T)$ and $(T_n)_n$ is equicontinuous (because T is Cesàro bounded), we get $T_n z \to 0$ for every $z \in \overline{\mathrm{Im}(I - T)}$ (recall Remark 2.11). This finally shows that 3 implies 1 and completes the proof. □

Given an operator $T \in L(E)$, we define the set

$$E_{me}(T) = \big\{x \in E \colon (T_n x)_n \text{ converges in } E\big\}\,.$$

Note that, if T is Cesàro bounded, then it is mean ergodic if and only if $E = E_{me}(T)$ (just by the equicontinuity of $(T_n)_n$, using Remark 2.10).

Theorem 5.5 (Yosida) *Let $T \in L(E)$ be Cesàro bounded satisfying*

$$\frac{1}{n} T^n x \to 0 \text{ as } n \to \infty \text{ for every } x \in E . \tag{5.16}$$

Then

$$E_{me}(T) = \ker(I - T) \oplus \overline{\mathrm{Im}(I - T)} . \tag{5.17}$$

The mapping $P : E_{me}(T) \to E$ defined by $Px = \lim_n T_n x$ for $x \in E_{me}(T)$ is the projection of $E_{me}(T)$ into $\ker(I - T)$, and $P = P^2 = TP = PT$.

In particular, T is mean ergodic if and only if

$$E = \ker(I - T) \oplus \overline{\mathrm{Im}(I - T)} .$$

Proof We prove (5.17) in three steps. We begin by showing that

$$\ker(I - T) \cap \overline{\mathrm{Im}(I - T)} = \{0\} .$$

Choose some z in the intersection and fix $\varepsilon > 0$ and a seminorm q on E. Being $(T_n)_n$ equicontinuous (because T is Cesàro bounded), we can find a seminorm p so that $q(T_n x) \leq p(x)$ for every x and n. First, $z \in \overline{\mathrm{Im}(I - T)}$; hence, there exists some $y \in E$ with $p(z - (I - T)y) < \varepsilon$ and then

$$q\big(T_n(z - (I - T)y)\big) < \varepsilon .$$

Second, z is in $\ker(I - T)$, and this implies that $T_n z = z$ for every n. Third, by Remark 5.2 and (5.16), we have $T_n(y - Ty) \to 0$ as $n \to \infty$, and there is some m for which

$$q\big(T_m(y - Ty)\big) < \varepsilon .$$

Joining these three facts together, we have

$$q(z) = q(T_m z) \leq q\big(T_m(z - (I - T)y)\big) + q\big(T_m(y - Ty)\big) < 2\varepsilon .$$

Since this holds for arbitrary ε and q, we have $z = 0$.

As a second step, we show that

$$E_{me}(T) \subseteq \ker(I - T) + \overline{\mathrm{Im}(I - T)} .$$

To do this, we pick some $x \in E_{me}(T)$ and define $Px = \lim_n T_n x$. By Eberlein's Theorem 5.4, $TPx = Px$, which gives $Px \in \ker(I - T)$. This also implies that

$T_n Px = Px$ for every n, and letting $z = x - Px$, we have

$$T_n z = T_n x - T_n Px = T_n x - Px ,$$

and this clearly tends to 0 as $n \to \infty$. Lemma 5.3 shows that $z \in \overline{\mathrm{Im}(I-T)}$ and yields our claim.

We see now that

$$\ker(I-T) + \overline{\mathrm{Im}(I-T)} \subseteq E_{me}(T) ,$$

which will complete the proof of (5.17). Suppose that $z = a + b$ with $a = Ta$ (hence $T_n a = a$ for every n) and $b \in \overline{\mathrm{Im}(I-T)}$ (and, then, $\lim_n T_n b = 0$, by Lemma 5.3). Thus $\lim_n T_n z = a$ and $z \in E_{me}(T)$. This completes the proof of (5.17).

To finish the proof, just observe that we have just seen that Eberlein's theorem implies that $TP = P$, and then $Px \in \ker(I-T)$ for every $x \in E_{me}(T)$. Also, $T_n Px = Px$ for every n and x. Then

$$P^2 x = \lim_n T_n Px = Px$$

for every $x \in E_{me}(T)$. Finally, since $T_n x - T_n Tx \to 0$ as $n \to \infty$ and $E_{me}(T)$ is invariant by T, we immediately have $Px - PTx = 0$ for every $x \in E_{me}(T)$. □

Corollary 5.8 *Let $T \in L(E)$ be Cesàro bounded satisfying*

$$\frac{1}{n} T^n x \to 0 \text{ as } n \to \infty \text{ for every } x \in E .$$

Then T is mean ergodic if and only if $\ker(I-T)$ *separates points of* $\ker(I - {}^tT)$.

Proof Let us note in first place that, if $u \in \ker(I - {}^tT)$, then $u(x) = u(Tx)$ for every $x \in E$, and a straightforward computation shows that $u(x) = u(T_n x)$ for every $x \in E$ and $n \in \mathbb{N}$. If T is mean ergodic, this implies that $u(Px) = u(x)$ for every x. Choose now $u_1, u_2 \in \ker(I - {}^tT)$ with $u_1 \neq u_2$. Then there is some $x_0 \in E$ so that $u_1(x_0) \neq u_2(x_0)$. By what we have just seen, $u_1(Px_0) \neq u_2(Px_0)$, and by Yosida's Theorem 5.5, $Px_0 \in \ker(I-T)$. This shows that indeed $\ker(I-T)$ separates points of $\ker(I - {}^tT)$.

Let us suppose now that T is not mean ergodic and take $y \notin \ker(I-T) \oplus \overline{\mathrm{Im}(I-T)}$ (which we know that exists by Yosida's theorem). By the Hahn–Banach theorem, there is $u \in E'$ such that $u(y) = 1$ and $u(x) = 0$ for every $x \in \ker(I-T) \oplus \overline{\mathrm{Im}(I-T)}$. This implies that $u((I-T)x) = 0$ for every $x \in E$ and all $u \in \ker(I - {}^tT)$. By assumption $\ker(I-T)$ separates u and 0, so there is some $x_0 \in \ker(I-T)$ with $u(x_0) \neq 0$. This gives a contradiction and completes the proof. □

As a straightforward consequence, we have the following.

Corollary 5.9 *Let $T \in L(E)$ be Cesàro bounded satisfying*

$$\frac{1}{n} T^n x \to 0 \text{ as } n \to \infty \text{ for every } x \in E ,$$

and $\ker(I - T) = \{0\}$. *Then T is mean ergodic if and only if* $\ker(I - {}^tT) = \{0\}$.

An operator $T : E \to F$ between (Hausdorff) locally convex spaces is called *Montel* if it maps bounded subsets of E into relatively compact subsets of F. A space E is called *semi-Montel* if the identity is a Montel operator (that is, every bounded subset of E is relatively compact). Let us point out that every semi-Montel space is semi-reflexive.

Proposition 5.10 *Let $T \in L(E)$ be a Cesàro bounded Montel operator such that*

$$\frac{1}{n} T^n \to 0 \text{ as } n \to \infty \text{ uniformly on the bounded subsets of } E . \tag{5.18}$$

Then T is uniformly mean ergodic.

Proof By Lemma 5.7, the operator T is mean ergodic, and there is $P \in L(E)$ such that $T_n x \to Px$ for every $x \in E$. Moreover, Theorem 5.5 shows that $P = P^2 = PT = TP$, from which we immediately have $P = PT_n = T_n P$ for every n. We fix now some continuous seminorm p on E, some $\varepsilon > 0$, and take a bounded set $B \subseteq E$. Since T is Cesàro bounded, the sequence $(T_n)_n$ is equicontinuous, and we can find a continuous seminorm q and a constant $C > 0$ so that

$$p((T_n - P)x) \leq Cq(x)$$

for every $x \in E$ and $n \in \mathbb{N}$. Now, the set $T(B)$ is relatively compact (because T is Montel); hence, there exist $z_1, \dots, z_m \in T(B)$ so that for each $y \in T(B)$ there is some $i = 1, \dots, m$ with

$$q(y - z_i) < \frac{\varepsilon}{2C} . \tag{5.19}$$

With all this, given $x \in B$, we choose z_i satisfying (5.19) for Tx, and for each $n \in \mathbb{N}$, we have

$$\begin{aligned} p(T_n Tx - Px) = p\big((T_n - P)Tx\big) &\leq p\big((T_n - P)(Tx - z_i)\big) + p\big((T_n - P)z_i\big) \\ &\leq Cq(Tx - z_i) + p\big((T_n - P)z_i\big) < \frac{\varepsilon}{2} + p\big((T_n - P)z_i\big) . \end{aligned}$$

Therefore,

$$\sup_{x \in B} p(T_n Tx - Px) \leq \frac{\varepsilon}{2} + \max_{i=1,\dots,m} p\big((T_n - P)z_i\big)$$

for every $n \in \mathbb{N}$. The fact that T is mean ergodic gives $\lim_n \max_{i=1,\dots,m} p\big((T_n - P)z_i\big) = 0$. This and the fact that $\varepsilon > 0$ was arbitrary yield

$$\lim_n \sup_{x \in B} p(T_n T x - P x) = 0\,,$$

and $T_n T \to P$ as $n \to \infty$ uniformly on the bounded subsets of E. Now we have

$$T_n - P = T_n(I - T) + (T_n T - P)\,.$$

We have just seen that the second term of the sum tends to 0 uniformly on the bounded subsets of E and, by Remark 5.2, so also does the first term. This altogether shows that $T_n \to P$ uniformly on bounded sets, and T is uniformly mean ergodic.

□

Note that every operator on a semi-Montel space is Montel. Recall also that every power bounded operator is Cesàro bounded (see (5.9)) and satisfies (5.8). Then we can apply Proposition 5.10 to have the following result.

Proposition 5.11 *Every power bounded operator on a semi-Montel locally convex space is uniformly mean ergodic.*

Theorem 5.6 (Lin) *Let E be a Fréchet space and $T \in L(E)$ such that*

$$\frac{1}{n} T^n \to 0 \text{ as } n \to \infty \text{ uniformly on the bounded subsets of } E\,, \tag{5.20}$$

and $\ker(I - T) = \{0\}$. *Consider the following statements:*

1. *$I - T_n$ is surjective for some $n \in \mathbb{N}$.*
2. *$I - T$ is surjective.*
3. *$T_n \to 0$ in $L_\beta(E)$ as $n \to \infty$.*

Then 1 ⇒ 2 ⇒ 3. If E is Banach, then also 3 ⇒ 1.

Proof It is easy to check that

$$I - T_n = (I - T)\Big(\frac{1}{n} \sum_{k=1}^{n-1} \sum_{j=0}^{k} T^j\Big)$$

for every n. Then, if $I - T_n$ is surjective for some n, so also is $I - T$. Let us suppose now that $I - T \in L(E)$ is surjective. Since, by assumption, it is also injective, by the Open Mapping Theorem 3.22, it is a surjective linear isomorphism. Given a bounded set $B \subseteq E$, the set $C = (I - T)^{-1}(B)$ is also bounded and $(I - T)(C) = B$. Take now any continuous seminorm q in E. We have

$$q_B(T_n) = \sup_{z \in B} q(T_n z) = \sup_{z \in C} q((I - T)T_n z) = q_C((I - T)T_n)\,,$$

and this tends to 0 as $n \to \infty$, since (5.20) holds (recall Remark 5.2). This shows that $T_n \to 0$ in $L_\beta(E)$, and 2 $\Rightarrow$ 3. Let us finally suppose that E is a Banach space and that 3 holds. Then there is some $n \in \mathbb{N}$ such that $\|T_n\| < 1$. A straightforward computation shows that

$$S = \sum_{j=0}^{\infty} (T_n)^j$$

is a continuous linear inverse of $I - T_n$, and then 1 holds. □

Let us see that the implication 3 $\Rightarrow$ 1 does not hold in general for non-normable Fréchet spaces. To this end, consider the Köthe echelon space $\lambda_1(A)$ defined by the matrix $a_n(i) = i^n$, that is,

$$\lambda_1(A) = \Big\{x = (x_i)_i \in \mathbb{C}^{\mathbb{N}} \colon q_n(x) = \sum_{i=1}^{\infty} i^n |x_i| < \infty \text{ for every } n \in \mathbb{N}\Big\},$$

known as the space of *rapidly decreasing sequences*, and often denoted by s.

Example 5.7 The operator $T \in L(s)$ given by

$$T(x) = \Big(\big(1 - \tfrac{1}{2^i}\big)\Big)_i \text{ for } x \in s$$

satisfies:

1. $\ker(I - T) = \{0\}$.
2. $\frac{1}{n} T^n \to 0$ uniformly on bounded sets.
3. $T_n \to 0$ in $L_\beta(s)$.
4. $I - T$ is not surjective.

Note that the first two ones are the hypothesis of Proposition 5.6, and the remaining two are the proof of our claim that the implication 3 $\Rightarrow$ 1 does not hold in general.

It is straightforward that $\ker(I - T) = \{0\}$. On the other hand, for each m, we have

$$T^m x = \Big(\big(1 - \tfrac{1}{2^i}\big)^m x_i\Big)$$

for every $x \in s$. Then,

$$q_n(T^m x) = \sum_{i=1}^{\infty} i^n \big(1 - \tfrac{1}{2^i}\big)^m |x_i| \leq \sum_{i=1}^{\infty} i^n |x_i| = q_n(x)\,.$$

From this, we deduce that $\frac{1}{n} T^n \to 0$ uniformly on bounded sets (that is, 2 holds) and that T is power bounded. By Proposition 2.12, the space s is Montel and then

Proposition 5.11 gives that T is uniformly mean ergodic. Then there is an operator $P \in L(s)$ so that $T_k \to P$ in $L_\beta(s)$ as $k \to \infty$. Let us see that, in fact, $P = 0$. We begin with the canonical basis vectors. Given $j, k \in \mathbb{N}$, we have

$$T^j e_k = \Big(1 - \frac{1}{2^i}\Big)^j e_k ,$$

and this tends to 0 (for each fixed k) as $j \to \infty$. Then

$$\frac{1}{n}\sum_{j=1}^{n} T^j e_k \longrightarrow 0 \text{ as } n \to \infty,$$

and therefore, $Pe_k = 0$ for each fixed k. As $\operatorname{span}\{e_k : k \in \mathbb{N}\}$ is dense in s (since the finite truncations of each sequence $x \in s$ converge to x) and $P \in L(s)$, we conclude that $P = 0$, giving our claim. Finally, a simple computation shows that $(2^{-i})_i \notin \operatorname{Im}(I - T)$, which gives that $I - T$ is not surjective. □

Theorem 5.8 (Lin) *Let E be a Banach space and $T \in L(E)$ such that*

$$\frac{1}{n}\|T^n\| \to 0 \ \text{as } n \to \infty . \tag{5.21}$$

The following are equivalent:

1. *T is uniformly mean ergodic.*
2. *$\operatorname{Im}(I - T)$ is closed and $E = \ker(I - T) \oplus \operatorname{Im}(I - T)$.*
3. *$\operatorname{Im}(I - T)$ is closed.*

Proof In order to keep the notation as clear as possible, all through the proof we write $(I - T)(E) = \operatorname{Im}(I - T)$, $F = \overline{\operatorname{Im}(I - T)}$, and $S = T|_F$ (the restriction of T to F). Suppose that T is uniformly mean ergodic. By Yosida's Theorem 5.5, $E = \ker(I - T) \oplus F$. We have

$$T(F) = T\big(\overline{(I - T)(E)}\big) \subseteq \overline{T\big((I - T)(E)\big)} \subseteq \overline{F} = F ;$$

thus $S(F) \subseteq F$ and $S \in L(F)$.

There is some $P \in L(E)$ so that $\lim_n \|T_n - P\| = 0$. We know that (recall again Yosida's theorem) $\operatorname{Im} P = \ker(I - T)$ and then $F = \ker P$, from which we have

$$\|S_n\| = \sup_{\substack{x\in F\\ \|x\|\le 1}} \|S_n x\| \le \sup_{\substack{x\in F\\ \|x\|\le 1}} \|S_n x - Px\| + \sup_{\substack{x\in F\\ \|x\|\le 1}} \|Px\| = \|S_n - P\| ,$$

and this clearly implies that $\lim_n \|S_n\| = 0$ and then $\frac{1}{n}\|S_n\| \to 0$. We also have $\ker(I - S) = \{0\}$, so that S fulfils all conditions in Proposition 5.6–3 and $I - S$ is surjective. Then

$$\overline{(I-T)(E)} = F = (I-S)(F) = (I-T)(F) \subseteq (I-T)(E) \subseteq F = \overline{(I-T)(E)},$$

and $(I - T)(E)$ is closed. This shows that 1 implies 3.

Let us suppose that $\mathrm{Im}(I - T)$ is closed (hence Banach). Note that $\mathrm{Im}(I - T) = F$, and the operator $I - T : E \to F$ is continuous and surjective so, by the Open Mapping Theorem 3.22, $I - T$ is open and there is $K > 0$ such that for every $y \in E$ there is $z \in E$ so that $(I - T)z = y$ and $\|z\| \leq K\|y\|$. We fix some $y \in E$ and select z as above to get (recall (5.11))

$$\|T_n y\| = \|T_n(I-T)z\| \leq \frac{1}{n}\|T - T^{n+1}\|\,\|z\| \leq \frac{K}{n}\big(\|T\| + \|T^{n+1}\|\big)\|y\|\,.$$

The restriction $S = T|_F : F \to F$ is well defined, linear, and continuous. Then, taking the supremum over $\|y\| \leq 1$ in the previous expression, and using (5.21), we have $\lim_n \|S_n\| = 0$, and then $\frac{1}{n}\|S_n\| \to 0$. On the other hand, proceeding as in the proof of Yosida's Theorem 5.5, we get $E = \ker(I - T) \cap \mathrm{Im}(I - T) = \{0\}$, from which $\ker(I - S) = \{0\}$ follows. Then, by Proposition 5.6, $I - S : F \to F$ is surjective (hence, an isomorphism onto). Denote by $J : F \to E$ the canonical inclusion and define

$$P = I - J(I-S)^{-1}(I-T) \in L(E)\,.$$

Let us finally see that $\lim_n \|T_n - P\| = 0$, which will give that T is uniformly mean ergodic and complete the proof. For $x \in E$, let $y = (I - S)^{-1}(I - T)x$, and note that $(I - T)y = (I - S)y = (I - T)x$, which gives $T(x - y) = x - y$, and hence, $T_n(x - y) = x - y$ for every n. Note also that $Px = x - y$. With this, we have (recall again (5.11))

$$\begin{aligned}\|T_n x - Px\| &= \|T_n x-(x-y)\| = \|T_n x - T_n(x-y)\| = \|T_n y\| = \|T_n(I-S)^{-1}(I-T)x\| \\ &\leq \|(I-S)^{-1}\|\,\|T_n(I-T)x\| \leq \frac{1}{n}\|(I-S)^{-1}\|\big(\|T\| + \|T^{n+1}\|\big)\|x\|\,.\end{aligned}$$

Then

$$\|T_n - S\| \leq \frac{1}{n}\|(I-S)^{-1}\|\big(\|T\| + \|T^{n+1}\|\big)$$

for every $n \in \mathbb{N}$, and since the latter tends to 0 as $n \to \infty$, this yields the claim and completes the proof. □

Remark 5.4 We finish this section by noting that, as a straightforward consequence of the Banach–Steinhaus theorem, an operator $T \in L(E)$ on a Fréchet space is power bounded if and only if the orbit $\text{Orb}(x, T)$ of each $x \in E$ is bounded. Then Corollary 5.2 gives that a power bounded operator cannot be hypercyclic. Also, if T is an operator on a Banach space, then it is power bounded if and only if $\sup_{n\in\mathbb{N}} \|T^n\| < \infty$. Hence, a hypercyclic operator T on a Banach space must satisfy $\|T\| > 1$.

5.3 Examples

We look now at some concrete operators and study when they share the properties that we have been looking at.

5.3.1 *The Backward Shift*

We start by considering operators defined on sequence spaces. For the moment, let X denote either ℓ_p with $1 \leq p < \infty$ (recall Example 2.5) or c_0 (recall (2.9)). We consider the backward shift operator $B : X \to X$, defined as $B\big((x_n)_{n=1}^{\infty}\big) = (x_{n+1})_{n=1}^{\infty}$, that is,

$$B(x_1, x_2, x_3, \dots) = (x_2, x_3, \dots)$$

(recall (3.6)). Clearly, this defines a continuous linear operator (i.e., $B \in L(X)$), and moreover, $\|B\| \leq 1$ (in fact, $\|B\| = 1$, since $B(e_2) = e_1$). Our aim is to study the properties (transitivity and mean ergodicity) of the operator

$$T = \lambda B \text{ for } \lambda \in \mathbb{C}$$

in terms of λ. For $\lambda = 0$, there is not much to say, so we assume, from now on, $\lambda \neq 0$. We collect all the information in the following result.

Theorem 5.9 *Let $T = \lambda B$ with $\lambda \neq 0$:*

1. *If $|\lambda| > 1$, then T is transitive, hypercyclic, and chaotic.*
2. *If $|\lambda| \leq 1$, then T is power bounded, and:*

   a. *If $|\lambda| < 1$, then T is uniformly mean ergodic.*
   b. *If $|\lambda| = 1$, then T is mean ergodic but not uniformly mean ergodic.*

Proof Let us suppose in first place that $|\lambda| > 1$. We use Corollary 5.1–(2) to see that the operator is transitive. Let X_0 be the subspace of X consisting of those sequences with only finitely many non-zero coordinates, which is clearly dense in X. Define an operator $S : X_0 \to X$ by

$$S(y_1, y_2, y_3, \dots) = \frac{1}{\lambda}(0, y_1, y_2, y_3, \dots) .$$

Obviously, $(T \circ S)(y) = y$ for every $y \in X_0$. Also, given $y \in X_0$, we have $\lim_n T^n y = 0$ (because y has only a finite number of non-zero coordinates). Finally, if $y \in X_0$, then

$$\|S^n y\| = \frac{1}{\lambda^n}\|y\|$$

for each n, and $\lim_n \|S^n y\| = 0$. Then Corollary 5.1–(2) gives that T is transitive. The space X is separable, and then by Birkhoff's Theorem 5.2, T is also hypercyclic. Let us see that it is also chaotic. First of all, it is easy to see that $x \in X$ is periodic if and only if it is of the form

$$x = \big(x_1, \dots, x_N, \lambda^{-N}x_1, \dots, \lambda^{-N}x_N, \lambda^{-2N}x_1, \dots, \lambda^{-2N}x_N, \dots\big).$$

Since X_0 is dense in X, in order to see that the set of periodic points is dense in X, it suffices to see that every $y \in X_0$ can be approximated by periodic points. Take, then, some $y = (y_1, \dots, y_m, 0, 0, \dots)$ in X_0. Now, for each $N \geq m$, we define x_N as

$$x_N = \big(\underbrace{y_1, \dots, y_m, 0, \dots, 0}_{N}, \underbrace{\lambda^{-N}y_1, \dots, \lambda^{-N}y_m, 0, \dots, 0}_{N}, \dots\big),$$

each one periodic. If $X = c_0$, then

$$\|x_N - y\|_\infty = \frac{1}{\lambda^N}\|y\| \xrightarrow{N\to\infty} 0.$$

On the other hand, if $X = \ell_p$ for $1 \leq p < \infty$, we have

$$\begin{aligned}\|x_N - y\|_p &= \Big(\sum_{j=1}^\infty \sum_{k=1}^m |\lambda^{-Nj}|^p |y_k|^p\Big)^{\frac{1}{p}} = \Big(\sum_{j=1}^\infty |\lambda^{-Nj}|^p \|y\|^p\Big)^{\frac{1}{p}} \\ &\leq \sum_{j=1}^\infty |\lambda^{-Nj}| \|y\| = \frac{|\lambda|^{-N}}{1-|\lambda|^{-N}} \xrightarrow{N\to\infty} 0.\end{aligned}$$

So, in each case, we have a sequence of periodic points converging to y. This gives our claim and completes the proof of 1.

The proof of 2 begins by noting that, since $\|B\| = 1$, we have $\|T\| = |\lambda| \leq 1$, and T is power bounded. Suppose now that $|\lambda| < 1$, then

$$\frac{1}{n}\Big\|\sum_{j=1}^n T^j\Big\| \leq \frac{1}{n}\sum_{j=1}^n \|\lambda^j B^j\| \leq \frac{1}{n}\sum_{j=1}^n |\lambda|^j \|B\| \leq \frac{1}{n}\frac{1}{1-|\lambda|} \xrightarrow{n\to\infty} 0.$$

This shows that the Cesàro means T_n converge (in the operator norm) to the operator $P = 0$. Hence, T is uniformly mean ergodic (showing 2a).

We complete the proof by showing that if $|\lambda| = 1$, then T is mean ergodic but not uniformly mean ergodic. Since $B^j x \to 0$ for every $x \in X$, we get $T^j x \to 0$ and then

$$\frac{1}{n}\sum_{j=1}^{n} T^j x \longrightarrow 0$$

for every $x \in X$. This yields that T is mean ergodic. Observe that, if $X = \ell_p$ with $1 < p < \infty$, the space is reflexive, and the result follows also from Corollary 5.7.

Let us see now that T is not uniformly mean ergodic. We consider two cases. Assume first that $X = \ell_p$ with $1 \leq p < \infty$. Fix $n \in \mathbb{N}$ and observe that

$$\Big\| \frac{1}{n}\sum_{j=1}^{n} T^j (e_{n+1}) \Big\|_p = \frac{1}{n}\Big(\sum_{j=1}^{n} |\lambda|^{jp}\Big)^{\frac{1}{p}} = 1\,.$$

Since the canonical vectors have norm 1, this implies that

$$\frac{1}{n}\Big\|\sum_{j=1}^{n} T^j\Big\|_p \not\to 0\,,$$

and T is not uniformly mean ergodic.

The case of $X = c_0$ is slightly more involved. To begin with, from the fact that $\|B\| = 1$ and $|\lambda| = 1$, we have $\lim_n \frac{1}{n}\|T^n\| = 0$, and by Lin's Theorem 5.6, to see that T is not uniformly mean ergodic, it suffices to see that

$$\mathrm{Im}(I - T) = (I - \lambda B)(c_0)$$

is not closed in c_0. We see in first place that $I - T$ is injective. If this were not the case, there would be some non-zero $x \in c_0$ so that $x = \lambda B x$. This forces x to be of the form $x = x_1\big(1, 1/\lambda, 1/\lambda^2, 1/\lambda^3 \ldots\big)$, but the sequence $\big(1/\lambda^j\big)_{j=0}^{\infty}$ does not belong to c_0. This gives a contradiction and yields our claim. In the second place, we observe that

$$(I - \lambda B)(e_1) = e_1, \text{ and } (I - \lambda B)(e_n) = e_n - \lambda e_{n-1} \text{ for } n \geq 2\,,$$

which clearly implies

$$X_0 = \mathrm{span}\{e_n : n \in \mathbb{N}\} \subseteq \mathrm{Im}(I - T)\,.$$

This gives that $\mathrm{Im}(I - T)$ is dense in c_0 (because X_0 is so). Therefore, if $\mathrm{Im}(I - T)$ were closed, we would have

$$\mathrm{Im}(I - T) = c_0$$

and, by the Open Mapping Theorem 3.22, $I - T : c_0 \to c_0$ would be an isomorphism. Then its bi-transpose

$$^{tt}(I - T) : \ell_\infty \longrightarrow \ell_\infty$$

would be also an isomorphism. It is easy to see that $^{tt}(I - T) = (I - \lambda B) : \ell_\infty \to \ell_\infty$, where B is the backward shift defined by (3.6) on ℓ_∞. However,

$$(I - \lambda B)\big(1, \tfrac{1}{\lambda}, \tfrac{1}{\lambda^2}, \dots\big) = 0\,,$$

and $\big(1, \frac{1}{\lambda}, \frac{1}{\lambda^2}, \dots\big) \in \ell_\infty$ (since $|\lambda| = 1$). This shows that $I - \lambda B$ is not injective on ℓ_∞, and henceforth not an isomorphism. Then $\mathrm{Im}\, I - T$ is not closed in c_0 and this completes the proof. □

Remark 5.5 The case of $T = \lambda B : \ell_\infty \to \ell_\infty$ is more delicate. To begin with, it is easy to see that $\|T^n\| = |\lambda|^n$ and, then, λB is power bounded if and only if $|\lambda| \leq 1$. On the other hand, if $|\lambda| < 1$, the same argument as in Theorem 5.9 shows that $\lambda B \in L(\ell_\infty)$ is uniformly mean ergodic.

Suppose that $|\lambda| > 1$. Since the space is not separable, λB cannot be hypercyclic in any case (actually, no operator on ℓ_∞ can be hypercyclic). Let us see that neither can it be transitive. It is enough to see that $^tT \in L(\ell'_\infty)$ has eigenvalues (recall Proposition 5.5). Let $L : \ell_\infty \to \mathbb{C}$ be a Banach limit (recall Sect. 3.2.4). By Theorem 3.7, $L \in \ell'_\infty$ and $L(Bx) = L(x)$ for every $x \in \ell_\infty$, which implies

$$\big({}^tT\big)(L)(x) = {}^t(\lambda B)(L)(x) = L(\lambda Bx) = \lambda L(x)$$

for every $x \in \ell_\infty$, that is, $^tT(L) = \lambda L$, and L is an eigenvector of ^tT. This shows that T is not transitive. As a matter of fact, Bermúdez and Kalton showed in [2, Theorem 3.4] that there cannot be any transitive operator on ℓ_∞.

It remains to study the behaviour of the operator for $|\lambda| = 1$. In this case, the operator λB is not mean ergodic (hence, obviously, neither is it uniformly mean ergodic). This relies on some known results. We just indicate how to use them to get the conclusion. First of all, by [8, Theorem 5], an operator on ℓ_∞ is mean ergodic if and only if it is uniformly mean ergodic, so it suffices to see that the latter fails for λB. Now, by [7, Chapter 2, Theorem 2.7, p.90], if S is any operator on a Banach space and $I - S$ is invertible, then S is uniformly mean ergodic. But, on the other hand, from [4, Chapter VII, Corollary 6.6, p. 214], we get that for $|\lambda| = 1$, the operator $I - \lambda B$ is not invertible. Hence, λB is not uniformly mean ergodic, and this gives our claim.

5.3.2 Composition Operators

We look now at composition operators defined on $H(\mathbb{D})$. Let us recall (see Example 4.16) that the composition operator $C_\varphi : H(\mathbb{D}) \to H(\mathbb{D})$ is always well defined and continuous for every holomorphic self-map $\varphi : \mathbb{D} \to \mathbb{D}$. Our aim now is to study when these operators are power bounded, mean ergodic, or hypercyclic. We look for conditions on φ that ensure that the composition operator C_φ has these properties. We answer these questions in Theorems 5.12 and 5.13.

Given a self-map $\varphi : \mathbb{D} \to \mathbb{D}$ and $n \in \mathbb{N}$, we denote, as usual, φ^n for the n-iterate of φ. That is, we denote $\varphi^0 = \mathrm{id}_\mathbb{D}$ (the identity mapping on $\mathbb{D}$), and $\varphi^n = \varphi^{n-1} \circ \varphi$ for $n \geq 1$. We clearly have that $(C_\varphi)^n(f) = f \circ \varphi^n$ for every $f \in H(\mathbb{D})$. In other words, $C_\varphi^n = C_{\varphi^n}$ for every n. Hence, if we want to study the behaviour of C_φ^n as $n \to \infty$, then the behaviour of the sequence of iterates $(\varphi^n)_n$ has to appear in one way or another.

We begin by looking at power boundedness and mean ergodicity. Let us point out in first place that by Montel's Theorem 4.9 and Proposition 5.11, every power bounded operator on $H(\mathbb{D})$ is uniformly mean ergodic. Whether or not C_φ has some of these properties depends on the shape of symbol φ. We use some very classical results on complex analysis. We state them without proof. The first one is the Schwarz lemma, the proof of which can be found in [9, Theorem 12.2].

Theorem 5.10 (Schwarz Lemma) *Let $\varphi : \mathbb{D} \to \mathbb{D}$ be holomorphic such that $\varphi(0) = 0$. Then:*

1. $|\varphi(z)| \leq |z|$ *for every* $z \in \mathbb{D}$. *If* $|\varphi(z_0)| = |z_0|$ *for some* $0 \neq z_0 \in \mathbb{D}$, *then there exists* $\lambda \in \mathbb{C}$ *with* $|\lambda| = 1$ *such that* $\varphi(z) = \lambda z$ *for every* $z \in \mathbb{D}$.
2. $|\varphi'(0)| \leq 1$. *Moreover,* $|\varphi'(0)| = 1$ *if and only if there exists* $\lambda \in \mathbb{C}$ *with* $|\lambda| = 1$ *such that* $\varphi(z) = \lambda z$ *for every* $z \in \mathbb{D}$.

We begin by looking at a particular class of symbols. A self-map $T : \mathbb{D} \to \mathbb{D}$ is an *automorphism* of $\mathbb{D}$ if it is holomorphic and bijective (note that, by the open mapping theorem, T^{-1} is also holomorphic). The set of all automorphisms of $\mathbb{D}$ clearly is a group and is denoted by $\mathrm{Aut}(\mathbb{D})$. Automorphisms can be described in a very precise way. For $a \in \mathbb{D}$, we define $T_a : \mathbb{D} \to \mathbb{D}$ by

$$T_a(z) = \frac{a - z}{1 - \overline{a}z},$$

which is clearly holomorphic. Moreover, it is easy to see that $T_a(\mathbb{D}) = \mathbb{D}$ and that it is invertible and $T_a^{-1} = T_a$. So, $T_a \in \mathrm{Aut}(\mathbb{D})$ and (this is also straightforward) $T_a(0) = a$ and $T_a(a) = 0$. Also, some simple computations show that $z_0 = a\frac{1-\sqrt{1-|a|^2}}{|a|^2}$ is the only fixed point of T_a in $\mathbb{D}$. Every automorphism is the composition of one of this type and a rotation. This is the content of the following result (see [9, Theorem 12.6]).

Proposition 5.12 *Let $T \in \mathrm{Aut}(\mathbb{D})$ with $T(a) = 0$. Then there exists $\lambda \in \mathbb{C}$ with $|\lambda| = 1$ so that $T(z) = \lambda T_a(z)$ for every $z \in \mathbb{D}$.*

As a consequence, every automorphism of $\mathbb{D}$ can be continuously extended to the closed disc $\overline{\mathbb{D}}$. From Proposition 5.12, we can study the fixed points of automorphisms. The proof of the following result is not difficult and can be found in [3, Lemma 1.8.1].

Lemma 5.4 *Let $T \in \mathrm{Aut}(\mathbb{D})$ be different from the identity. Then T has at least one fixed point in $\overline{\mathbb{D}}$. Moreover, if T does not have fixed points in $\mathbb{D}$, then it has at most two fixed points $\tau, \sigma \in \partial\mathbb{D}$ (that may be equal) that satisfy $T'(\sigma)T'(\tau) = 1$.*

Taking this into account, we say that an automorphism T of $\mathbb{D}$ is:

- *Elliptic* if it has a (then unique) fixed point in $\mathbb{D}$.
- *Parabolic* if it has a unique fixed point in $\partial\mathbb{D}$.
- *Hyperbolic* if it has two fixed points in $\partial\mathbb{D}$.

For example, $T(z) = iz$ is elliptic (with $\tau = 0$ as unique fixed point), $T(z) = \frac{(1+i)z-i}{iz+1-i}$ is parabolic (with fixed point $\tau = 1$), and $T(z) = \frac{z+1/2}{1+z/2}$ is hyperbolic (with fixed points $\tau = 1$ and $\sigma = -1$).

We begin by studying composition operators defined by elliptic automorphisms. As we said before, the iterates of the symbol play an important rôle. In this case (thanks to Proposition 5.12), they have a very convenient shape.

Lemma 5.5 *Let T be an elliptic automorphism with fixed point $\tau \in \mathbb{D}$. Then there is $\lambda \in \mathbb{C}$ with $|\lambda| = 1$ such that*

$$T^n(z) = T_\tau\big(\lambda^n T_\tau(z)\big) \tag{5.22}$$

for every $z \in \mathbb{D}$ and $n \in \mathbb{N}$.

Proof Consider $S = T_\tau \circ T \circ T_\tau \in \mathrm{Aut}(\mathbb{D})$, which satisfies $S(0) = 0$. By Proposition 5.12, there is λ with $|\lambda| = 1$ so that $S(z) = \lambda z$ for every $z \in \mathbb{D}$. Now, using the fact that $T_\tau \circ T_\tau = \mathrm{id}_{\mathbb{D}}$, we have

$$\lambda^n z = S^n(z) = \big(T_\tau \circ T \circ T_\tau\big)^n(z) = T_\tau \circ T^n \circ T_\tau(z)$$

for every n and z. This yields the conclusion. □

Proposition 5.13 *Let T be an elliptic automorphism. Then the composition operator $C_T : H(\mathbb{D}) \to H(\mathbb{D})$ is power bounded and uniformly mean ergodic.*

Proof We just have to show that the composition operator is power bounded since, as we have already noticed, this (by Theorem 4.9 and Proposition 5.11) automatically gives that the operator is also uniformly mean ergodic. Let $\tau \in \mathbb{D}$ be the fixed point of T and take λ as in Lemma 5.5. Now, T_τ is continuous, and then, for each $0 < \rho < 1$, there is some $0 < R(\rho) < 1$ so that $|T_\tau(z)| < R(\rho)$ for every $|z| < \rho$. Fix now $0 < r < 1$ and write $s = R(r)$ and $R = R(s)$. Then, for every $f \in H(\mathbb{D})$ and all $|z| < r$ and $n \in \mathbb{N}$, we have, using (5.22),

$$|f(T^n z)| = \left|f\big(T_\tau(\lambda^n T_\tau z)\big)\right| \leq \sup_{|w|\leq s} \left|f\big(T_\tau(w)\big)\right| \leq \sup_{|\zeta|\leq R} |f(\zeta)|\,.$$

This shows that C_T is power bounded and completes the proof. □

This describes completely the composition operators defined by elliptic automorphisms, showing that they are power bounded and, then, uniformly mean ergodic. What can be said for composition operators defined by arbitrary holomorphic self-maps? When are they power bounded? As we have seen, having a fixed point plays some rôle. A holomorphic self-map of the disc may (or may not) have a fixed point in $\mathbb{D}$, but if it does have one, then it has only one (unless the mapping is the identity).

Proposition 5.14 *Let $\varphi : \mathbb{D} \to \mathbb{D}$ be holomorphic and different from the identity. Then there exists at most one point $z_0 \in \mathbb{D}$ such that $\varphi(z_0) = z_0$.*

Proof Suppose that there are $a \neq b$ so that $\varphi(a) = a$ and $\varphi(b) = b$. We define $\psi = T_a \circ \varphi \circ T_a \in H(\mathbb{D})$ that satisfies $\psi(0) = 0$ and take $w \in \mathbb{D}$ so that $T_a(w) = b$. Then

$$\psi(w) = T_a \circ \varphi(b) = T_a(b) = w\,. \tag{5.23}$$

By Theorem 5.10, there exists some $\lambda \in \mathbb{C}$ with $|\lambda| = 1$ such that $\psi(z) = \lambda z$ for every $z \in \mathbb{D}$. But (5.23) forces $\lambda = 1$, and, in fact, $\psi = \mathrm{id}_{\mathbb{D}}$. Finally, the fact that $T_a \circ T_a = \mathrm{id}_{\mathbb{D}}$ yields that φ is the identity. □

Our next step is to show that, if a holomorphic self-map (not an automorphism) has a fixed point τ, then the iterates converge uniformly on the compact subsets of $\mathbb{D}$ to the constant function given by this point. We denote this function by c_τ; that is, $c_\tau(z) = \tau$ for every $z \in \mathbb{D}$.

Proposition 5.15 *Let $\varphi : \mathbb{D} \to \mathbb{D}$ be holomorphic and not an automorphism such that there is $\tau \in \mathbb{D}$ with $\varphi(\tau) = \tau$. Then the sequence of iterates $(\varphi^n)_n$ converges to c_τ uniformly on the compact subsets of $\mathbb{D}$.*

Proof Let us suppose first that $\tau = 0$ (i.e., $\varphi(0) = 0$). Since φ is not an automorphism, from Schwarz's Lemma (Theorem 5.10), we have

$$|\varphi(z)| < |z| \tag{5.24}$$

for every $0 \neq z \in \mathbb{D}$. For each $0 < r < 1$, we define

$$M(\varphi, r) = \max_{|z|\leq r} |\varphi(z)| \text{ and } \delta(r) = \frac{M(\varphi, r)}{r}\,.$$

From (5.24), we have $\delta(r) < 1$ for every $0 < r < 1$. Fix one such r and consider the function given by

$$\psi_r(z) = \frac{\varphi(rz)}{M(\varphi, r)}$$

for $z \in \mathbb{D}$. Clearly, it is holomorphic and $\psi_r(\mathbb{D}) \subseteq \mathbb{D}$. Even more, it can be extended to a continuous function on $\overline{\mathbb{D}}$ and $\psi_r(0) = 0$. Then, Schwarz's Lemma again gives

$$|\psi_r(z)| \leq |z|$$

for every $z \in \overline{\mathbb{D}}$. With this, for $|z| \leq r$, we have

$$|\varphi(z)| = M(\varphi, r)\big|\psi_r\big(z/r\big)\big| \leq M(\varphi, r)\Big|\frac{z}{r}\Big| = \delta(r)|z| < \delta(r) .$$

Iterating this, we get

$$|\varphi^n(z)| \leq \delta(r)^n$$

for every $z \in \overline{\mathbb{D}(0, r)}$ and all n. Since $\delta(r) < 1$, this shows that (φ^n) converges to 0 uniformly on the compact sets, completing the proof if $\tau = 0$.

Now, if $\tau \neq 0$, we define $\psi = T_\tau \circ \varphi \circ T_\tau$. This is a holomorphic self-map that fixes 0 and that is not an automorphism. By the previous case, the iterates $(\psi^n)_n$ converge to 0 uniformly on compact sets. But, since clearly $\varphi^n = T_\tau \circ \psi^n \circ T_\tau$ for every n, we have the conclusion. □

We are now in the position to give an analogue of Proposition 5.13 for holomorphic self-maps with a fixed point.

Proposition 5.16 *Let $\varphi : \mathbb{D} \to \mathbb{D}$ be holomorphic, not an automorphism and with a fixed point $\tau \in \mathbb{D}$. Then the composition operator $C_\varphi : H(\mathbb{D}) \to H(\mathbb{D})$ is power bounded and uniformly mean ergodic.*

Proof As in Proposition 5.13, it suffices to see that C_φ is power bounded. We know from Proposition 5.15 that the iterates of φ converge to the constant function c_τ uniformly on the compact subsets of $\mathbb{D}$. Our first step in the proof is to see that for every compact $K \subseteq \mathbb{D}$ there is some compact $L \subseteq \mathbb{D}$ such that

$$\varphi^n(K) \subseteq L \text{ for every } n \in \mathbb{N} . \tag{5.25}$$

Fix some compact set $K \subseteq \mathbb{C}$ and, for $\varepsilon = \frac{1-|\tau|}{2} > 0$, choose $n_0 \in \mathbb{N}$ so that

$$|\varphi^n(z) - \tau| < \varepsilon ,$$

for every $z \in K$ and all $n \geq n_0$ we have. Then, for every such z and n, we have

$$|\varphi^n(z)| < \frac{1+|\tau|}{2}.$$

Define now the set

$$L = \bigcup_{j=1}^{n_0-1} \varphi^j(K) \cup \overline{\mathbb{D}(0, \tfrac{1+|\tau|}{2})},$$

which is clearly compact and satisfies (5.25).

Let us see now that the operator is power bounded. Given some compact set $K \subseteq \mathbb{D}$, we take some compact set L as in (5.25). Then, for each $f \in \mathrm{H}(\mathbb{D})$, we have

$$\sup_{z\in K} |(C_\varphi)^n(f)(z)| = \sup_{z\in K} |f(\varphi^n(z))| \leq \sup_{w\in L} |f(w)|.$$

This shows that $\big((C_\varphi)^n\big)_n$ is equicontinuous and completes the proof. □

The key property to get that the composition operator is power bounded is (5.25). We isolate this property and say that a function $\varphi : \mathbb{D} \to \mathbb{D}$ has *stable orbits* if for every compact $K \subseteq \mathbb{D}$ there is some compact $L \subseteq \mathbb{D}$ so that

$$\varphi^n(K) \subseteq L \text{ for every } n \in \mathbb{N}.$$

It is easy to see that every elliptic automorphism has stable orbits. Also, the proof of Proposition 5.16 shows that every holomorphic function $\varphi : \mathbb{D} \to \mathbb{D}$, which is not an automorphism and has a fixed point, also has stable orbits. That is, every holomorphic self-map of $\mathbb{D}$ with a fixed point in $\mathbb{D}$ has stable orbits. As a matter of fact, both things are equivalent. To see this, we need the following deep result, the proof of which can be found, e.g., in [3, Theorem 1.8.4] or [10, Sections 5.3 and 5.4].

Theorem 5.11 (Denjoy–Wolff) *Let $\varphi : \mathbb{D} \to \mathbb{D}$ be a holomorphic function without fixed points in $\mathbb{D}$. Then there is a unique fixed point $\tau \in \partial\mathbb{D}$ such that the sequence of iterates $(\varphi^n)_n$ converges to the constant function c_τ uniformly on compact subsets of $\mathbb{D}$.*

The point τ in the theorem is called the *Denjoy–Wolff point of* φ. It has many important properties (see the references mentioned above).

Proposition 5.17 *Let $\varphi : \mathbb{D} \to \mathbb{D}$ be holomorphic and without fixed points in $\mathbb{D}$. Then for every compact $K \subseteq \mathbb{D}$, there exists some $n \in \mathbb{N}$ so that $\varphi^n(K) \cap K = \varnothing$. Moreover, φ does not have stable orbits.*

Proof We are going to see that for any two compact sets $K \subseteq L \subseteq \mathbb{D}$, there is some $n \in \mathbb{N}$ so that

$$\varphi^n(K) \cap L = \varnothing . \tag{5.26}$$

This clearly yields both statements. Take, then, two such compact sets and fix $0 < r < 1$ so that $L \subseteq \overline{\mathbb{D}(0,r)}$. Let τ be the Denjoy–Wolff point of φ that we obtain from Theorem 5.11. Then there is $n_0 \in \mathbb{N}$ so that $|\varphi^n(z) - \tau| < 1 - r$ for every $n \geq n_0$ and $|z| \leq r$. This gives

$$1 - |\varphi^n(z)| = |\tau| - |\varphi^n(z)| = \big||\tau| - |\varphi^n(z)|\big| \leq |\tau - \varphi^n(z)| < 1 - r ,$$

and then $|\varphi^n(z)| > r$ for every $n \geq n_0$ and $|z| \leq r$ (in particular for $z \in K$). This gives (5.26) and completes the proof. □

As a straightforward consequence, we have the following:

Corollary 5.10 *Let $\varphi : \mathbb{D} \to \mathbb{D}$ be holomorphic. Then φ has stable orbits if and only if it has a fixed point in $\mathbb{D}$.*

We have finally everything at hand to characterize mean ergodicity for composition operators.

Theorem 5.12 *Let $\varphi : \mathbb{D} \to \mathbb{D}$ be holomorphic. The following statements are equivalent:*

1. *φ has stable orbits.*
2. *φ has a fixed point in $\mathbb{D}$.*
3. *$C_\varphi : H(\mathbb{D}) \to H(\mathbb{D})$ is power bounded.*
4. *$C_\varphi : H(\mathbb{D}) \to H(\mathbb{D})$ is uniformly mean ergodic.*
5. *$C_\varphi : H(\mathbb{D}) \to H(\mathbb{D})$ is mean ergodic.*
6. *$\frac{1}{n}(C_\varphi)^n \to 0$ in $L_s\big(H(\mathbb{D})\big)$.*

Proof The equivalence between 1 and 2 has been already established in Corollary 5.10. That 2 implies 3 follows from Proposition 5.13 (if φ is an automorphism) and Proposition 5.16 (if it is not). The implication 3$\Rightarrow$ 4 holds for any arbitrary operator on H($\mathbb{D}$) (recall Proposition 5.11). Obviously, 4 implies 5, and this implies 6 (by Proposition 5.9). So, in order to complete the proof, only the implication 6 $\Rightarrow$ 1 is left. Suppose, then that $\frac{1}{n}(C_\varphi)^n \to 0$ in $L_s\big(\mathrm{H}(\mathbb{D})\big)$. This implies that $\big(\frac{1}{n}C_{\varphi^n}\big)_n$ is equicontinuous and, in particular, for each compact set $K \subseteq \mathbb{D}$ there are $0 < s < 1$ and $M > 0$ so that

$$\sup_{z \in K} \frac{1}{n}\big|f\big(\varphi^n(z)\big)\big| \leq M \sup_{|w| \leq s} |f(w)| , \tag{5.27}$$

for every $n \in \mathbb{N}$ and $f \in \mathrm{H}(\mathbb{D})$. Our aim now is to see that

$$\varphi^n(K) \subseteq \overline{\mathbb{D}(0,s)}$$

for every n, which clearly will give the conclusion. Suppose that this does not hold; then there are $a \in K$ and $m \in \mathbb{N}$ so that, if $\alpha = \varphi^m(a)$, then $|\alpha| > s$. Take $1 < R < \frac{|\alpha|}{s}$ and define the function $g : \mathbb{D} \to \mathbb{C}$ by

$$g(z) = \frac{R\overline{\alpha}z}{|\alpha|^2},$$

which is clearly holomorphic. Then $g(\alpha) = R$ and, for $|w| \leq s$, we have

$$|g(w)| = \frac{R|\overline{\alpha}w|}{|\alpha|^2} \leq \frac{Rs}{|\alpha|} < 1,$$

from which we deduce that $g^k(\varphi^m(a)) = g(\alpha)^k = R^k$ and $\sup_{|w|\leq s} |g^k(w)| < 1$ for every k. Now, using (5.27) with $f = g^k$ and $n = m$, we have

$$\frac{1}{m}R^k = \frac{1}{m}\big|g^k\big(\varphi^m(a)\big)\big| \leq \sup_{z\in K}\frac{1}{m}\big|g^k\big(\varphi^m(z)\big)\big| \leq M \sup_{|w|\leq s}|g^k(w)| < M.$$

This gives $R^k \leq Mm$ for every $k \in \mathbb{N}$, but this contradicts the fact that $R > 1$. □

We have clarified completely when a composition operator is mean ergodic. Let us see now when is it hypercyclic. Given a holomorphic self-map, $\varphi : \mathbb{D} \to \mathbb{D}$, we say that the sequence of iterates $(\varphi^n)_n$ *runs away* if for every compact $K \subseteq \mathbb{D}$ there is some $n \in \mathbb{N}$ so that

$$\varphi^n(K) \cap K = \varnothing.$$

Proposition 5.18 *Let $\varphi : \mathbb{D} \to \mathbb{D}$ be holomorphic. Then the iterates run away if and only if φ does not have fixed points in $\mathbb{D}$.*

Proof Let us suppose that φ does has a fixed point. By Corollary 5.10, it has stable orbits and, given the compact set $K = \{0\}$, there exists $L \subseteq \mathbb{D}$ compact such that $\varphi^n(0) \subseteq L$ for every n. Using again that φ has stable orbits, we can find another compact $M \subseteq \mathbb{D}$ such that $\varphi^n(L) \subseteq M$ for every n. Now, the set

$$\tilde{L} = \overline{L \cup \Big(\bigcup_{n=1}^{\infty} \varphi^n(L)\Big)} \subseteq L \cup M$$

is compact and satisfies $\varphi^n(\tilde{L}) \subseteq \tilde{L}$ for every $n \in \mathbb{N}$. Hence, $(\varphi^n)_n$ do not run away. This gives one implication. The converse one follows from Proposition 5.17. □

With this, we can describe the hypercyclic composition operators.

Theorem 5.13 *Let $\varphi : \mathbb{D} \to \mathbb{D}$ be holomorphic. The following statements are equivalent:*

1. *φ is injective and $(\varphi^n)_n$ runs away.*
2. *$C_\varphi : H(\mathbb{D}) \to H(\mathbb{D})$ is hypercyclic.*
3. *$C_\varphi : H(\mathbb{D}) \to H(\mathbb{D})$ is transitive.*

Proof The equivalence between 2 and 3 follows from Birkhoff's Theorem 5.2, since H($\mathbb{D}$) is Fréchet (Theorem 4.4) and separable (Proposition 4.1). Assume now that C_φ is hypercyclic, and let us see that 1 holds. Suppose in first place that φ is not injective. Then we can find $a, b \in \mathbb{D}$ with $a \neq b$ so that $\varphi(a) = \varphi(b)$. Choose some $g \in$ H($\mathbb{D}$) such that $g(a) \neq g(b)$ and a hypercyclic vector f of C_φ. Since the orbit of f is τ_{co}-dense, there is a subsequence $(n_j)_j$ for which $(f \circ \varphi^{n_j})_j$ converges to g uniformly on compact subsets of $\mathbb{D}$ and, in particular, on every compact set that contains a and b. Since $f(\varphi^n(a)) = f(\varphi^n(b))$ for every n, we have $g(a) = g(b)$. This is a contradiction and shows that φ is injective.

Suppose now that $(\varphi^n)_n$ does not run away. Then there is a compact $K \subseteq \mathbb{D}$ and a sequence $(z_n)_n \subseteq K$ so that $\varphi^n(z_n) \subseteq K$. Now, for each $f \in$ H($\mathbb{D}$), set $M = \sup_{z\in K} |f(z)|$ and note that, for every $n \in \mathbb{N}$, we have

$$\inf_{z\in K} \big|(C_\varphi)^n(f)(z)\big| \leq \big|f(\varphi^n(z_n))\big| \leq M\,,$$

And this shows that $(f \circ \varphi^n)_n$ cannot approximate the constant function $M + 1$ uniformly on K. Hence, C_φ cannot have hypercyclic vectors. This completes the proof of 2 ⇒ 1.

The most difficult part of the proof is the remaining one: if 1 holds, then $C_\varphi \in L(\text{H}(\mathbb{D}))$ is transitive. Assume that φ is injective and $(\varphi^n)_n$ runs away, and fix two open non-empty sets $U, V \subseteq$ H($\mathbb{D}$). Choose $K \subseteq \mathbb{D}$ compact, $f, g \in$ H($\mathbb{D}$) and $\varepsilon > 0$ so that

$$\big\{h\in \text{H}(\mathbb{D}) : \sup_{z\in K} |h(z) - f(z)| < \varepsilon\big\} \subseteq U \text{ and } \big\{h \in \text{H}(\mathbb{D}) : \sup_{z\in K} |h(z) - g(z)| < \varepsilon\big\} \subseteq V\,.$$

Pick $0 < r < 1$ so that $K \subseteq \mathbb{D}(0, r)$, fix $r < s < 1$, and take $K_1 = \overline{\mathbb{D}(0, r)}$ and $K_2 = \overline{\mathbb{D}(0, s)}$ (both compact, and clearly $\mathring{K}_1 \subseteq K_2$). Since the iterates of φ run away, there is some $m \in \mathbb{N}$ for which $\varphi^m(K_2) \cap K_2 = \varnothing$. On the other hand, being φ injective, the complement of $K_1 \cup \varphi^m(K_1)$ has no connected relatively compact components. By the open mapping theorem for holomorphic functions [9, Theorem 10.32], the set $\varphi^m(\mathring{K}_2)$ is open, and $\psi = (\varphi^m)^{-1} : \varphi^m(\mathring{K}_2) \to \mathring{K}_2$ is holomorphic. We now apply Runge's approximation theorem [9, Theorem 13.6] to f on K_1 and $g \circ \psi$ on $\varphi^m(\mathring{K}_1)$ (which is contained in $\varphi^m(\mathring{K}_2)$) and find $h_0 \in$ H($\mathbb{D}$) so that

$$|h_0(z) - f(z)| < \varepsilon \text{ for all } z \in K_1\,, \tag{5.28}$$

$$|h_0(w) - g(\psi(w))| < \varepsilon \text{ for all } w \in \varphi^m(\mathring{K}_1)\,. \tag{5.29}$$

Note that, since $K \subseteq K_1$, from (5.28), we deduce that $h_0 \in U$. Also, (5.29) gives

$$|h_0(\varphi^m(z)) - g(z)| < \varepsilon \text{ for every } z \in K_1 ,$$

and, therefore, $\big(C_\varphi\big)^m(h_0) = h_0 \circ \varphi^m \in V$. Altogether we have shown that

$$\big(C_\varphi\big)^m(U) \cap V \neq \varnothing ,$$

and C_φ is transitive. □

As a straightforward consequence of Theorem 5.13 and Proposition 5.18, we have the following.

Corollary 5.11 *Let $\varphi : \mathbb{D} \to \mathbb{D}$ be holomorphic and injective. Then $C_\varphi : H(\mathbb{D}) \to H(\mathbb{D})$ is hypercyclic if and only if φ does not have fixed points.*

5.3.3 Multiplication and Integration Operators

We look now at the dynamical properties of multiplication and integration operators that we already defined in Examples 4.15 and 4.18. For the first one, let us recall that if $\Omega \subseteq \mathbb{C}$ is open and connected, and $g : \Omega \to \mathbb{C}$ is holomorphic, then the associated multiplication operator $M_g : \mathrm{H}(\Omega) \to \mathrm{H}(\Omega)$ is defined by $M_g(f) = fg$. Let us note that

$$(M_g)^n(f)(z) = g(z)^n f(z)$$

for every $z \in \Omega$ and $n \in \mathbb{N}$. Hypercyclicity is easily settled.

Proposition 5.19 *The multiplication operator M_g is never hypercyclic.*

Proof For each $z \in \Omega$, we consider the evaluation functional $\delta_z : \mathrm{H}(\Omega) \to \mathbb{C}$, given by $\delta_z(f) = f(z)$. It is clearly non-zero, and for each $f \in \mathrm{H}(\Omega)$, we have

$$\langle {}^t M_g(\delta_z), f \rangle = \langle \delta_z, M_g(f) \rangle = g(z)\delta_z(f) .$$

Then ${}^t M_g(\delta_z) = g(z)\delta_z$, and δ_z is an eigenvector of ${}^t M_g$. By Proposition 5.5, M_g is not transitive, and hence (recall Remark 5.1), neither is it hypercyclic. □

We see now when the operator is mean ergodic.

Proposition 5.20 *Let $\Omega \subseteq \mathbb{C}$ be open and connected, and $g : \Omega \to \mathbb{C}$ holomorphic. The following statements are equivalent:*

1. *$M_g : H(\Omega) \to H(\Omega)$ is power bounded.*
2. *$M_g : H(\Omega) \to H(\Omega)$ is uniformly mean ergodic.*
3. *$M_g : H(\Omega) \to H(\Omega)$ is mean ergodic.*

4. $\frac{1}{n}(M_g)^n f \to 0$ *for every* $f \in H(\Omega)$.
5. $|g(z)| \leq 1$ *for every* $z \in \Omega$.

Proof The implications 1 ⇒ 2 ⇒ 3 ⇒ 4 follow from Proposition 5.11 (recall that H(Ω) is a Montel space) and Proposition 5.9. Suppose now that $\frac{1}{n}(M_g)^n f \to 0$ for every $f \in \mathrm{H}(\Omega)$, and take f to be the constant function equal to 1 for every $z \in \Omega$. Then $(M_g)^n f(z) = g(z)^n$ for every $z \in \Omega$, and then $\frac{1}{n}g(z)^n \to 0$, and this implies that $|g(z)| \leq 1$ for every z. Let us finally assume that 5 holds and fix some compact $K \subseteq \Omega$. Then

$$\sup_{z\in K} \left|(M_g)^n(f)(z)\right| = \sup_{z\in K} |g(z)^n f(z)| \leq \sup_{z\in K} |f(z)|$$

for every f and n, and M_g is power bounded. □

We now move on to the study of the integration operator $J : \mathrm{H}(\mathbb{C}) \to \mathrm{H}(\mathbb{C})$ defined (recall Example 4.18) as

$$Jf(z) = \int_{[0,z]} f(w)dw\,,$$

which we know is continuous.

Proposition 5.21 *The integration operator* $J : H(\mathbb{C}) \to H(\mathbb{C})$ *is power bounded, uniformly mean ergodic, and not transitive.*

Proof It is enough to see that the operator is power bounded. The uniform mean ergodicity follows from Proposition 5.11 and the fact that H(ℂ) is a Montel space (Theorem 4.9). Corollary 5.2 will give that the operator is not transitive. In fact, what we are going to see is that $(J^n f)_n$ tends to 0 uniformly on compact sets for every $f \in \mathrm{H}(\mathbb{C})$. Take, then, an entire function $f(z) = \sum_{k=0}^{\infty} a_k z^k$. By the Cauchy inequalities (4.1), for each $R > 0$, there exists $c_R > 0$ so that $|a_k| \leq \frac{c_R}{R^k}$ for every k. Now, given some compact set $K \subseteq \mathbb{C}$, let $r > 0$ be such that $K \subseteq \mathbb{D}(0, r)$, and fix $R > r$. Note that, given $z \in K$, the power series converges uniformly on $[0, z]$ and, then,

$$\begin{aligned} Jf(z) = \int_{[0,z]} \sum_{k=0}^{\infty} a_k w^k dw = \sum_{k=0}^{\infty} a_k \int_{[0,z]} w^k dw \\ = \sum_{k=0}^{\infty} \frac{a_k}{k+1} z^{k+1} = z\Big(\sum_{k=0}^{\infty} \frac{a_k}{k+1} z^k\Big) \end{aligned}$$

and, iterating this,

$$J^n f(z) = z^n\Big(\sum_{k=0}^{\infty} \frac{a_k}{(k+1)\cdots(k+n)} z^k\Big)\,,$$

for every $n \in \mathbb{N}$. Then

$$|J^n f(z)| \leq |z^n| \sum_{k=0}^{\infty} \frac{|a_k|}{(k+1)\cdots(k+n)} |z^k| \leq \frac{|z^n|}{n!} \sum_{k=0}^{\infty} |a_k||z^k|$$

$$\leq \frac{r^n}{n!} \sum_{k=0}^{\infty} \frac{c_R}{R^k} r^k = \frac{r^n}{n!} c_R \sum_{k=0}^{\infty} \left(\frac{r}{R}\right)^k .$$

The latter tends to 0 as $n \to \infty$ and does not depend on z. This yields the claim and completes the proof. □

With a similar argument, it can be shown that the integration operator $J : \mathrm{H}(\mathbb{D}) \to \mathrm{H}(\mathbb{D})$ is power bounded, uniformly mean ergodic but not transitive.

5.3.4 Differential Operators

Let $\Omega \subseteq \mathbb{C}$ be open. Let us recall that the differentiation operator $D : \mathrm{H}(\Omega) \to \mathrm{H}(\Omega)$ is defined by $D(f) = f'$. An obvious modification of Example 4.17 shows that this operator is continuous. Given $m \geq 1$ and a polynomial $P(z) = \sum_{k=0}^{m} a_k z^k$, we consider the operator $P(D) = \sum_{k=0}^{m} a_k D^k$. That is, $P(D) : \mathrm{H}(\Omega) \to \mathrm{H}(\Omega)$ is given by

$$P(D)(f) = \sum_{k=0}^{m} a_k f^{(k)} .$$

Our aim now is to study whether or not is this operator hypercyclic, chaotic, power bounded, or uniformly mean ergodic. For the remainder of this section, we establish the following notation: for $\lambda \in \mathbb{C}$, we consider the function $e_\lambda \in \mathrm{H}(\mathbb{C})$ given by $e_\lambda(z) = e^{\lambda z}$. Note that, for each such function (and every operator $P(D)$), we have

$$P(D)(e_\lambda) = \sum_{k=0}^{m} a_k e_\lambda^{(k)} = \sum_{k=0}^{m} a_k \lambda^k e_\lambda = P(\lambda) e_\lambda . \tag{5.30}$$

Lemma 5.6 *If $\Lambda \subseteq \mathbb{C}$ has an accumulation point, then* $\operatorname{span}\{e_\lambda : \lambda \in \Lambda\}$ *is dense in $H(\mathbb{C})$.*

Proof Let us recall from Theorem 4.12 that, if $\mu \in \mathrm{H}(\mathbb{C})'$, then its Fourier–Borel transform $\widehat{\mu}$ given by $\widehat{\mu}(z) = \mu_w(e^{zw}) = \mu(e_z)$ is a holomorphic function of exponential type. Suppose now that $\mu \in \mathrm{H}(\mathbb{C})'$ is such that $\mu(e_z) = 0$ for every $\lambda \in \Lambda$. Then $\widehat{\mu}(\lambda) = 0$ for every $\lambda \in \Lambda$, and since $\widehat{\mu}$ is entire and Λ has an accumulation point, the identity principle (Theorem 4.7) yields that $\widehat{\mu} = 0$ and, therefore, $\mu = 0$. Corollary 3.3 gives the claim. □

We give our first step by describing the dynamical properties of the differential operators defined on the space of entire functions.

Theorem 5.14 *Let $P(z) = \sum_{k=0}^{m} a_k z^k$ be a non-constant polynomial of degree $m \geq 1$. The operator $P(D) : H(\mathbb{C}) \to H(\mathbb{C})$ is hypercyclic and chaotic, and neither power bounded, nor mean ergodic.*

Proof Let us note in first place that the set $\{\lambda \in \mathbb{C} : |P(\lambda)| < 1\}$ is open and non-empty and hence has an accumulation point. By Lemma 5.6, $\{e_\lambda : |P(\lambda)| < 1\}$ is dense in $\mathrm{H}(\mathbb{C})$. With this and (5.30), we have that

$$E_0 = \operatorname{span}\{f \in \mathrm{H}(\mathbb{C}) \colon P(D)f = \mu f \text{ for some } \mu \in \mathbb{C},\ |\mu| < 1\}$$

is dense in $\mathrm{H}(\mathbb{C})$. A similar argument shows that

$$F_0 = \operatorname{span}\{f \in \mathrm{H}(\mathbb{C}) \colon P(D)f = \mu f \text{ for some } \mu \in \mathbb{C},\ |\mu| > 1\}$$

is also dense and, by Proposition 5.3, the operator $P(D)$ is hypercyclic (hence not power bounded; see Remark 5.4).

Let us see now that the operator is also chaotic. By Proposition 5.7, it is enough to show that

$$G_0 = \operatorname{span}\{f \in \mathrm{H}(\mathbb{C}) \colon \text{ there exists } \mu \in \mathbb{C} \text{ with } \mu^n = 1 \\ \text{for some } n \in \mathbb{N} \text{ so that } P(D)f = \mu f\}$$

is dense in $\mathrm{H}(\mathbb{C})$. Note that, by (5.30), the set

$$\{e_\lambda \colon P(\lambda)^n = 1 \text{ for some } n\} \tag{5.31}$$

is contained in G_0. Note now that for every $w \in \mathbb{C}$ with $|w| = 1$, there is some λ with $P(\lambda) = w$. Also, the set $\{\lambda \in \mathbb{C} \colon P(\lambda)^n = 1 \text{ for some } n\}$ consists of pre-images through P of roots of unity, and since these are dense in $\mathbb{T}$ (because $e^{q\pi}$ is a root of 1 for every $q \in \mathbb{Q}$ and $\mathbb{Q}$ is dense in $\mathbb{R}$), this set has an accumulation point. Lemma 5.6 gives that the set in (5.31) is dense in $\mathrm{H}(\mathbb{C})$ and then so also is G_0. This yields the claim.

Finally, take λ_0 such that $P(\lambda_0) = 2$ to have, by (5.30), that $P(D)e_{\lambda_0} = 2e_{\lambda_0}$. Remark 5.3 gives that $P(D)$ is not mean ergodic. □

We move now to operators defined on spaces of holomorphic functions on some open set.

Theorem 5.15 (Shapiro) *Let $\Omega \subseteq \mathbb{C}$ be open and connected, and $P(z) = \sum_{k=0}^{m} a_k z^k$ be a non-constant polynomial of degree $m \geq 1$. The following statements are equivalent:*

1. *$P(D) : H(\Omega) \to H(\Omega)$ is chaotic.*
2. *$P(D) : H(\Omega) \to H(\Omega)$ is hypercyclic.*
3. *Ω is simply connected.*

Proof If the operator is chaotic, it is transitive, and then, by Birkhoff's Theorem 5.2, hypercyclic. Let us suppose now that Ω is not simply connected. Then we can find $a \in \mathbb{C} \setminus \Omega$ and a closed piecewise C^1 curve γ contained in Ω so that

$$\int_\gamma \frac{1}{z-a} dz \neq 0 . \tag{5.32}$$

We define now $u : \mathrm{H}(\Omega) \to \mathbb{C}$ by $u(f) = \int_\gamma f(z)dz$. It is easy to check that $u \in \mathrm{H}(\Omega)'$ and (5.32) gives $u \neq 0$. Moreover, $u(f') = 0$ for every $f \in \mathrm{H}(\Omega)$. We now write the polynomial as

$$P(z) = P(0) + \sum_{k=1}^{m} a_k z^k$$

and, for each $f \in \mathrm{H}(\Omega)$, we have

$$\langle {}^t P(D)u, f \rangle = \langle u, P(D)f \rangle = \langle u, P(0)f \rangle + \sum_{k=1}^{m} a_k \langle u, f^{(k)} \rangle = P(0)u(f) .$$

Hence, ${}^t P(D)u = P(0)u$, and $u \neq 0$ is an eigenvector of ${}^t P(D)$, which, by Proposition 5.5, cannot be transitive. Remark 5.1 gives that it cannot be hypercyclic.

Let us finally assume that Ω is simply connected. The polynomials are dense in $\mathrm{H}(\Omega)$ (see, e.g., [9, Theorem 13.11] for a proof of this fact) and, therefore, $\mathrm{H}(\mathbb{C})$ is dense in $\mathrm{H}(\Omega)$. From Theorem 5.14, we know that $P(D) : \mathrm{H}(\mathbb{C}) \to \mathrm{H}(\mathbb{C})$ is chaotic. Corollary 5.5 yields 1 and completes the proof. □

5.4 Exercises

5.1 Let T be an invertible transitive operator on a locally convex space. Show that $T^{-1} \in L(E)$ is also transitive.

5.2 Let $H \subseteq L(E)$ be equicontinuous. Show that there is a fundamental system of seminorms $\mathcal{P}$ defining the topology of E such that

$$p(Tx) \leq p(x)$$

for every $T \in H$, every $x \in E$, and $p \in \mathcal{P}$.
Hint: Given a fundamental system of seminorms Q of E, we know that for each $q \in Q$ there are $q \in Q$ and $C > 0$ such that $q(Tx) \leq Cq'(x)$ for every $T \in H$ and $x \in E$. Set, for a given $q \in Q$,

$$\tilde{q}(x) = \max \big(q(x), \sup_{T \in H} q(Tx) \big) .$$

This defines a continuous seminorm on E that satisfies $q \leq \tilde{q} \leq Cq'$. Taking $\mathcal{P} = \{\tilde{q} : q \in \mathcal{Q}\}$ gives the conclusion.

5.3 The resolvent $\rho(T)$ of an operator $T \in L(E)$ is defined as

$$\rho(T) = \{\lambda \in \mathbb{C} : \lambda I - T \text{ is invertible}\},$$

and the spectrum as $\sigma(T) = \mathbb{C} \setminus \rho(T)$. Prove that, if T is such that $\left(\frac{1}{n}T^n\right)_n$ is equicontinuous in $L(E)$, then $\sigma(T) \subseteq \{\lambda \in \mathbb{C} : |\lambda| \leq 1\}$.
Hint: By the previous exercise, there is a fundamental system of seminorms $\mathcal{P}$ such that $p(T^n x) \leq np(x)$ for every $x \in E$ and $n \in \mathbb{N}$. If $\lambda \in \mathbb{C}$ satisfies $|\lambda| > 1$, then the series $\sum_{n=0}^{\infty} \frac{T^n x}{\lambda^n}$ converges in E and so defines a continuous operator R, since $\sum_{n=1}^{\infty} \frac{n}{\lambda^n} < \infty$. Now $(\lambda I - T)R = R(\lambda I - T) = \lambda I$, as can be easily seen. The series $\sum_{n=0}^{\infty} \frac{T^n}{\lambda^n}$ is called the *von Neumann series*.

5.4 Prove that, if $T \in L(E)$ is power bounded or mean ergodic on a Fréchet space, then $\sigma(T) \subseteq \{\lambda \in \mathbb{C} : |\lambda| \leq 1\}$.
Hint: Recall that if E is a Fréchet space and $\frac{1}{n}T^n \to 0$ in $L_s(E)$, then $\left(\frac{1}{n}T^n\right)_n$ is equicontinuous.

5.5 Show that the operator $T : c_0 \to c_0$ given by $T(x_1, x_2, x_3, \ldots) = (x_1, x_1, x_2, x_3, \ldots)$ is power bounded but not mean ergodic.

5.6 Show that the operator $T : \ell_1 \to \ell_1$ given by $T(x_1, x_2, x_3, \ldots) = (0, x_1, x_2, x_3, \ldots)$ is power bounded but not mean ergodic.

5.7 Let E and F be Hausdorff locally convex spaces. Assume that every $\beta(F', F)$-bounded subset of F' is equicontinuous. Show that the linear map $\phi : L_\beta(E, F) \to L_\beta(F'_\beta, E'_\beta)$ defined as $\phi(T) = {}^tT$ is continuous. Prove that, as a consequence, if E satisfies that every bounded subset of E'_β is equicontinuous and $T_n \to T$ in $L_\beta(E)$, then ${}^tT_n \to {}^tT$ in $L_\beta(E'_\beta)$.

5.8 Let E be a Hausdorff locally convex space and $T \in L(E)$. Prove:

1. If T is uniformly mean ergodic, then ${}^tT \in L(E'_\beta)$ is mean ergodic.
2. If every bounded subset of E'_β is equicontinuous and T is uniformly mean ergodic, then ${}^tT \in L(E'_\beta)$ is uniformly mean ergodic.

5.9 A linear operator $T : E \to F$ between Hausdorff locally convex spaces is called *compact* if there is a neighbourhood U in E such that $T(U)$ is relatively compact in F:

1. Let E be a complex Banach space. Show that if $T \in L(E)$ is compact, then T cannot be hypercyclic.
   Hint: The proof that we suggest relies on several results on spectral theory. The transpose ${}^tT : E' \to E'$ is also compact. By Proposition 5.5, ${}^tT$ has no eigenvalues. By Riesz's spectral theory for compact operators, the spectrum $\sigma({}^tT)$ reduces to $\{0\}$ (see, e.g., [4, Theorem 7.7.1]). Then $\sigma(T) = \sigma({}^tT) = \{0\}$.

By the spectral radius formula ([4, Proposition 7.3.8]), $\lim_n \|T^n\| = 0$. This implies that T is power bounded, which contradicts Remark 5.4.

2. Show that no compact operator on a Hausdorff locally convex space is hypercyclic.
Hint: Let $T \in L(E)$ be compact and U be a closed absolutely convex set so that $T(U)$ is relatively compact. Let p_U be the Minkowski functional of U and E_U the normed space $(E/\ker p_U, \widehat{p}_U)$, where $\widehat{p}_U(x + \ker U) = p_U(x)$. The map $\tilde{T} : E_U \to E_U$ defined as $\tilde{T}(x + \ker p_U) = Tx + p_U(x)$ is well defined and continuous. Denote also by $\tilde{T}$ the extension to the completion. Show that if T is hypercyclic, then so also is $\tilde{T}$.

5.10 Let A be a Köthe matrix such that $a_1(i) > 0$ for every $i \in \mathbb{N}$. Let $1 \leq p \leq \infty$ or $p = 0$. For $\lambda = (\lambda_1, \lambda_2, \ldots)$, define the diagonal operator by $D_\lambda x = (\lambda_1 x_1, \lambda_2 x_2, \ldots)$:

1. When is D_λ continuous on $\lambda_p(A)$?
2. Show that D_λ is never hypercyclic on $\lambda_p(A)$.
3. Can D_λ have periodic points other than $(0, 0, 0, \ldots)$?
4. When is D_λ power bounded on $\lambda_p(A)$?
5. When is D_λ mean ergodic on $\lambda_p(A)$?
6. When is D_λ uniformly mean ergodic on $\lambda_p(A)$?
7. When is $\mathrm{Im}(I - D_\lambda)$ closed in $\lambda_p(A)$?

5.11 Consider the Banach space $C[0, 1]$ of all complex continuous functions on $[0, 1]$. For each $f \in C[0, 1]$, define the Volterra integration operator

$$Vf(x) = \int_0^x f(t)dt$$

for $0 \leq x \leq 1$.

1. Show that V is continuous and $\|V\| \leq 1$.
2. Show by induction that

$$V^n f(x) = \frac{1}{(n-1)!} \int_0^x (x-t)^{n-1} f(t)dt$$

for every $n \in \mathbb{N}$ (and $0 \leq x \leq 1$).

3. Compute the norm of V^n for each $n \in \mathbb{N}$.
4. Is V power bounded, hypercyclic, or mean ergodic?
5. Show that $\sigma(V) = \{0\}$.

Further Reading
[1, 3, 5–7, 10]

References

1. Bayart, F., Matheron, E.: Dynamics of Linear Operators. Cambridge Tracts in Mathematics, vol. 179. Cambridge University Press, Cambridge (2009). https://doi.org/10.1017/CBO9780511581113
2. Bermúdez, T., Kalton, N.J.: The range of operators on von Neumann algebras. Proc. Am. Math. Soc. **130**(5), 1447–1455 (2002). https://doi.org/10.1090/S0002-9939-01-06292-X
3. Bracci, F., Contreras, M.D., Díaz-Madrigal, S.: Continuous Semigroups of Holomorphic Self-Maps of the Unit Disc. Springer Monographs in Mathematics. Springer, Cham (2020). https://doi.org/10.1007/978-3-030-36782-4
4. Conway, J.B.: A Course in Functional Analysis. Graduate Texts in Mathematics, vol. 96. Springer, New York (1985). https://doi.org/10.1007/978-1-4757-3828-5
5. Eisner, T., Farkas, B., Haase, M., Nagel, R.: Operator Theoretic Aspects of Ergodic Theory. Graduate Texts in Mathematics, vol. 272. Springer, Cham (2015). https://doi.org/10.1007/978-3-319-16898-2
6. Grosse-Erdmann, K.G., Peris Manguillot, A.: Linear Chaos. Universitext. Springer, London (2011). https://doi.org/10.1007/978-1-4471-2170-1
7. Krengel, U.: Ergodic Theorems. De Gruyter Studies in Mathematics, vol. 6. Walter de Gruyter & Co., Berlin (1985). https://doi.org/10.1515/9783110844641. With a supplement by Antoine Brunel
8. Lotz, H.P.: Uniform convergence of operators on L^∞ and similar spaces. Math. Z. **190**(2), 207–220 (1985). https://doi.org/10.1007/BF01160459
9. Rudin, W.: Real and Complex Analysis, 3rd edn. McGraw-Hill Book Co., New York (1987)
10. Shapiro, J.H.: Composition Operators and Classical Function Theory. Universitext: Tracts in Mathematics. Springer, New York (1993). https://doi.org/10.1007/978-1-4612-0887-7

Chapter 6
Schwartz Distributions and Linear Partial Differential Operators

6.1 Test Functions and Distributions

6.1.1 *Definition and Examples*

If Ω is some open subset of $\mathbb{R}^N$ (with $N \in \mathbb{N}$), we want to deal with functions in $C^\infty(\Omega)$ that are 0 except on a relatively compact subset of Ω. To achieve this, we define the *support* of an arbitrary function $f : \mathbb{R}^N \to \mathbb{K}$ as the set

$$\operatorname{supp}(f) = \overline{\{x \in \mathbb{R}^N : f(x) \neq 0\}}\,.$$

If $K \subseteq \mathbb{R}^N$ is a compact subset, we consider the space of C^∞-functions whose support is contained in K; that is,

$$\mathscr{D}(K) = \{f \in C^\infty(\mathbb{R}^N) : \operatorname{supp}(f) \subseteq K\}\,.$$

If we endow it the induced topology from $C^\infty(\mathbb{R}^N)$, this is clearly a closed subspace of $C^\infty(\mathbb{R}^N)$. Then, in view of Theorem 4.13, $\mathscr{D}(K)$ is a Fréchet space. Note that the family of seminorms (recall (4.8)), which are in fact norms, defining the topology of $\mathscr{D}(K)$ are

$$p_{K,n}(f) = \sup_{|\alpha| \leq n} \sup_{x \in K} \left| \partial^\alpha f(x) \right|,$$

for $n \in \mathbb{N}_0$. If K is a subset of the open set $\Omega \subseteq \mathbb{R}^N$, then $\mathscr{D}(K)$ is also a closed subspace of $C^\infty(\Omega)$ for the induced topology.

J. Bonet et al., *Function Spaces and Operators between them*,
RSME Springer Series 11, https://doi.org/10.1007/978-3-031-41602-6_6

Example 6.1 Consider the function $g : \mathbb{R} \to \mathbb{R}$ given by

$$g(x) = \begin{cases} e^{-1/x} & \text{if } x > 0 \\ 0 & \text{if } x \le 0 \end{cases}.$$

For each n, there is a polynomial P_n of degree n so that $g^{(n)}(x) = P_n(1/x)e^{-1/x}$ for every $x > 0$. Then it is easy to see that $g \in C^\infty(\mathbb{R})$ and, moreover, $g^{(n)}(0) = 0$ for every $n \in \mathbb{N}$.

Define now a function $\rho : \mathbb{R}^N \to \mathbb{R}$ by $\rho(x) = Cg(1 - |x|^2)$, where $C > 0$ is chosen in such a way that $\int_{\mathbb{R}^N} \rho(x)dx = 1$. Then $\rho \in C^\infty(\mathbb{R}^N)$ and $\operatorname{supp} \rho = \overline{B(0, 1)}$ (that is, $\rho \in \mathscr{D}(\overline{B(0, 1)})$).
Finally, for a given $\varepsilon > 0$, we define $\rho_\varepsilon \in C^\infty(\mathbb{R}^N)$ by

$$\rho_\varepsilon(x) = \frac{1}{\varepsilon^N} \rho\left(\tfrac{x}{\varepsilon}\right), \tag{6.1}$$

for $x \in \mathbb{R}^N$. This clearly satisfies $\operatorname{supp} \rho_\varepsilon = \overline{B(0, \varepsilon)}$, $0 \le \rho_\varepsilon \le 1$, and $\int_{\mathbb{R}^N} \rho_\varepsilon(x)dx = 1$ for each $\varepsilon > 0$. □

Given any two measurable functions $f, g : \mathbb{R}^N \to \mathbb{K}$, their *convolution* is defined, for $x \in \mathbb{R}^N$, as

$$(f * g)(x) = \int_{\mathbb{R}^N} f(y)g(x - y)dy, \tag{6.2}$$

whenever the integral exists. It satisfies the following properties.

Proposition 6.1 *Let* $f, g \in L_1(\mathbb{R}^N)$*; then:*

1. $f * g \in L_1(\mathbb{R}^N)$ *and* $\|f * g\|_{L_1(\mathbb{R}^N)} \le \|f\|_{L_1(\mathbb{R}^N)} \|g\|_{L_1(\mathbb{R}^N)}$.
2. $f * g = g * f$ *almost everywhere.*
3. *If* $\operatorname{supp} f$ *is compact, then* $\operatorname{supp}(f * g) \subseteq \operatorname{supp} f + \operatorname{supp} g$.
4. *If* $g \in C^1(\mathbb{R}^N) \cap L_1(\mathbb{R}^N)$, *then* $f * g \in C^1(\mathbb{R}^N) \cap L_1(\mathbb{R}^N)$, *and*

$$\frac{\partial}{\partial x_k}(f * g) = f * \frac{\partial g}{\partial x_k},$$

for $k = 1, \ldots, N$.

A detailed study of the convolution of functions (including the proof of these properties) can be found, e.g., in [2, pages 53–54] or [12].

If one of the functions is in L_1, then we know that we can calculate its convolution with any other function in L_1. However, if our function has 'better' properties (namely, it is continuous and has compact support), then we can pick functions in a larger space to convolve it with. Let us be more precise.

For an open set $\Omega \subseteq \mathbb{R}^N$, a measurable function $f : \Omega \to \mathbb{K}$ is said to be *locally integrable* if $f \in L_1(K)$ for every compact set $K \subseteq \Omega$. We write $L_{1,\text{loc}}(\Omega)$ for the space of all locally integrable functions on Ω. If $f \in L_{1,\text{loc}}(\Omega)$ and $K \subseteq \Omega$ is compact, then clearly $f\chi_K \in L_1(\Omega)$.

If φ is continuous and has compact support (that is, $\operatorname{supp}\varphi \subseteq \mathbb{R}^N$ is compact), and $f \in L_{1,\text{loc}}(\mathbb{R}^N)$, then we have

$$(\varphi * f)(x) = \int_{\mathbb{R}^N} \varphi(y) f(x-y)dy = \int_{\operatorname{supp}\varphi} \varphi(y) f(x-y)dy\,.$$

In particular, if $\varphi \in \mathscr{D}(K)$ for some compact K, then we can convolute it with any $f \in L_{1,\text{loc}}(\mathbb{R}^N)$.

Given a set $A \subseteq \mathbb{R}^N$, it is convenient to have a function that takes value 1 on A and 0 elsewhere. Obviously, the characteristic function does this job, but it has the drawback that it is not continuous. If we go down to one variable, it is very easy to see that, given $A < B$, one may construct a function that is 1 on $[-A, A]$ and 0 on $\mathbb{R} \setminus [-B, B]$. It is slightly more involved (but still easy) to see that, if $K, \Omega \subseteq \mathbb{R}^N$ are, respectively, compact and open, and $K \subseteq \Omega$, then a continuous function can be constructed taking value 1 on K and 0 outside Ω. Or next result shows that, in fact, the function can be taken to be C^∞.

Proposition 6.2 *Let $\Omega \subsetneq \mathbb{R}^N$ be open, and $K \subseteq \Omega$ compact. Then there exists $\psi \in C^\infty(\mathbb{R}^N)$ such that:*

1. $\operatorname{supp}\psi$ *is a compact subset of* Ω.
2. $0 \leq \psi(x) \leq 1$ *for every* $x \in \mathbb{R}^N$.
3. $\psi(x) = 1$ *for every* $x \in K$.

Proof Select $\delta > 0$ such that $2\delta < d(K, \mathbb{R}^N \setminus \Omega)$ and set

$$L = K + \overline{B(0,\delta)} = \{x \in \mathbb{R}^N \;:\; d(x, K) \leq \delta\} \subseteq \Omega\,,$$

which is again compact. By Example 6.1, we can find $\varphi \in \mathscr{D}(\overline{B(0,\delta)})$ satisfying $0 \leq \varphi \leq 1$ and $\int_{\mathbb{R}^N} \varphi(x)dx = 1$. We then define $\psi = \varphi * \chi_L$. To begin with, since $\varphi \in C^\infty(\mathbb{R}^N)$, Proposition 6.1 gives $\psi \in C^\infty(\mathbb{R}^N)$ and (since φ has compact support) $\operatorname{supp}\psi \subseteq K + \operatorname{supp}\varphi \subseteq L$ and, then, it is also compact.
Finally, since $\varphi \geq 0$ and $0 \leq \chi_L \leq 1$, we have

$$0 \leq \psi(x) = \int_{\mathbb{R}^N} \varphi(y)\chi_L(x-y)dy \leq \int_{\mathbb{R}^N} \varphi(y)dy \leq 1\,,$$

for every $x \in \mathbb{R}^N$. On the other hand, if $x \in K$, note that $x - y \in K + \overline{B(0, \delta)} \subseteq L$ for each $y \in \operatorname{supp} \varphi$. Therefore,

$$\psi(x) = (\varphi * \chi_L)(x) = \int_{\mathbb{R}^N} \varphi(y)\chi_L(x-y)dy$$

$$= \int_{\operatorname{supp}\varphi} \varphi(y)\chi_L(x-y)dy = \int_{\operatorname{supp}\varphi} \varphi(y)dy = 1\,.$$

□

Functions that are C^∞ and have compact support are called *test functions*. If $\Omega \subseteq \mathbb{R}^N$ is open and not empty, we denote by $\mathscr{D}(\Omega)$ the space of all test functions belonging to $C^\infty(\Omega)$ (that is, whose support is compact in Ω). Note that

$$\mathscr{D}(\Omega) = \bigcup \{\mathscr{D}(K) : K \subseteq \Omega,\ K \text{ compact in } \Omega\}\,.$$

Note that if $(K_n)_n$ is a fundamental sequence of compact subsets in Ω, then

$$\mathscr{D}(\Omega) = \bigcup_{n=1}^{\infty} \mathscr{D}(K_n)\,.$$

Proposition 6.3 *Let $\Omega \subseteq \mathbb{R}^N$ be open. Then the space $\mathscr{D}(\Omega)$ is dense in $C^\infty(\Omega)$.*

Proof Let $(K_m)_m$ a fundamental sequence of compact subsets of Ω such that $K_m \subseteq \overset{\circ}{K}_{m+1}$ for each $m \in \mathbb{N}$ and $\Omega = \bigcup_{m=1}^{\infty} K_m$. We apply Proposition 6.2 to find, for each m, a test function $\varphi_m \in \mathscr{D}(\Omega)$ such that $\varphi_m(x) = 1$ for each $x \in K_m$, $\operatorname{supp} \varphi_m \subseteq \overset{\circ}{K}_{m+1}$, and $0 \leq \varphi_m \leq 1$.

Given $f \in C^\infty(\Omega)$, we consider the sequence $(\varphi_m f)_m \subseteq \mathscr{D}(\Omega)$. Let us see that this sequence converges to f in $C^\infty(\Omega)$ (recall in (4.8) the seminorms defining the topology). Note that $\varphi_m f = f$ on K_m for every m. Now, for each fixed $n \in \mathbb{N}$, if $m \geq n+1$, we have $K_n \subseteq \overset{\circ}{K}_{n+1} \subseteq K_m$, and then

$$p_n(\varphi_m f - f) = \sup_{|\alpha| \leq n} \sup_{x \in K_n} |\partial^\alpha (\varphi_m f - f)(x)| = 0\,.$$

This gives our claim and completes the proof. □

Proposition 6.4 *Let $\Omega_1, \ldots, \Omega_m$ be open proper subsets of $\mathbb{R}^N$ and $K \subseteq \bigcup_{j=1}^{m} \Omega_j$ be compact. Then there are $\varphi_1, \ldots, \varphi_m$ with $\varphi_j \in \mathscr{D}(\Omega_j)$ for $j = 1, \ldots, m$, such that:*

1. $0 \leq \varphi_j(x) \leq 1$ *for every* $x \in \mathbb{R}^N$ *and* $j = 1, \ldots, m$.
2. $\sum_{j=1}^{m} \varphi_j(x) = 1$ *for every* $x \in K$.

The family $\{\varphi_1, \dots, \varphi_m\}$ is often called a *partition of unity* on $\Omega_1, \dots, \Omega_m$.

Proof Given $x \in K$, there is some $j \in \{1, \dots, m\}$ such that $x \in \Omega_j$, and since this set is open, we can find $r(x) > 0$ with $\overline{B(x, r(x))} \subseteq \Omega_j$. This builds an open cover of K, which is compact. Then there are $x_1, \dots, x_s \in K$ such that

$$K \subseteq B(x_1, r(x_1)) \cup \dots \cup B(x_s, r(x_s)) .$$

Now, for each $i = 1, \dots, s$, there is $j = j(i) \in \{1, \dots, m\}$ such that $\overline{B(x_i, r(x_i))} \subseteq \Omega_{j(i)}$. Then we define

$$K_j = \bigcup \big\{ \overline{B(x_i, r(x_i))} \ : \ j(i) = j \big\} ,$$

which is clearly compact and $K_j \subseteq \Omega_j$. Note also that $K \subseteq \bigcup_{j=1}^{m} K_j$. Now, for each j, we apply Proposition 6.2 to find some $\psi_j \in C^\infty(\mathbb{R}^N)$ with compact support contained in Ω_j (i.e., $\psi_j \in \mathscr{D}(\Omega_j)$) with $0 \leq \psi_j \leq 1$ on $\mathbb{R}^N$ and $\psi_j(x) = 1$ for every $x \in K_j$.

Now we define

$$\begin{aligned} \varphi_1 &= \psi_1 \\ \varphi_k &= (1 - \psi_1) \cdots (1 - \psi_{k-1}) \psi_k \quad \text{for } k = 2, \dots, m . \end{aligned}$$

Clearly, $\operatorname{supp} \varphi_j \subseteq \operatorname{supp} \psi_j \subseteq K_j \subseteq \Omega_j$ and $0 \leq \varphi_j \leq 1$ for each $j = 1, \dots, m$. Note also that

$$\varphi_1 + \varphi_2 = \psi_1 + (1 - \psi_1)\psi_2 = \psi_1 + \psi_2 - \psi_1 \psi_2 = 1 - (1 - \psi_1)(1 - \psi_2)$$

and, with the same argument,

$$\varphi_1 + \dots + \varphi_k = 1 - (1 - \psi_1) \cdots (1 - \psi_k) ,$$

for each $k = 2, \dots, m$. If $x \in K$, then $x \in K_j$ for some $1 \leq j \leq m$ and, so, $\psi_j(x) = 1$. Hence, $\varphi_1(x) + \dots + \varphi_k(x) = 1$ for each $j \leq k \leq m$. This gives 2 and completes the proof. □

A *distribution* on an open subset Ω of $\mathbb{R}^N$ is a linear map $T : \mathscr{D}(\Omega) \to \mathbb{K}$ such that for each compact $K \subseteq \Omega$ there are $c > 0$ and $n \in \mathbb{N}_0$ so that

$$|T(\varphi)| \leq c \sup_{|\alpha| \leq n} \sup_{x \in K} |\partial^\alpha \varphi(x)| = c\, p_{K,n}(\varphi), \tag{6.3}$$

for every $\varphi \in \mathscr{D}(K)$. The space of all distributions on Ω is denoted by $\mathscr{D}'(\Omega)$. Observe that, if T is linear, then $T \in \mathscr{D}'(\Omega)$ if and only if, for each $K \subseteq \Omega$ compact, the restriction $T|_{\mathscr{D}(K)}$ of T to $\mathscr{D}(K)$ is continuous on $\mathscr{D}(K)$.

We give now some examples of distributions that will play a fundamental rôle for now on.

Example 6.2 The Dirac 'delta' function $\delta : \mathscr{D}(\mathbb{R}^N) \to \mathbb{K}$, defined by $\delta(\varphi) = \varphi(0)$, is clearly linear. Also, for each compact $K \subseteq \mathbb{R}^N$ and $\varphi \in \mathscr{D}(K)$, we have

$$|\delta(\varphi)| = |\varphi(0)| \leq \sup_{x\in K} |\varphi(x)| = p_{K,0}(\varphi) .$$

Then $\delta \in \mathscr{D}'(\mathbb{R}^N)$ (that is, δ is a distribution). Note that $\varphi(0) = 0$ whenever $0 \notin K$. With the same argument, for each fixed $a \in \mathbb{R}^N$, the linear map $\delta_a : \mathscr{D}(\mathbb{R}^N) \to \mathbb{K}$, defined by $\delta_a(\varphi) = \varphi(a)$, is a distribution. □

Example 6.3 If $\Omega \subseteq \mathbb{R}^N$ is open, then each locally integrable function on Ω defines a distribution in the following way. Given $f \in L_{1,\mathrm{loc}}(\Omega)$, we define the map $T_f : \mathscr{D}(\Omega) \to \mathbb{K}$ as

$$T_f(\varphi) = \int_{\mathbb{R}^N} f(x)\varphi(x)dx = \int_{\mathrm{supp}(\varphi)} f(x)\varphi(x)dx . \tag{6.4}$$

This is linear and, for each compact subset $K \subseteq \Omega$ and every $\varphi \in \mathscr{D}(K)$, we get

$$\left|T_f(\varphi)\right| = \left|\int_K f(x)\varphi(x)dx\right| \leq \left(\int_K |f(x)|dx\right) \sup_{x\in K} |\varphi(x)| .$$

Hence, $T_f \in \mathscr{D}'(\Omega)$. □

Example 6.4 The Heaviside step function, defined as

$$H(x) = \begin{cases} 0 & \text{if } x < 0 \\ 1 & \text{if } x \geq 0, \end{cases}$$

is a locally integrable function (which is not in $L_1(\mathbb{R})$). We denote also by H the associated distribution $H \in \mathscr{D}'(\mathbb{R})$, that is,

$$H(\varphi) = \int_{\mathbb{R}} \varphi(x)H(x)dx = \int_0^\infty \varphi(x)dx ,$$

for each $\varphi \in \mathscr{D}(\mathbb{R})$. For $a \in \mathbb{R}$, the translated Heaviside function is defined as $H_a(x) = H(x - a)$ for $x \in \mathbb{R}$. It is locally integrable and, therefore, defines a distribution. □

Example 6.5 Define a mapping $T : \mathscr{D}(\mathbb{R}) \longrightarrow \mathbb{K}$ by

$$T(\varphi) = \sum_{n=1}^{\infty} \left(\varphi'\left(\tfrac{1}{n^2}\right) - \varphi'(0)\right) .$$

Let us see that it is well defined and that $T \in \mathscr{D}'(\mathbb{R})$. Once it is well defined, it is straightforward to check that it is linear.
Take $A > 0$ and $\varphi \in \mathscr{D}([-A, A])$. For each $n \in \mathbb{N}$, there is $c_n \in]0, 1/n^2[$ such that

$$\varphi'\big(\tfrac{1}{n^2}\big) - \varphi'(0) = \frac{1}{n^2}\varphi''(c_n).$$

This gives

$$\begin{aligned} |T(\varphi)| &\leq \sum_{n=1}^{\infty} |\varphi'(1/n^2) - \varphi'(0)| = \sum_{n=1}^{\infty} \frac{1}{n^2} |\varphi''(c_n)| \\ &\leq \Big(\sum_{n=1}^{\infty} \frac{1}{n^2}\Big) \sup_{x\in[-A,A]} |\varphi''(x)| \leq \frac{\pi^2}{6} \sup_{x\in[-A,A]} |\varphi''(x)|, \end{aligned}$$

which (recall (6.3)) shows that $T \in \mathscr{D}'(\mathbb{R})$. □

Example 6.6 We define the Cauchy principal value of $\frac{1}{x}$ as the mapping $PV\frac{1}{x} : \mathscr{D}(\mathbb{R}) \to \mathbb{K}$ given by

$$PV\tfrac{1}{x}(\varphi) = \lim_{\varepsilon\to 0+} \int_{|x|>\varepsilon} \frac{\varphi(x)}{x} dx\,.$$

We show that $PV\frac{1}{x}$ is well defined and that it belongs to $\mathscr{D}'(\mathbb{R})$. Fix $A > 0$ and take some $\varphi \in \mathscr{D}([-A, A])$. Since $\varphi(A) = \varphi(-A) = 0$ (because $\operatorname{supp}\varphi \subseteq [-A, A]$), we get, integrating by parts

$$\begin{aligned} PV\tfrac{1}{x}(\varphi) &= \lim_{\varepsilon\to 0+} \int_{\varepsilon<|x|<A} \frac{\varphi(x)}{x} dx = \lim_{\varepsilon\to 0+} \left(\int_{\varepsilon}^{A} \frac{\varphi(x)}{x} dx + \int_{-A}^{-\varepsilon} \frac{\varphi(x)}{x} dx \right) \\ &= -\lim_{\varepsilon\to 0+} \left(\big(\varphi(\varepsilon) - \varphi(-\varepsilon)\big) \log\varepsilon + \int_{\varepsilon<|x|<A} \varphi'(x) \log|x| dx \right). \end{aligned}$$

Since $\big|\log|x|\big| \leq \frac{M}{|x|^{1/2}}$ for each $0 \neq x \in [-A, A]$, we get $\log|x| \in L_1([-A, A])$ and

$$\lim_{\varepsilon\to 0+} \int_{\varepsilon<|x|<A} \varphi'(x) \log|x| dx = \int_{-A}^{A} \varphi'(x) \log|x| dx.$$

On the other hand, by the mean value theorem, we also have

$$\big|(\varphi(\varepsilon) - \varphi(-\varepsilon)) \log\varepsilon\big| \leq 2\varepsilon \log\varepsilon \max_{x\in[-A,A]} |\varphi'(x)|,$$

which tends to 0 as $\varepsilon \to 0+$. All this altogether yields

$$\left|PV\tfrac{1}{x}(\varphi)\right| \leq \left(\int_{-A}^{A} \log|x|dx\right) \max_{x\in[-A,A]} |\varphi'(x)|,$$

and therefore (see again (6.3)), $PV\frac{1}{x} \in \mathscr{D}'(\mathbb{R})$. Note that, in fact, $\log|x| \in L_{1,\mathrm{loc}}(\mathbb{R})$ and

$$PV\tfrac{1}{x}(\varphi) = -\int_{-A}^{A} \varphi'(x)\log|x|dx$$

for all $\varphi \in \mathscr{D}([-A, A])$. This implies

$$PV\tfrac{1}{x}(\varphi) = -T_{\log|x|}(\varphi') \tag{6.5}$$

for every $\varphi \in \mathscr{D}(\mathbb{R})$. □

Lemma 6.1 (du Bois-Reymond) *Let $\Omega \subseteq \mathbb{R}^n$ be open and $f \in L_{1,\mathrm{loc}}(\Omega)$ satisfy $\int_\Omega f(x)\varphi(x)dx = 0$ for each $\varphi \in \mathscr{D}(\Omega)$. Then $f = 0$ almost everywhere in Ω.*

Proof We may assume that f is a real-valued function. Given any $a \in \Omega$ and $r > 0$ such that $B(a, r) \subset \Omega$, we show that $f = 0$ almost everywhere in $B(a, r)$. We set $A = \{x \in B(a, r) : f(x) \geq 0\}$. The set A is measurable. If m_N denotes the Lebesgue measure on $\mathbb{R}^N$, for each $m \in \mathbb{N}$, we may choose a compact set K_m and an open set G_m such that $K_m \subseteq A \subseteq G_m \subseteq B(a, r)$ and $m_N(G_m \setminus K_m) < 1/m$ (see, e.g., [12, Theorem 2.20]). By Proposition 6.2, we may find $\varphi_m \in \mathscr{D}(G_m)$ such that $0 \leq \varphi_m \leq 1$ and $\varphi_m(x) = 1$ for every $x \in K_m$. Then $\lim_{m\to\infty} \varphi_m(x) = \chi_A(x)$ for almost every $x \in B(a, r)$. This yields

$$\int_A \max(f(x), 0)dx = \lim_{m\to\infty} \int_{\mathbb{R}^N} f(x)\varphi_m(x)dx = 0.$$

Hence, $\max(f, 0) = 0$ on $B(a, r)$. The same argument shows that $\max(-f, 0) = 0$ on $B(a, r)$ and completes the proof. □

We have not defined any topology on $\mathscr{D}(\Omega)$. We define now instead a notion of convergence of sequences and say that a sequence $(\varphi_j)_j \subseteq \mathscr{D}(\Omega)$ converges to 0 if the following two conditions are satisfied:

1. There is some compact set $K \subseteq \Omega$ such that $\operatorname{supp}\varphi_j \subseteq K$ for each $j \in \mathbb{N}$.
2. $(\partial^\alpha \varphi_j)_j$ converges to 0 uniformly on Ω for each multi-index α.

Our aim now is to characterize when a linear mapping is a distribution in terms of this convergence.

Proposition 6.5 *Let $T : \mathscr{D}(\Omega) \to \mathbb{K}$ be linear. Then the following statements are equivalent:*

1. $T \in \mathscr{D}'(\Omega)$.
2. *If* $(\varphi_j)_j \subseteq \mathscr{D}(\Omega)$ *is such that* $\lim_{j\to\infty} \varphi_j = 0$ *in* $\mathscr{D}(\Omega)$, *then* $\lim_{j\to\infty} T(\varphi_j) = 0$.

Proof Assume first that $T \in \mathscr{D}'(\Omega)$. If $(\varphi_j)_j \subseteq \mathscr{D}(\Omega)$ converges to 0 in $\mathscr{D}(\Omega)$, then there is some compact $K \subseteq \Omega$ such that $\operatorname{supp} \varphi_j \subseteq K$ for every j. By (6.3), we can find $c > 0$ and $n \in \mathbb{N}$ with

$$|T(\varphi)| \leq c \sup_{|\alpha|\leq n} \sup_{x\in K} |\partial^\alpha \varphi(x)|$$

for every $\varphi \in \mathscr{D}(K)$. Since $\partial^\alpha \varphi_j \to 0$ as $j \to \infty$ uniformly on K for each $|\alpha| \leq n$, we immediately deduce that $\lim_{j\to\infty} T(\varphi_j) = 0$.

To prove the converse, assume that there is some compact $K \subseteq \Omega$ such that, for each $j \in \mathbb{N}$, there is $\psi_j \in \mathscr{D}(K)$ with

$$|T(\psi_j)| > j \sup_{|\alpha|\leq j} \sup_{x\in K} |\partial^\alpha \psi_j(x)| .$$

Now, for each $j \in \mathbb{N}$, we define $\varphi_j = \frac{\psi_j}{T(\psi_j)}$. Clearly, $|T(\varphi_j)| = 1$ and

$$\begin{aligned} \frac{1}{j} = \frac{1}{j}|T(\varphi_j)| &= \frac{1}{j}\frac{1}{|T(\psi_j)|}|T(\psi_j)| \\ &> \frac{1}{|T(\psi_j)|} \sup_{|\alpha|\leq j} \sup_{x\in K} |\partial^\alpha \psi_j(x)| = \sup_{|\alpha|\leq j} \sup_{x\in K} |\partial^\alpha \varphi_j(x)|. \end{aligned}$$

Hence, we find a sequence $(\varphi_j)_j \subseteq \mathscr{D}(K)$ such that $|T(\varphi_j)| = 1$ and $\sup_{x\in K} |\partial^\alpha \varphi_j(x)| < 1/j$ for every $j \in \mathbb{N}$ and $|\alpha| \leq j$. Then $(\varphi_j)_j$ tends to 0 in $\mathscr{D}(\Omega)$, but $(T(\varphi_j))_j$ does not tend to 0. □

A distribution $T \in \mathscr{D}'(\Omega)$ is called of *finite order* if there is $s \in \mathbb{N}$ such that for all compact subset K of Ω there is $c > 0$ such that

$$|T(\varphi)| \leq c \sup_{|\alpha|\leq s} \sup_{x\in K} |\partial^\alpha \varphi(x)| ,$$

for all $\varphi \in \mathscr{D}(K)$. The distributions in Examples 6.2, 6.3, 6.5, and 6.6 are of finite order. There are, however, distributions that are not of finite order.

Example 6.7 Define the linear map $T : \mathscr{D}(\mathbb{R}) \to \mathbb{K}$ by

$$T(\varphi) = \sum_{n=1}^{\infty} \varphi^{(n)}(n) .$$

To each $A > 0$, select $m \in \mathbb{N}$ such that $A \leq m$. For each $\varphi \in \mathscr{D}([-A, A])$, we have

$$|T(\varphi)| \leq \sum_{n=1}^{m} |\varphi^{(n)}(n)| \leq m \sup_{k \leq m} \sup_{x \in [-A,A]} |\varphi^{(k)}(x)|,$$

and $T \in \mathscr{D}'(\mathbb{R})$. However, this distribution is not of finite order. Otherwise, there is $k \in \mathbb{N}$ such that for each compact set $K \subseteq \mathbb{R}$ there is $c > 0$ with

$$|T(f)| \leq c \sup_{n \leq k} \sup_{x \in K} |f^{(n)}(x)|,$$

for every $f \in \mathscr{D}(K)$. Take $\psi \in \mathscr{D}([-1, 1])$ with $\psi^{(k+1)}(0) = 1$ and define, for each $\varepsilon > 0$, the function $\varphi_\varepsilon(x) = \psi(\frac{x-k-1}{\varepsilon})$. If $\varepsilon < 1$, we have

$$\operatorname{supp} \varphi_\varepsilon \subseteq [k + 1 - \varepsilon, k + 1 + \varepsilon] \subseteq [k, k + 2],$$

and

$$T(\varphi_\varepsilon) = \sum_{n=0}^{\infty} \varphi_\varepsilon^{(n)}(n) = \varphi_\varepsilon^{(k+1)}(k + 1) = \frac{1}{\varepsilon^{k+1}}.$$

On the other hand, for each $n = 1, \dots, k$, we get

$$\sup_{x \in [k,k+2]} |\varphi_\varepsilon^{(n)}(x)| \leq \frac{1}{\varepsilon^n} \sup_{y \in \mathbb{R}} |\psi^{(n)}(y)| \leq \frac{C_1}{\varepsilon^k},$$

for some $C_1 > 0$ independent of $\varepsilon \in]0, 1[$. Hence,

$$\frac{1}{\varepsilon^{k+1}} = |T(\varphi_\varepsilon)| \leq C \sup_{n \leq k} \sup_{x \in L} |\varphi_\varepsilon^{(n)}(x)| \leq \frac{CC_1}{\varepsilon^k},$$

and $1/\varepsilon < CC_1$ for each $0 < \varepsilon < 1$. This is a contradiction and yields our claim. □

Remark 6.1 The pair $(\mathscr{D}(\Omega), \mathscr{D}'(\Omega))$ is a dual pair (with the bilinear mapping $\langle T, \varphi \rangle = T(\varphi)$). Indeed, clearly, if $T \in \mathscr{D}'(\Omega)$ and $T \neq 0$, then there is $\varphi \in \mathscr{D}(\Omega)$ with $T(\varphi) \neq 0$. On the other hand, if $\varphi \in \mathscr{D}(\Omega)$ is non-zero, then $T_{\overline{\varphi}}(\varphi) = \int_{\mathbb{R}^N} \overline{\varphi(x)}\varphi(x)dx = \int_{\mathbb{R}^N} |\varphi(x)|^2 dx > 0$.

6.1.2 *Differentiation of Distributions*

Our aim now is to define the derivative of a distribution. We would expect to do it in such a way that, when we consider the distribution defined by a differentiable function of one variable, its derivative coincides with the distribution defined by the derivative of the function. That is, if $f \in C^1(\mathbb{R})$, we have that $f' \in C(\mathbb{R}) \subseteq L_{1,\text{loc}}(\mathbb{R})$, and what we want is to define a derivative in such a way that $\left(T_f\right)' = T_{f'}$.

If $\varphi \in \mathscr{D}(\mathbb{R})$ is so that $\operatorname{supp}\varphi \subseteq [-A, A]$ (note that then $\varphi(A) = \varphi(-A) = 0$), we can integrate by parts to have

$$T_{f'}\varphi = \int f'(x)\varphi(x)dx = \int_{[-A,A]} f'(x)\varphi(x)dx$$
$$= -\int_{[-A,A]} f(x)\varphi'(x)dx = -\int f(x)\varphi'(x)dx\,.$$

This suggests to define the derivative of T_f as $(T_f)'(\varphi) = -T_f(\varphi')$ for $\varphi \in \mathscr{D}(\mathbb{R})$. We generalize this to arbitrary distributions.

If $\Omega \subseteq \mathbb{R}^N$ is an open set and $\alpha \in \mathbb{N}_0^N$, then the *$\alpha$th derivative* of a distribution $T \in \mathscr{D}(\Omega)$ is defined as the mapping $\partial^\alpha T : \mathscr{D}(\Omega) \longrightarrow \mathbb{K}$ given by

$$\partial^\alpha T(\varphi) = (-1)^{|\alpha|} T(\partial^\alpha \varphi)\,.$$

The first natural question we have to answer is if $\partial^\alpha T$ defines a distribution. It is clearly linear, so we only have to check that (6.3) holds or, equivalently, that for each compact $K \subseteq \Omega$ the restriction $\partial^\alpha T|_{\mathscr{D}(K)} : \mathscr{D}(K) \to \mathbb{K}$ is continuous.

We know that the operator $\partial^\alpha : C^\infty(\Omega) \to C^\infty(\Omega)$ is continuous. If K is compact and $\operatorname{supp}\varphi \subseteq K$, then $\operatorname{supp}\partial^\alpha\varphi \subseteq K$. This (since the topology in $\mathscr{D}(K)$ is induced by $C^\infty(\Omega)$) implies that the operator $\partial^\alpha : \mathscr{D}(K) \to \mathscr{D}(K)$ is well defined and continuous. Then we can see $\partial^\alpha T|_{\mathscr{D}(K)}$ as the composition of two continuous operators

$$\partial^\alpha T|_{\mathscr{D}(K)} : \mathscr{D}(K) \xrightarrow{\partial^\alpha} \mathscr{D}(K) \xrightarrow{(-1)^{|\alpha|}T} \mathbb{K}$$

and it is continuous.

We compute now the derivative of some of the basic distributions.

Example 6.8

1. The derivative of the Dirac function $\delta_a : \mathscr{D}(\mathbb{R}) \to \mathbb{K}$ is easily computed:

$$\delta_a'(\varphi) = -\varphi'(a)\,.$$

2. For the Heaviside function, we have, for each $\varphi \in \mathscr{D}(\mathbb{R})$ with $\operatorname{supp}\varphi \subseteq [-A, A]$,

$$(T_H)'(\varphi) = -\int H(x)\varphi'(x)dx = -\int_0^A \varphi'(x)dx = \varphi(0) = \delta(\varphi)\,.$$

That is,

$$H' = \delta\,. \tag{6.6}$$

Similarly, we have $H_a' = \delta_a$.

3. We look finally at the derivative of Cauchy's principal value of $\frac{1}{x}$. If $\varphi \in \mathscr{D}(\mathbb{R})$, we have (recall (6.5))

$$(T_{\log|x|})'(\varphi) = -T_{\log|x|}(\varphi') = PV\frac{1}{x}(\varphi)\,.$$

Hence, $(T_{\log|x|})' = PV\frac{1}{x}$.

□

6.1.3 Multiplication of a Distribution by a C^∞-Function

What we want to do now is to define a product of distributions and C^∞-functions. As we did before, we check first how it should work for distributions defined by locally integrable functions of one variable. If $f \in L_{1,\mathrm{loc}}(\mathbb{R})$ and $g \in C^\infty(\mathbb{R})$, then $gf \in L_{1,\mathrm{loc}}(\mathbb{R})$, and what we want is to have that $gT_f = T_{fg}$. But this is

$$T_{fg}(\varphi) = \int f(x)g(x)\varphi(x)dx = \int f(x)\big(g\varphi\big)(x)dx = T_f(g\varphi)\,.$$

So we define $gT_f(\varphi) = T_f(g\varphi)$, and we want to follow this idea for arbitrary distributions.

Given $g \in C^\infty(\Omega)$, note that for every $\varphi \in \mathscr{D}(\Omega)$, the product is obviously C^∞ and has compact support, that is, $g\varphi \in \mathscr{D}(\Omega)$. With this in mind, given $T \in \mathscr{D}'(\Omega)$, we define the mapping $gT : \mathscr{D}(\Omega) \to \mathbb{K}$ by

$$\big(gT\big)(\varphi) = T(g\varphi)\,, \tag{6.7}$$

which is clearly linear. To see that it is a distribution, we have to check that, for each compact set $K \subseteq \Omega$, the restriction to $\mathscr{D}(K)$ is continuous. To do that we have to see first that multiplying test functions and C^∞-functions is continuous. For multi-indices $\alpha, \beta \in \mathbb{N}_0^n$, we write $\beta \leq \alpha$ whenever $\beta_k \leq \alpha_k$ for $k = 1, \ldots, n$. Also $\alpha! = \alpha_1! \cdots \alpha_n!$ and $\binom{\alpha}{\beta} = \frac{\alpha!}{\beta!(\alpha-\beta)!}$.

Lemma 6.2 *Let $g \in C^\infty(\Omega)$; then the function $M_g : C^\infty(\Omega) \longrightarrow C^\infty(\Omega)$ given by $M_g(f) = fg$ is continuous.*

Proof First of all, by Leibniz rule, we have $\partial^\alpha(fg) = \sum_{\beta\le\alpha}\binom{\alpha}{\beta}\partial^\beta f\cdot\partial^{\alpha-\beta}g$. Then, for each $K\subseteq\Omega$ compact and each $n\in\mathbb{N}$, we have

$$\begin{aligned}\sup_{x\in K}\sup_{|\alpha|\le n}|\partial^\alpha(fg)(x)| &\le \sup_{x\in K}\sup_{|\alpha|\le n}\sum_{\beta\le\alpha}\binom{\alpha}{\beta}|\partial^\beta f\;\partial^{\alpha-\beta}g| \\ &\le \Big(\sup_{|\alpha|\le n}\sum_{\beta\le\alpha}\binom{\alpha}{\beta}\Big)\Big(\sup_{x\in K}\sup_{|\gamma|\le n}|\partial^\gamma g(x)|\Big)\Big(\sup_{x\in K}\sup_{|\eta|\le n}|\partial^\eta f(x)|\Big).\end{aligned}$$

Taking $c = \Big(\sup_{|\alpha|\le n}\sum_{\beta\le\alpha}\binom{\alpha}{\beta}\Big)\Big(\sup_{x\in K}\sup_{|\gamma|\le n}|\partial^\gamma g(x)|\Big)$, this shows that M_g is continuous. □

We can now see that the linear mapping gT defined in (6.7) is actually a distribution. Note first of all that if K is compact and $\operatorname{supp}\varphi\subseteq K$, then $\operatorname{supp} g\varphi\subseteq K$. So, by Lemma 6.2, the mapping $M_g|_{\mathscr{D}(K)}:\mathscr{D}(K)\longrightarrow\mathscr{D}(K)$ is continuous. Therefore, $(gT)|_{\mathscr{D}(K)} = T|_{\mathscr{D}(K)}\circ M_g|_{\mathscr{D}(K)}$ is also continuous. Since this holds for every compact $K\subseteq\Omega$, we have that gT is a distribution.

Example 6.9 Let us see some examples of products of functions and distributions. To begin with, if $g\in C^\infty(\mathbb{R})$, then, for each $\varphi\in\mathbb{R}$, we have

$$(g\delta)(\varphi) = \delta(g\varphi) = g(0)\varphi(0) = g(0)\delta(\varphi),$$

which gives $g(x)\delta = g(0)\delta$. In particular, $x\delta = 0$. The same argument yields $g\delta_a = g(a)\delta_a$.

Let us multiply the function $g(x) = x$ with the distribution δ' (the derivative of the Dirac function). If $\varphi\in\mathbb{R}$, then

$$(x\delta')(\varphi) = \delta'(x\varphi) = -\delta((x\varphi)') = -\delta(\varphi + x\varphi') = -\delta(\varphi) - (x\delta)(\varphi') = -\delta(\varphi).$$

Hence, $x\delta' = -\delta$. The same argument shows that $\delta_a = (a-x)\delta_a'$ for each fixed $a\in\mathbb{R}$.

Finally, taking Cauchy's principal value of $\frac{1}{x}$, we have

$$\begin{aligned}\big(x(PV\tfrac{1}{x})\big)(\varphi) = (PV\frac{1}{x})(x\varphi) &= \lim_{\varepsilon\to 0+}\int_{\varepsilon<|x|} x\frac{1}{x}\varphi(x)dx \\ &= \lim_{\varepsilon\to 0+}\int_{\varepsilon<|x|}\varphi(x)dx = \int 1\cdot\varphi(x)dx = \mathbb{1}(\varphi),\end{aligned}$$

for every $\varphi\in\mathscr{D}(\mathbb{R})$, where $\mathbb{1}$ denotes the distribution defined by the constant function equal to 1. Thus

$$x(PV\tfrac{1}{x}) = \mathbb{1}.$$

□

Remark 6.2 Let us note that this product is not associative. On the one hand, we have $(x\delta)(PV\frac{1}{x}) = 0$, but on the other hand $\delta(x PV\frac{1}{x}) = \delta \cdot \mathbb{1} = \delta$.

Example 6.10 Distributional derivatives and products allow us to give a meaning to the derivative for functions that, in the classical sense, are not differentiable like, for example, the absolute value. If $f(x) = |x| = 2xH(x) - x$, then (note that here we are identifying the functions and the distributions that they define) $f'(x) = 2H(x) + 2x\delta(x) - 1 = 2H(x) - 1$. □

We can use these ideas to solve differential equations.

Example 6.11 Let us consider the differential equation

$$y' + 2y = \delta_1 .$$

Multiplying by e^{2x}, we have $e^{2x}(y' + 2y) = e^{2x}\delta_1$, and this gives $\left(e^{2x}y\right)' = e^2\delta_1$. Hence, we have

$$y(x) = e^{-2x}(e^2 H(x-1) + C).$$

□

Example 6.12 We take now the differential equation

$$y'' + 4y = \delta, \tag{6.8}$$

whose homogeneous part has solution

$$y_h(x) = A\cos(2x) + B\sin(2x) .$$

We look now for a particular solution of the form $y_p(x) = u(x)H(x)$. Then

$$y_p' = u'H + u\delta = u'H + u(0)\delta$$

and

$$y_p'' = u''H + u'\delta + u(0)\delta' = u''H + u'(0)\delta + u(0)\delta' .$$

Therefore,

$$y_p'' + 4y_p = (u'' + 4u)H + u'(0)\delta + u(0)\delta' .$$

So, if we want the right-hand side to be δ, we need u to satisfy

$$\begin{cases} u'' + 4u = 0 \\ u(0) = 0 \\ u'(0) = 1 \end{cases} .$$

Solving this, we have $u(x) = \frac{1}{2}\sin(2x)$. So, the solution to (6.8) is given by

$$y(x) = A\cos(2x) + \big(B + \frac{1}{2}H(x)\big)\sin(2x)\,.$$

□

6.1.4 *Support of a Distribution and Distributions with Compact Support*

We want to define now the support of a distribution. The first step is to give a meaning to a distribution vanishing on a certain subset. Once again, we take distributions defined by functions as starting point. If f is locally integrable and vanishes at some open set V, then clearly (just have a look at (6.4)) $T_f(\varphi) = 0$ for every test function φ with $\operatorname{supp}\varphi \subseteq V$ (i.e., $\varphi \in \mathscr{D}(V)$). With this idea in mind, we say that a distribution $T \in \mathscr{D}'(\Omega)$ (being $\Omega \subseteq \mathbb{R}^N$ open) *vanishes* on the open set $V \subseteq \Omega$ if $T(\varphi) = 0$ for every $\varphi \in \mathscr{D}(V)$. We say that two distributions $S, T \in \mathscr{D}'(\Omega)$ *coincide* on V if $T - S$ vanishes on V.

Lemma 6.3 *Let $\Omega \subseteq \mathbb{R}^N$ be open, and $(V_i)_{i\in I}$ be a family of open subsets of Ω. If $T \in \mathscr{D}'(\Omega)$ vanishes on V_i for each $i \in I$, then T vanishes on $V = \bigcup_{i\in I} V_i$.*

Proof Take some $\varphi \in \mathscr{D}(V)$. Since the support of φ is compact, there are $V_{i(1)}, \dots, V_{i(m)}$ such that $\operatorname{supp}\varphi \subseteq \bigcup_{j=1}^m V_{i(j)}$. We apply Proposition 6.4 to find $\psi_j \in \mathscr{D}(V_{i(j)})$ such that $0 \leq \psi_j \leq 1$ and $\sum_{j=1}^m \psi_j(x) = 1$ for all $x \in \operatorname{supp}\varphi$. Then $\varphi = \sum_{j=1}^m \varphi\psi_j$. Since $\varphi\psi_j \in \mathscr{D}(V_{i(j)})$ and T vanishes on $V_{i(j)}$ for each $j = 1, \dots, m$, we conclude

$$T(\varphi) = \sum_{j=1}^m T(\varphi\psi_j) = 0\,.$$

□

As a consequence of Lemma 6.3, given $T \in \mathscr{D}(\Omega)$, the set

$$V_T = \bigcup \{V : V \subseteq \Omega \text{ open, and } T \text{ vanishes on } V\}$$

is the largest open set where T vanishes. With this, we define the *support* of a distribution $T \in \mathscr{D}'(\Omega)$ as

$$\operatorname{supp} T = \Omega \setminus V_T\,,$$

which is clearly closed.

Example 6.13 The support of the Dirac function $\delta \in \mathscr{D}'(\mathbb{R})$ is $\{0\}$. □

As we did with functions, we are interested in distributions that have compact support, the space of which is denoted by $\mathscr{E}'(\Omega)$. Let us note that a distribution $T \in \mathscr{D}'(\Omega)$ satisfies $T \in \mathscr{E}'(\Omega)$ if and only if there are $K \subseteq \Omega$ compact, $C > 0$, and $m \in \mathbb{N}$ with

$$|T(\varphi)| \leq C \sup_{|\alpha| \leq m} \sup_{x \in K} |\partial^\alpha \varphi(x)| ,$$

for $\varphi \in \mathscr{D}(\Omega)$. However, it is not possible to replace in general K by $\operatorname{supp} T$. See [20, Chapter 2, Exercise 10]. Our next result describes distributions with compact support in a very precise way.

Theorem 6.14 *Let $\Omega \subseteq \mathbb{R}^N$ be open, and $T \in \mathscr{D}'(\Omega)$. Then* $\operatorname{supp} T$ *is compact if and only if T is the restriction to $\mathscr{D}(\Omega)$ of some $S \in C^\infty(\Omega)'$, which is uniquely determined by T.*

Proof Take $S \in C^\infty(\Omega)'$, denote by T its restriction to $\mathscr{D}(\Omega)$, and let us see that, in fact, T is a distribution with compact support. Given any compact set $K \subseteq \Omega$, the inclusion $\mathscr{D}(K) \subseteq C^\infty(\Omega)$ is continuous. Then $T : \mathscr{D}(K) \to \mathbb{K}$ is continuous for every compact K. This shows that $T \in \mathscr{D}'(\Omega)$. Let us see now that it has compact support. Since S is continuous, we can find $L \subseteq \Omega$ compact, $m \in \mathbb{N}$, and $C > 0$ so that

$$|S(f)| \leq C \sup_{|\alpha| \leq m} \sup_{x \in L} |\partial^\alpha f(x)| ,$$

for every $f \in C^\infty(\Omega)$. If $\varphi \in \mathscr{D}(\Omega)$ is such that $\operatorname{supp} \varphi \subseteq \Omega \setminus L$, we have $S(\varphi) = 0$. This implies $\operatorname{supp} T \subseteq L$, and then T has compact support.

Note also that, since $\mathscr{D}(\Omega)$ is dense in $C^\infty(\Omega)$ (recall Proposition 6.3), $S \in C^\infty(\Omega)'$ is uniquely determined by its restriction to $\mathscr{D}(\Omega)$.

Suppose now that $T \in \mathscr{D}'(\Omega)$ has compact support $K = \operatorname{supp} T$. Take some relatively compact $V \subseteq \Omega$ such that $K \subseteq V \subseteq \overline{V} \subseteq \Omega$, and use Proposition 6.2 to find a function $\psi \in \mathscr{D}(\Omega)$ with $\psi(x) = 1$ for each $x \in V$. We define $S : C^\infty(\Omega) \to \mathbb{K}$ by $S(f) = T(\psi f)$. We first prove that $S \in C^\infty(\Omega)'$. It is clearly linear, so it is only left to see that it is continuous. The set $L = \operatorname{supp} \psi$ is compact in Ω; then (see (6.3)), there are $C > 0$ and $m \in \mathbb{N}$ with

$$|T(\varphi)| \leq C \sup_{|\alpha| \leq m} \sup_{x \in L} |\partial^\alpha \varphi(x)| ,$$

for every $\varphi \in \mathscr{D}(L)$. If $f \in C^\infty(\Omega)$, then clearly $\psi f \in \mathscr{D}(L)$, and

$$|S(f)| \leq \sup_{|\alpha| \leq m} \sup_{x \in L} |\partial^\alpha (\psi f)(x)| .$$

Now, using Leibniz formula, we can find $D > 0$ such that

$$|S(f)| \leq D \sup_{|\alpha| \leq m} \sup_{x \in L} |\partial^\alpha f(x)|,$$

for every $f \in C^\infty(\Omega)$. This shows that $S : C^\infty(\Omega) \to \mathbb{K}$ is continuous. Finally, we show that T is the restriction of S to $\mathscr{D}(\Omega)$. Given $\varphi \in \mathscr{D}(\Omega)$, the function $\varphi - \psi\varphi \in \mathscr{D}(\Omega)$ vanishes on V. Thus $\text{supp}(\varphi - \psi\varphi) \subseteq \Omega \setminus V \subseteq \Omega \setminus K$ and, hence, $\varphi - \psi\varphi \in \mathscr{D}(\Omega \setminus K)$. Since T vanishes in $\Omega \setminus K$, $T(\varphi - \psi\varphi) = 0$, which gives $S(\varphi) = T(\psi\varphi) = T(\varphi)$. □

6.2 The Space of Rapidly Decreasing Functions

So far we have been working with the space $C^\infty(\mathbb{R}^N)$ of all C^∞-functions, and its subspace $\mathscr{D}(\mathbb{R}^N)$ of functions with compact support. We are going to consider now functions that, although not having compact support (and maybe even being non-zero everywhere), decrease to 0 very fast as x grows. Let us give a precise meaning to this idea and say that a function $f \in C^\infty(\mathbb{R}^N)$ is *rapidly decreasing* if

$$p_k(f) = \sup_{|\alpha| \leq k} \sup_{x \in \mathbb{R}^N} (1 + |x|^2)^k |\partial^\alpha f(x)| < +\infty, \tag{6.9}$$

for each $k \in \mathbb{N}$. The space of all rapidly decreasing functions (also known as the *Schwartz space*) is denoted by $S(\mathbb{R}^N)$ and is endowed with the metrizable locally convex topology defined by the increasing system of seminorms $(p_k)_k$ given in (6.9). Let us point out that if $f \in S(\mathbb{R}^N)$, then for each $k \in \mathbb{N}$ there is $C_k > 0$ such that $|\partial^\alpha f(x)| \leq C_k \frac{1}{(1+|x|^2)^{k/2}}$ for every x and all $|\alpha| \leq k$. In particular,

$$\lim_{|x| \to \infty} \partial^\alpha f(x) = 0, \tag{6.10}$$

for every α.

Remark 6.3 Every test function is clearly rapidly decreasing. In other words, we have $\mathscr{D}(\mathbb{R}^N) \subseteq S(\mathbb{R}^N)$. This inclusion is strict, since the function $f \in C^\infty(\mathbb{R}^N)$ given by $f(x) = e^{-\pi|x|^2}$ clearly belongs to $S(\mathbb{R}^N)$ but does not have compact support.

Remark 6.4 For each $\alpha, \beta \in \mathbb{N}_0^N$, we consider

$$q_{\alpha,\beta}(f) = \sup_{x \in \mathbb{R}^N} |x^\beta \partial^\alpha f(x)|.$$

It is a simple exercise to show that the family of seminorms $(q_{\alpha,\beta})_{\alpha,\beta}$ is equivalent to $(p_k)_k$ in (6.9) and, therefore, also defines the topology on $S(\mathbb{R}^N)$. As a straightforward consequence, we have the following.

Proposition 6.6 *A sequence $(f_j)_j \subseteq S(\mathbb{R}^N)$ converges to zero in $S(\mathbb{R}^N)$ if and only if, for each pair of multi-indices $\alpha, \beta \in \mathbb{N}_0^N$, we have*

$$\lim_{j\to\infty} x^\beta \partial^\alpha f_j(x) = 0$$

uniformly on $\mathbb{R}^N$.

Proposition 6.7 *For each compact $K \subseteq \mathbb{R}^N$, the inclusions $\mathscr{D}(K) \subseteq S(\mathbb{R}^N) \subseteq C^\infty(\mathbb{R}^N)$ are continuous. Moreover, $\mathscr{D}(\mathbb{R}^N)$ is dense in $S(\mathbb{R}^N)$, and $S(\mathbb{R}^N)$ is dense in $C^\infty(\mathbb{R}^N)$.*

Proof Fix some compact set $K \subseteq \mathbb{R}^N$. For each $k \in \mathbb{N}_0$ and every $f \in \mathscr{D}(K)$, we have

$$\sup_{|\alpha|\le k}\sup_{x\in\mathbb{R}^N}(1+|x|^2)^k|\partial^\alpha f(x)| \le \max_{x\in K}(1+|x|^2)^k \sup_{|\alpha|\le k}\sup_{x\in K}|\partial^\alpha f(x)|\,.$$

This shows that the inclusion $\mathscr{D}(K) \subseteq S(\mathbb{R}^N)$ is continuous.
To show the continuity of the inclusion $S(\mathbb{R}^N) \subseteq C^\infty(\mathbb{R}^N)$, we fix $K \subseteq \mathbb{R}^N$ compact and $k \in \mathbb{N}$. Since $(1+|x|^2)^k \ge 1$ for each $x \in K$, we have

$$\sup_{|\alpha|\le k}\sup_{x\in K}|\partial^\alpha f(x)| \le \sup_{|\alpha|\le k}\sup_{x\in\mathbb{R}^N}(1+|x|^2)^k|\partial^\alpha f(x)|\,,$$

for every $f \in S(\mathbb{R}^N)$.
The fact that $S(\mathbb{R}^N)$ is dense in $C^\infty(\mathbb{R}^N)$ is an immediate consequence of Remark 6.3 and Proposition 6.3. It is only left, then, to see that $\mathscr{D}(\mathbb{R}^N)$ is dense in $S(\mathbb{R}^N)$. To this end, we choose $\varphi \in \mathscr{D}(\mathbb{R}^N)$ such that $\varphi(x) = 1$ on an open neighbourhood of $\overline{B(0,1)}$. For each $j \in \mathbb{N}$, we consider the function $\varphi_j(x) = \varphi(x/j)$. Clearly, $\varphi_j \in \mathscr{D}(\mathbb{R}^N)$ and $\varphi_j(x) = 1$ for every x in some open neighbourhood of $\overline{B(0,j)}$. Given $f \in S(\mathbb{R}^N)$, we set $f_j = \varphi_j f \in \mathscr{D}(\mathbb{R}^N)$. Let us see that $(f_j)_j$ converges to f in $S(\mathbb{R}^N)$. First of all, for each $x \in \mathbb{R}^N$, using Leibniz formula, we have

$$\begin{aligned}(1+|x|^2)^k&|\partial^\alpha(f_j-f)(x)|\\ &\le \frac{1}{1+|x|^2}\sum_{\beta\le\alpha}\binom{\alpha}{\beta}|\partial^\beta(\varphi_j(x)-1)|\sup_{y\in\mathbb{R}^N}(1+|y|^2)^{k+1}|\partial^{\alpha-\beta}f(y)|\,.\end{aligned}$$

Note now that $|\partial^\beta(\varphi(x)-1)| \le \sup_{y\in\mathbb{R}^N}|\partial^\beta\varphi(y)|$ for every $\beta \ne 0$, and recall that $\varphi_j(x) = 1$ in some open neighbourhood of $\overline{B(0,j)}$. With this, we can find $C > 0$

such that

$$\sup_{x\in\mathbb{R}^N}(1+|x|^2)^k|\partial^\alpha(f_j-f)(x)|$$
$$\leq C\sup_{|x|\geq j}\frac{1}{1+|x|^2}\sum_{\beta\leq\alpha}\binom{\alpha}{\beta}\sup_{x\in\mathbb{R}^N}(1+|x|^2)^{k+1}|\partial^{\alpha-\beta}f(x)|\,.$$

Since $f\in S(\mathbb{R}^N)$, we conclude

$$\lim_{j\to\infty}\sup_{x\in\mathbb{R}^N}(1+|x|^2)^k|\partial^\alpha(f_j-f)(x)|=0,$$

which proves that $\mathscr{D}(\mathbb{R}^N)$ is dense in $S(\mathbb{R}^N)$. □

Theorem 6.15 *The space $S(\mathbb{R}^N)$ is complete, and thus a Fréchet space.*

Proof Let $(f_j)_j$ be a Cauchy sequence in $S(\mathbb{R}^N)$. Since, by Proposition 6.7, the inclusion if S in C^∞ is continuous, the sequence is also Cauchy in $C^\infty(\mathbb{R}^N)$, which is complete (see Theorem 4.13). Then there is some $f\in C^\infty(\mathbb{R}^N)$ such that $(f_j)_j$ converges to f in $C^\infty(\mathbb{R}^N)$. Let us see now that, in fact, $f\in S(\mathbb{R}^N)$ and $(f_j)_j$ converges to f also in $S(\mathbb{R}^N)$. Given $\varepsilon>0$, $k\in\mathbb{N}_0$, and $\alpha\in\mathbb{N}_0^N$, there is $j_0\in\mathbb{N}$ such that, for $j,l\geq j_0$ and every $x\in\mathbb{R}^N$, we have

$$(1+|x|^2)^k|\partial^\alpha f_j(x)-\partial^\alpha f_l(x)|<\varepsilon\,.$$

Fixing $l=j_0$ and letting $j\to\infty$, it follows that $f\in S(\mathbb{R}^N)$. Now, for a fixed $l\geq j_0$ and letting $j\to\infty$, we conclude that $(f_j)_j$ converges to f in $S(\mathbb{R}^N)$. □

Proposition 6.8 *The space $S(\mathbb{R}^N)$ is contained in $L_1(\mathbb{R}^N)$ with continuous inclusion.*

Proof First note that

$$\int_{\mathbb{R}^N}\frac{1}{(1+|x|^2)^k}dx=c_N\int_0^\infty\frac{r^{N-1}}{(1+r^2)^k}dr<\infty\,,$$

if $N-1-2k<-1$. Then, for $k>N/2$ and $f\in S(\mathbb{R}^N)$, we get

$$\|f\|_{L_1}=\int_{\mathbb{R}^N}|f(x)|dx\leq Cp_k(f)\,, \tag{6.11}$$

for some $C>0$. This completes the proof. □

The following two results are straightforward consequences of the definitions, and their proof is left to the reader.

Proposition 6.9 *Fix $f \in S(\mathbb{R}^N)$ and $\alpha, \beta \in \mathbb{N}_0^N$. Then $\partial^\alpha f$ and $x^\beta f$ belong to $S(\mathbb{R}^N)$.*

Proposition 6.10 *For $a \in \mathbb{R}^N$ and $\lambda > 0$, the following operators*

$$\begin{aligned} T_a f(x) &= f(x-a) \\ E_a f(x) &= e^{2\pi i a\cdot x} f(x) \\ D_\lambda f(x) &= \lambda^{\frac{N}{2}} f(\lambda x) \\ \check{f}(x) &= f(-x) \end{aligned}$$

are well defined and continuous from $S(\mathbb{R}^N)$ into $S(\mathbb{R}^N)$.

6.3 Fourier Transform on $S(\mathbb{R}^N)$

Given a function $f \in L_1(\mathbb{R}^N)$, its *Fourier transform* is defined as

$$\mathcal{F}(w) = \widehat{f}(w) = \int_{\mathbb{R}^N} e^{-2\pi i x\cdot w} f(x)dx\,,$$

for $w \in \mathbb{R}^N$. The integral is well defined because $|e^{-2\pi i x\cdot w} f(x)| = |f(x)|$ for all $x, w \in \mathbb{R}^N$.

Proposition 6.11 *Let $f, g \in L_1(\mathbb{R}^N)$, $t \in \mathbb{R}^N$, and $\varepsilon > 0$. Then:*

1. *The Fourier transform $\widehat{f}$ is a bounded continuous function. Moreover,*

$$\sup_{w\in\mathbb{R}^N} |\widehat{f}(w)| \leq \int_{\mathbb{R}^N} |f(x)|dx\,. \tag{6.12}$$

2. *The following formula holds*

$$\int_{\mathbb{R}^N} \widehat{f}(w)g(w)dw = \int_{\mathbb{R}^n} f(w)\widehat{g}(w)dw\,.$$

3. $\widehat{f * g}(w) = \widehat{f}(w)\widehat{g}(w)$ *for every $w \in \mathbb{R}^N$.*
4. $\mathcal{F}\check{f}(w) = \widehat{f}(-w)$, *that is, $\widehat{\check{f}} = \check{\widehat{f}}$.*
5. *If $f(x) = g(x+t)$ for $x \in \mathbb{R}^N$, we have*

$$\widehat{f}(w) = \widehat{g}(w)e^{2\pi i t\cdot w}\,.$$

6. *If* $f(x) = g(\varepsilon x)$ *for* $x \in \mathbb{R}^N$, *we have*

$$\widehat{f}(w) = \varepsilon^{-N}\widehat{g}\big(\frac{w}{\varepsilon}\big).$$

Proof To see 1, we note in first place that the continuity of $\widehat{f}$ is a direct consequence of the dominated convergence theorem. On the other hand, if $w \in \mathbb{R}^N$, then

$$|\widehat{f}(w)| \le \int_{\mathbb{R}^N} |f(x)||e^{-2\pi i x\cdot w}|dx \le \int_{\mathbb{R}^N} |f(x)|dx .$$

For the proof of 2, define $h : \mathbb{R}^N \times \mathbb{R}^N \to \mathbb{K}$ by $h(x, w) = f(x)e^{-2\pi i x\cdot w}g(w)$, which is measurable and satisfies

$$\int_{\mathbb{R}^N}\Big(\int_{\mathbb{R}^N}|h(x, w)|dx\Big)dw < \infty .$$

Thus $h \in L_1(\mathbb{R}^N \times \mathbb{R}^N)$ and, by Fubini's theorem, we get

$$\begin{aligned}\int_{\mathbb{R}^N}\widehat{f}(w)g(w)dw &= \int_{\mathbb{R}^N}\Big(\int_{\mathbb{R}^N} f(x)e^{-2\pi i x\cdot w}dx\Big)g(w)dw \\ &= \int_{\mathbb{R}^N\times\mathbb{R}^N} h(x, w)dxdw = \int_{\mathbb{R}^n} f(x)\widehat{g}(x)dx.\end{aligned}$$

Fubini's theorem and a simple change of variable give 3:

$$\begin{aligned}\widehat{f*g}\,(w) &= \int_{\mathbb{R}^N} e^{-2\pi i x\cdot w}(f*g)(x)dx = \int_{\mathbb{R}^N} e^{-2\pi i x\cdot w}\int_{\mathbb{R}^N} f(y)g(x-y)dydx \\ &= \int_{\mathbb{R}^N}\int_{\mathbb{R}^N} e^{-2\pi i y\cdot w}f(y)e^{-2\pi i (x-y)\cdot w}g(x-y)dydx \\ &= \int_{\mathbb{R}^N} e^{-2\pi i y\cdot w}f(y)\Big(\int_{\mathbb{R}^N} e^{-2\pi i (x-y)\cdot w}g(x-y)dx\Big)dy \\ &= \int_{\mathbb{R}^N} e^{-2\pi i y\cdot w}f(y)\Big(\int_{\mathbb{R}^N} e^{-2\pi i u\cdot w}g(u)du\Big)dy \\ &= \Big(\int_{\mathbb{R}^N} e^{-2\pi i y\cdot w}f(y)dy\Big)\Big(\int_{\mathbb{R}^N} e^{-2\pi i u\cdot w}g(u)du\Big) = \widehat{f}(w)\widehat{g}(w) .\end{aligned}$$

Again a change of variable $-x = u$ yields 4

$$\begin{aligned}\widehat{\check{f}}(w) &= \int_{\mathbb{R}^N} e^{-2\pi i x\cdot w}\check{f}(x)dx = \int_{\mathbb{R}^N} e^{-2\pi i x\cdot w}f(-x)dx \\ &= \int_{\mathbb{R}^N} e^{2\pi i u\cdot w}f(u)du = \int_{\mathbb{R}^N} e^{-2\pi i u\cdot(-w)}f(u)du = \widehat{f}(-w) .\end{aligned}$$

For 5, we apply the definition of Fourier transform to obtain

$$\widehat{f}(w) = \int_{\mathbb{R}^N} g(x+t)e^{-2\pi ix\cdot w}dx = \int_{\mathbb{R}^N} g(z)e^{-2\pi i(z-t)\cdot w}dz = \widehat{g}(w)e^{2\pi it\cdot w}.$$

Finally, 6 again follows directly from the definition of Fourier transform by doing

$$\widehat{f}(w) = \int_{\mathbb{R}^N} g(\varepsilon x)e^{-2\pi ix\cdot w}dx = \frac{1}{\varepsilon^N}\int_{\mathbb{R}^N} g(z)e^{-2\pi iz\cdot\frac{w}{\varepsilon}}dz = \frac{1}{\varepsilon^N}\widehat{g}\Big(\frac{w}{\varepsilon}\Big).$$

□

The Fourier transform of a function in $L_1(\mathbb{R}^N)$ may not belong to $L_1(\mathbb{R}^N)$, as our next example shows.

Example 6.16 Fix $a > 0$ and take $f = \chi_{[-a/2,a/2]}$, the characteristic function of the interval $[-a/2, a/2]$, of length $a > 0$. Then

$$\begin{aligned}\widehat{f}(w) = \int_{\mathbb{R}} f(x)e^{-2\pi ix\cdot w}dx &= \int_{-a/2}^{a/2} e^{-2\pi ix\cdot\frac{w}{\varepsilon}}dx\\ &= \frac{1}{2\pi iw}\big(e^{i\pi aw} - e^{-i\pi aw}\big) = \frac{\sin(\pi aw)}{\pi w}.\end{aligned}$$

Observe that $\widehat{f}$ is continuous and bounded, but $\widehat{f} \notin L_1(\mathbb{R})$. □

Example 6.17 Let us compute the Fourier transform of the function $f(x) = e^{-|x|}$ for $x \in \mathbb{R}$. We have

$$\widehat{f}(w) = \int_{\mathbb{R}} e^{-2\pi ixw}e^{-|x|}dx = \int_{-\infty}^{0} e^{-2\pi ixw}e^{x}dx + \int_{0}^{+\infty} e^{-2\pi ixw}e^{-x}dx$$

$$\begin{aligned}&= \int_{0}^{+\infty} e^{2\pi ixw}e^{-x}dx + \int_{0}^{+\infty} e^{-2\pi ixw}e^{-x}dx = \int_{0}^{+\infty}\big(e^{2\pi ixw}+e^{-2\pi ixw}\big)e^{-x}dx\\ &= \int_{0}^{+\infty} 2\cos(2\pi ixw)e^{-x}dx.\end{aligned}$$

If $a \neq 0$, elementary computations show that

$$\int \cos(ax)e^{-x}dx = \frac{1}{1+a^2}e^{-x}\big(a\sin(ax) - \cos(ax)\big).$$

This altogether shows that

$$\widehat{f}(w) = \frac{2}{1+4\pi^2w^2},$$

for $w \in \mathbb{R}$. □

The following basic properties that relate the Fourier transform with differentiation are a direct consequence of derivation under the integral sign and integration by parts.

Proposition 6.12 *Fix $k \in \mathbb{N}$:*

1. *If $f \in L_1(\mathbb{R}^N)$ satisfies $x^\alpha f(x) \in L_1(\mathbb{R}^N)$ for every $|\alpha| \le k$, then $\widehat{f} \in C^k(\mathbb{R}^N)$ and*

$$\partial^\alpha \widehat{f}(w) = (-2\pi i)^{|\alpha|} \mathcal{F}\big(x^\alpha f(x)\big)(w),$$

for every $w \in \mathbb{R}^N$.

2. *If $f \in L_1(\mathbb{R}^N) \cap C^k(\mathbb{R}^N)$, and $\partial^\alpha f \in L_1(\mathbb{R}^N)$ for every $|\alpha| \le k$, then*

$$\widehat{\partial^\alpha f}(w) = \mathcal{F}\big(\partial^\alpha f\big)(w) = (2\pi i)^{|\alpha|} w^\alpha \widehat{f}(w),$$

for every $w \in \mathbb{R}^N$.

Proposition 6.12 shows that the Fourier transform has a nice behaviour with derivatives. This suggests that it can be a useful tool when dealing with differential equations. We will later give an example of how to use it (see Example 6.19). The general idea will be to apply the Fourier transform to the equation, to get an algebraic equation, and then solve it to get an expression for the Fourier transform of the solution. We need then a way to 'come back', recovering the function from its Fourier transform. We show in Theorem 6.18 how to do this, but we need some preliminary work.

Lemma 6.4 *We have*

$$\int_{\mathbb{R}} e^{-\pi t^2} dt = 1.$$

Proof Since

$$\int_{\mathbb{R}} e^{-\pi t^2} dt = \frac{1}{\sqrt{\pi}} \int_{\mathbb{R}} e^{-x^2} dx,$$

and $x \rightsquigarrow e^{-x^2}$ is an even function, it is enough to show that

$$\int_0^\infty e^{-x^2} dx = \frac{\sqrt{\pi}}{2}. \tag{6.13}$$

Consider the function $f(x, y) = e^{-x^2} e^{-y^2}$, for $(x, y) \in \mathbb{R}^2$, and the sets

$$\begin{aligned} S &= [0, R] \times [0, R], \\ D_R &= \overline{B(0, R)} \cap \{(x, y) \in \mathbb{R}^2 : x, y \ge 0\} \\ D_{\sqrt{2}R} &= \overline{B(0, \sqrt{2}R)} \cap \{(x, y) \in \mathbb{R}^2 : x, y \ge 0\}. \end{aligned}$$

Clearly, $D_R \subseteq S \subseteq D_{\sqrt{2}R}$; hence, since $f(x, y) > 0$ and it is continuous, we have

$$\int_{D_R} f(x, y)dxdy \le \int_S f(x, y)dxdy \le \int_{D_{\sqrt{2}R}} f(x, y)dxdy\,.$$

On the other hand, Fubini's theorem yields

$$\int_S f(x, y)dxdy = \Big(\int_0^R e^{-x^2}dx\Big)^2,$$

and with the change to polar coordinates $x = r\cos\theta$, $y = r\sin\theta$, for $\theta \in [0, \pi/2]$ and $r \in [0, R]$, we get

$$\frac{\pi}{4}\Big(1 - \frac{1}{e^{R^2}}\Big) \le \Big(\int_0^R e^{-x^2}dx\Big)^2 \le \frac{\pi}{4}\Big(1 - \frac{1}{e^{2R^2}}\Big)\,.$$

Letting $R \to \infty$, we conclude

$$\Big(\int_0^\infty e^{-x^2}dx\Big)^2 = \frac{\pi}{4},$$

which yields (6.13) and gives the conclusion. □

Lemma 6.5 *If $f(x) = e^{-\pi|x|^2}$ for $x \in \mathbb{R}^N$, then $\widehat{f}(w) = e^{-\pi|w|^2}$ for all $w \in \mathbb{R}^N$.*

Proof Since $f(x) = e^{-\pi x_1^2} \cdots e^{-\pi x_N^2}$ for all $x \in \mathbb{R}^N$, by Fubini's theorem, it is enough to show that $\widehat{g} = g$ for $g(x) = e^{-\pi x^2}$ for $x \in \mathbb{R}$. To do this, we observe that g is a solution of the differential equation $2\pi xy + y' = 0$ (this follows by straightforward computation). Suppose now that $h(x)$ is a differentiable function that is also a solution of the same differential equation. Then we have

$$(e^{\pi x^2}h)' = (2\pi xh + h')e^{\pi x^2} = 0\,.$$

So $e^{\pi x^2}h$ is constant, and $h(x) = Ce^{-\pi x^2}$ for some $C \in \mathbb{R}$ and all $x \in \mathbb{R}$.

Now, since $g' \in L_1(\mathbb{R})$ and $xg(x) \in L_1(\mathbb{R})$, we can apply Proposition 6.12 to conclude that $\widehat{g}$ is also differentiable and

$$\begin{aligned} 0 = \mathcal{F}(2\pi xg + g')(w) &= 2\pi\mathcal{F}(xg(x))(w) + \mathcal{F}(g')(w) \\ &= -2\pi\frac{1}{2\pi i}\widehat{g}'(w) + 2\pi iw\widehat{g}(w) = i\big(\widehat{g}'(w) + 2\pi iw\widehat{g}(w)\big). \end{aligned}$$

This implies that $\widehat{g}$ is a solution of the same differential equation, and, by what we have just observed, there is $C \in \mathbb{R}$ with $\widehat{g}(w) = Ce^{-\pi w^2}$. On the other hand, using Lemma 6.4, we have

$$C = \widehat{g}(0) = \int_{\mathbb{R}} e^{-\pi x^2} dx = 1 .$$

This shows that $g = \widehat{g}$ and completes the proof. □

Theorem 6.18 (Fourier inversion formula) *Let f be a bounded and continuous function in $L_1(\mathbb{R}^N)$. If $\widehat{f} \in L_1(\mathbb{R}^N)$, then*

$$f(x) = \int_{\mathbb{R}^N} \widehat{f}(w) e^{2\pi i x \cdot w} dw , \tag{6.14}$$

for every $x \in \mathbb{R}^N$.

Proof We set $g(x) = e^{-\pi |x|^2}$ and $g_\varepsilon(x) = g(\varepsilon x)$ for each $\varepsilon > 0$. For $x \in \mathbb{R}^N$ fixed, we consider

$$I_\varepsilon = \int_{\mathbb{R}^N} \widehat{f}(w) e^{2\pi i x \cdot w} g_\varepsilon(w) dw.$$

The function g_ε satisfies $|g_\varepsilon(w)| \leq 1$ for all $w \in \mathbb{R}$ and $g_\varepsilon \to 0$ pointwise as $\varepsilon \to 0+$, and then

$$|\widehat{f}(w) e^{2\pi i x \cdot w} g_\varepsilon(w)| \leq |\widehat{f}(w)| ,$$

for all $w, x \in \mathbb{R}^N$ and $\varepsilon > 0$. Since $\widehat{f} \in L_1(\mathbb{R}^N)$, we can apply the dominated convergence theorem to get

$$\lim_{\varepsilon \to 0+} I_\varepsilon = \int_{\mathbb{R}^N} \widehat{f}(w) e^{2\pi i x \cdot w} dw$$

for every $x \in \mathbb{R}^N$. On the other hand, by Proposition 6.11–2 and 5,

$$I_\varepsilon = \int_{\mathbb{R}^N} \widehat{f}(w) e^{2\pi i x \cdot w} g_\varepsilon(w) dw = \int_{\mathbb{R}^N} f(x+w) \widehat{g_\varepsilon}(w) dw . \tag{6.15}$$

From Propositions 6.11–6 and Lemma 6.5, we have

$$\widehat{g_\varepsilon}(w) = \int_{\mathbb{R}^N} g_\varepsilon(x) e^{-2\pi i x \cdot w} dx = \frac{1}{\varepsilon^N} \widehat{g}\big(\tfrac{w}{\varepsilon}\big) = \frac{1}{\varepsilon^N} g\big(\tfrac{w}{\varepsilon}\big) . \tag{6.16}$$

Joining (6.15) and (6.16), we obtain

$$I_\varepsilon = \frac{1}{\varepsilon^N} \int_{\mathbb{R}^N} f(x+w) g\big(\tfrac{w}{\varepsilon}\big) dw = \int_{\mathbb{R}^N} f(x+\varepsilon t) g(t) dt .$$

We let $\varepsilon \to 0^+$ and get that $f(x+\varepsilon t)g(t)$ converges pointwise to $f(x)g(t)$, and

$$|f(x+\varepsilon t)g(t)| \leq \sup_{z\in\mathbb{R}^N} |f(z)||g(t)|,$$

as f is continuous and bounded in $\mathbb{R}^N$. We apply again the dominated convergence theorem, which is possible since $g \in L_1(\mathbb{R}^N)$, to get

$$\lim_{\varepsilon\to 0+} I_\varepsilon = f(x)\int_{\mathbb{R}^N} g(t)dt = f(x)\int_{\mathbb{R}^N} e^{-\pi|t|^2}dt = f(x).$$

Hence,

$$f(x) = \int_{\mathbb{R}^N} \widehat{f}(w)e^{2\pi i x\cdot w}dw.$$

□

Corollary 6.1 *Let $f, g \in L_1(\mathbb{R}^N)$ be continuous and bounded. If $\widehat{f} = \widehat{g}$, then $f = g$.*

Proof Define $h(x) = f(x) - g(x)$ for $x \in \mathbb{R}^N$. Taking the Fourier transform, we have $\widehat{h} = \widehat{f-g} = \widehat{f} - \widehat{g} = 0$, which clearly belongs to $L_1(\mathbb{R}^N)$. Then (6.14) gives $h(x) = 0$ for every $x \in \mathbb{R}^N$, which gives the conclusion. □

Example 6.19 Take $f \in L_1(\mathbb{R})$ and consider the differential equation on $\mathbb{R}$ given by

$$y'' - y = f(x).$$

If we apply Fourier transform to the equation and use Proposition 6.12, it becomes

$$\widehat{f} = \mathcal{F}(y''-y) = (2\pi i)^2\widehat{y}(w) - \widehat{y}(w) = -(4\pi^2w^2+1)\widehat{y}(w),$$

where $\widehat{f}$ is continuous and bounded. Taking Example 6.17 and Proposition 6.11–3 into account, we have

$$\widehat{y}(w) = -\frac{1}{4\pi^2w^2+1}\widehat{f}(w) = \mathcal{F}\big(-\tfrac{1}{2}e^{-|x|}\big)(w)\mathcal{F}(f)(w) = \mathcal{F}\big(-\tfrac{1}{2}e^{-|x|} * f\big)(w).$$

Corollary 6.1 yields

$$y(x) = -\frac{1}{2}\int_{\mathbb{R}} e^{-|x-y|}f(y)dy.$$

□

The action $f \rightsquigarrow \widehat{f} = \mathcal{F}(f)$ defines an operator (say $\mathcal{F}$) on $L_1(\mathbb{R}^N)$. But taking $L_1(\mathbb{R}^N)$ as the domain has certain flaws. First of all, as we have seen in Example 6.16, the Fourier transform of an L_1-function may not belong to $L_1(\mathbb{R}^N)$ (that is, we cannot define the operator $\mathcal{F} : L_1(\mathbb{R}^N) \to L_1(\mathbb{R}^N)$). Also, we need to be in a setting where we are able to use the inversion formula to recover a function from its Fourier transform. We need, then, to find a subspace of $L_1(\mathbb{R}^N)$ that is closed under the Fourier transform, and where we can safely play this game of going forth and back. The Schwartz space is the one. We recall from the notation of Proposition 6.10 that, given any $f \in L_1(\mathbb{R}^N)$, we denote $\check{f}(x) = f(-x)$ for $x \in \mathbb{R}^N$.

Theorem 6.20 *The Fourier transform $\mathcal{F} : S(\mathbb{R}^N) \to S(\mathbb{R}^N)$ is a topological isomorphism. Also, if $f \in S(\mathbb{R}^N)$, then*

$$f(x) = \int_{\mathbb{R}^N} \widehat{f}(w) e^{2\pi i x \cdot w} dw ,$$

for every $x \in \mathbb{R}^N$. In particular, $f = \mathcal{F}\big(\check{\widehat{f}}\big)$ for every $f \in S(\mathbb{R}^N)$.

Proof Let us see in first place that it is well defined and continuous. Take $f \in S(\mathbb{R}^N)$, and recall that, by Proposition 6.9, the function $\partial^\beta\big(x^\alpha f(x)\big)$ belongs to $S(\mathbb{R}^N)$ for all $\alpha, \beta \in \mathbb{N}_0^N$. Since $S(\mathbb{R}^N) \subseteq L_1(\mathbb{R}^N)$ (recall Proposition 6.8), we can apply Proposition 6.12 to get

$$\begin{aligned} w^\beta \partial^\alpha \widehat{f}(w) &= w^\beta (-2\pi i)^{|\alpha|} \mathcal{F}\big(x^\alpha f(x)\big)(w) \\ &= (-2\pi i)^{|\alpha|} w^\beta \mathcal{F}\big(x^\alpha f(x)\big)(w) = \frac{(-2\pi i)^{|\alpha|}}{(2\pi i)^{|\beta|}} \mathcal{F}\big(\partial^\beta (x^\alpha f(x))\big)(w) . \end{aligned}$$

Once we have this, (6.12) gives

$$\sup_{w \in \mathbb{R}^N} |w^\beta \partial^\alpha \widehat{f}(w)| \le \frac{(-2\pi i)^{|\alpha|}}{(2\pi i)^{|\beta|}} \int_{\mathbb{R}^N} |\partial^\beta (x^\alpha f(x))| dx < \infty .$$

Since this holds for every α and β, we have $\widehat{f} \in S(\mathbb{R}^N)$ (recall Remark 6.4). This shows that $\mathcal{F} : S(\mathbb{R}^N) \to S(\mathbb{R}^N)$ is well defined, and (6.11) gives that it is continuous.

We already saw in Corollary 6.1 that $\mathcal{F}$ is injective. In order to see that it is surjective, take some $f \in S(\mathbb{R}^N)$. Since $\widehat{f} \in S(\mathbb{R}^N) \subseteq L_1(\mathbb{R}^N)$, we can use the inversion formula in Theorem 6.18 and a change of variable $w = -t$ to get

$$f(x) = \int_{\mathbb{R}^N} \widehat{f}(w) e^{2\pi i x \cdot w} dw = \int_{\mathbb{R}^N} \widehat{f}(-t) e^{-2\pi i x \cdot t} dt$$

for every $x \in \mathbb{R}^N$. Hence,

$$f = \mathcal{F}\big(\check{\widehat{f}}\big), \tag{6.17}$$

which, since $\check{\check{f}} \in S(\mathbb{R}^N)$, shows that $\mathcal{F}$ is surjective. Also, (6.17) immediately gives that $\mathcal{F}^{-1}$ is continuous and completes the proof. □

Corollary 6.2 *For each $f \in S(\mathbb{R}^N)$, we have $\widehat{\widehat{f}} = \check{f}$.*

Proof We use the Fourier inversion formula to get

$$\check{f}(x) = f(-x) = \int_{\mathbb{R}^N} \widehat{f}(w) e^{2\pi i (-x)\cdot w} dw = \int_{\mathbb{R}^N} \widehat{f}(w) e^{-2\pi i x\cdot w} dw = \widehat{\widehat{f}}(x)\,.$$

□

We know from Proposition 6.11 that the Fourier transform of an L_1-function is continuous and bounded. We can now say more.

Corollary 6.3 (Riemann–Lebesgue lemma) *If $f \in L_1(\mathbb{R}^N)$, then $\widehat{f}$ is a continuous function vanishing at infinity, that is,*

$$\lim_{|x|\to\infty} \widehat{f}(x) = 0\,.$$

Proof Fix $f \in L_1(\mathbb{R}^N)$, and for each $n \in \mathbb{N}$, consider the function $\rho_{1/n}$ defined in (6.1). Each one of these belongs to $C^\infty(\mathbb{R}^N)$ (see Example 6.1) and, then, by Proposition 6.1, we have $\rho_{1/n} * f \in C^\infty(\mathbb{R}^N)$ for every n, and this sequence converges to f in $L_1(\mathbb{R}^N)$ (see [2, Theorem 6.2] for a proof of this fact). This shows that $C^\infty(\mathbb{R}^N)$ is dense in $L_1(\mathbb{R}^N)$, and this implies (see Proposition 6.3) that $\mathcal{D}(\mathbb{R}^N)$ is dense in $L_1(\mathbb{R}^N)$. So, we can find a sequence $(\varphi_n)_n \subseteq \mathcal{D}(\mathbb{R}^N)$ such that $\varphi_n \to f$ in $L_1(\mathbb{R}^N)$. Now, Proposition 6.11–1 shows that the Fourier transform defines a continuous mapping from $L_1(\mathbb{R}^N)$ into the space of bounded continuous functions. This implies that $\widehat{\varphi_n} \to \widehat{f}$ uniformly on $\mathbb{R}^N$. By Theorem 6.20, each $\widehat{\varphi_n} \in S(\mathbb{R}^N)$ and this (recall (6.10)) gives the conclusion. □

Proposition 6.13 *If $f, g \in S(\mathbb{R}^N)$, then the convolution $f * g \in S(\mathbb{R}^N)$.*

Proof By Proposition 6.11–3, as $f, g \in L_1(\mathbb{R}^N)$, we have $\widehat{f * g} = \widehat{f}\,\widehat{g}$, and, by Theorem 6.20, $\widehat{f}, \widehat{g} \in S(\mathbb{R}^N)$. Leibniz formula easily implies that $\widehat{f}\,\widehat{g} \in S(\mathbb{R}^N)$. An application of Theorem 6.20 again shows that $f * g \in S(\mathbb{R}^N)$. □

6.4 Tempered Distributions and the Fourier Transform

We look now at the dual of the space $S(\mathbb{R}^N)$, denoted $S'(\mathbb{R}^N)$. This consists, then, of the continuous linear maps $T : S(\mathbb{R}^N) \to \mathbb{K}$, which are called *tempered distributions*. Note that this means that there are $C > 0$ and $m \in \mathbb{N}$ such that

$$|T(\varphi)| \leq C \sup_{|\alpha|\leq m} \sup_{x\in\mathbb{R}^N} (1 + |x|^2)^m |\partial^\alpha \varphi(x)|\,,$$

for all $\varphi \in S(\mathbb{R}^N)$.

Remark 6.5 The first question that we have to address at this point is why do we call them 'distributions'. The reason is easy: if $T \in S'(\mathbb{R}^N)$, then its restriction $T|_{\mathscr{D}(\mathbb{R}^N)}$ defines a distribution, since for each compact set $K \subseteq \mathbb{R}^N$, the inclusion $\mathscr{D}(K) \hookrightarrow S(\mathbb{R}^N)$ is continuous (recall Proposition 6.7). Even more, if $T|_{\mathscr{D}(\mathbb{R}^N)}$ vanishes, the density of $\mathscr{D}(\mathbb{R}^N)$ in $S(\mathbb{R}^N)$ (see again Proposition 6.8) implies that $T(\varphi) = 0$ for each $\varphi \in S(\mathbb{R}^N)$. Hence, $T \in S'(\mathbb{R}^N)$ is uniquely determined by its restriction to $\mathscr{D}(\mathbb{R}^N)$ or, in other words, that each $T \in S'(\mathbb{R}^N)$ defines, by restricting it to $\mathscr{D}(\mathbb{R}^N)$, a unique distribution in $\mathscr{D}'(\mathbb{R}^N)$. Summarizing, the restriction map $J : S'(\mathbb{R}^N) \to \mathscr{D}'(\mathbb{R}^N)$ given by $J(T) = T|_{\mathscr{D}(\mathbb{R}^N)}$, is well defined and injective.

Example 6.21 Not every distribution is a tempered distribution. Consider the C^∞-function given by $f(x) = e^x$ for $x \in \mathbb{R}$, and its associated distribution, defined as

$$T_f(\varphi) = \int_{\mathbb{R}} e^x \varphi(x)dx\,,$$

for $\varphi \in \mathscr{D}(\mathbb{R})$. Choose some $\phi \in C^\infty(\mathbb{R})$ such that $\phi(x) = 1$ for $x \geq 0$, and $\phi(x) = 0$ for $x \leq -1$. Consider now the function $\psi \in C^\infty(\mathbb{R})$ given by $\psi(x) = \phi(x)e^{-x}$, which clearly satisfies

$$\lim_{x \to +\infty} x^\beta |\partial^\alpha \psi(x)| = 0\,,$$

for all $\beta, \alpha \in \mathbb{N}_0$. Hence, $\psi \in S(\mathbb{R})$. If $T_f \in S'(\mathbb{R})$, then

$$T_f(\psi) = \int_{\mathbb{R}} e^x \psi(x)dx \in \mathbb{K}.$$

But $e^x \psi(x) = 1$ for all $x > 0$ and so $e^x \psi(x) \notin L_1(\mathbb{R})$. □

Proposition 6.14

1. *$L_p(\mathbb{R}^N) \subseteq S'(\mathbb{R}^N)$ for every $1 \leq p \leq \infty$. In particular, $S(\mathbb{R}^N) \subseteq S'(\mathbb{R}^N)$.*
2. *If $f \in L_{1,\text{loc}}(\mathbb{R}^N)$ satisfies $|f(x)| \leq C(1 + |x|^2)^{n/2}$ for some $n \in \mathbb{N}_0$ and $C > 0$, we have $T_f \in S'(\mathbb{R}^N)$. In particular, every polynomial defines a tempered distribution.*

Proof If $f \in L_p(\mathbb{R}^N)$, then it belongs to $L_{1,\text{loc}}(\mathbb{R}^N)$, and we have by the definition of the associated distribution (6.4) and Hölder's inequality

$$|T_f(\varphi)| \leq \int |f(x)\varphi(x)|dx \leq \|f\|_p \|\varphi\|_{p'}\,.$$

Then, to show that T_f is a tempered distribution, we just have to bound $\|\varphi\|_{p'}$ by the action on φ of some convenient element of the fundamental system of seminorms from (6.9). We distinguish two cases. First of all, if $p' = \infty$, then it is obvious since

$$\|\varphi\|_\infty = \sup_{x\in\mathbb{R}^N} |\varphi(x)| = p_0(\varphi) .$$

Now, if $1 \le p' < \infty$, we choose $m > N/p'$ and have

$$\|\varphi\|_{p'} = \Big(\int |\varphi(x)|^{p'} dx\Big)^{\frac{1}{p'}} = \Big(\int \frac{(1+|x|^2)^{\frac{m}{2}}}{(1+|x|^2)^{\frac{m}{2}}} |\varphi(x)|^{p'} dx\Big)^{\frac{1}{p'}}$$

$$\le p_m(\varphi)\Big(\int \frac{1}{(1+|x|^2)^{\frac{mp'}{2}}} dx\Big)^{\frac{1}{p'}} .$$

The last integral converges because of the choice of m. This completes the proof of the first statement.

The argument for the second one is rather similar. Take some $\varphi \in S(\mathbb{R}^N)$ and estimate

$$|T_f(\varphi)| \le \int |f(x)\varphi(x)|dx \le C \int \frac{(1+|x|^2)^{\frac{N+1}{2}}}{(1+|x|^2)^{\frac{N+1}{2}}} (1+|x|^2)^{\frac{n}{2}} |\varphi(x)|dx$$

$$\le C\Big(\int \frac{1}{(1+|x|^2)^{\frac{N+1}{2}}} dx\Big) \sup_{x\in\mathbb{R}^N} (1+|x|^2)^{\frac{N+1+n}{2}} |\varphi(x)| \le C_1 p_{N+1+n}(\varphi) .$$

□

As we have just seen, every polynomial defines a tempered distribution. Nevertheless, there are tempered distributions defined by functions other than polynomials, as our next example shows.

Example 6.22 Consider the function $f(x) = e^x \cos e^x$ for $x \in \mathbb{R}$. If $\varphi \in S(\mathbb{R})$, then we have

$$T_f(\varphi) = \int_{\mathbb{R}} e^x \cos(e^x)\varphi(x)dx = -\int_{\mathbb{R}} \sin(e^x)\varphi'(x)dx ,$$

and

$$|T_f(\varphi)| \le \int_{\mathbb{R}} |\sin(e^x)||\varphi'(x)|dx \le C \sup_{x\in\mathbb{R}}(1+x^2)|\varphi'(x)| ,$$

which shows that $T_f \in S'(\mathbb{R}^N)$.

However, there is no polynomial $P(x)$ with $|f(x)| \le |P(x)|$ for every $x \in \mathbb{R}$. If this were the case, we would have

$$|\cos(e^x)| \le \frac{|P(x)|}{e^x},$$

for every $x \in \mathbb{R}$, which implies

$$\lim_{x\to+\infty} \cos(e^x) = 0.$$

But this is not true, since $\cos(e^{\log 2k\pi}) = 1$ for each $k \in \mathbb{N}$. □

Proposition 6.15 *Every distribution with compact support is tempered. That is,* $\mathscr{E}'(\mathbb{R}^N) \subseteq S'(\mathbb{R}^N)$.

Proof If $T \in \mathscr{E}'(\mathbb{R}^N)$, then by Theorem 6.14 there is a unique continuous linear map $S : C^\infty(\mathbb{R}^N) \to \mathbb{K}$ such that $T = S|_{\mathscr{D}(\mathbb{R}^N)}$. Now, by Theorem 6.7, the inclusion $S(\mathbb{R}^N) \subseteq C^\infty(\mathbb{R}^N)$ is continuous and has dense range. Then the restriction $S|_{S(\mathbb{R}^N)}$ is a uniquely determined distribution that coincides with T on $\mathscr{D}(\mathbb{R}^N)$. □

We have already seen how some of the basic operations on functions (such as differentiation or multiplication by a C^∞-function) can be extended also to distributions. Our aim now is to extend the definition of the Fourier transform to tempered distributions. As in previous occasions, we look first at distributions defined by functions, in order to get some ideas. If $f \in L_1(\mathbb{R}^N)$ is such that $\widehat{f} \in L_1(\mathbb{R}^N)$, we would expect that the Fourier transform of T_f coincides with $T_{\widehat{f}}$. If $\varphi \in S(\mathbb{R}^N)$, we can use Propositions 6.11–2 to get

$$T_{\widehat{f}}(\varphi) = \int_{\mathbb{R}^N} \widehat{f}(x)\varphi(x)dx = \int_{\mathbb{R}^N} f(x)\widehat{\varphi}(x)dx = T_f(\widehat{\varphi}).$$

This suggests to define the Fourier transform of T_f as $\widehat{T_f}(\varphi) = T_f(\widehat{\varphi})$ for $\varphi \in S(\mathbb{R}^N)$.

With this idea, we define the *Fourier transform* of a tempered distribution $u \in S'(\mathbb{R}^N)$ as the mapping $\widehat{u} : S(\mathbb{R}^N) \to \mathbb{K}$ defined as

$$\widehat{u}(\varphi) = u(\widehat{\varphi}).$$

Let us note that $\widehat{u} = u \circ \mathcal{F}$ for every $u \in S'(\mathbb{R}^N)$, where $\mathcal{F} : S(\mathbb{R}^N) \to S(\mathbb{R}^N)$. Since we know that $\mathcal{F}$ is continuous (see Theorem 6.20), we immediately have $\widehat{u} \in S'(\mathbb{R}^N)$ for every $u \in S'(\mathbb{R}^N)$.

Example 6.23 The Dirac function δ is a tempered distribution, since $\delta \in \mathscr{E}'(\mathbb{R}^N)$. Let us compute its Fourier transform. For $\varphi \in S(\mathbb{R}^N)$, we have

$$\widehat{\delta}(\varphi) = \delta(\widehat{\varphi}) = \widehat{\varphi}(0) = \int e^{-2\pi i(0\cdot w)}\varphi(w)dw = \int \varphi(w)dw = \mathbb{1}(\varphi),$$

which shows that $\widehat{\delta}$ is the constant function $\mathbb{1}$.
With the same idea, for every $a \in \mathbb{R}^N$, we have $\delta_a \in S'(\mathbb{R}^N)$, and

$$\widehat{\delta_a}(\varphi) = \delta_a(\widehat{\varphi}) = \widehat{\varphi}(a) = \int_{\mathbb{R}^N} \varphi(x)e^{-2\pi i(a\cdot x)}dx,$$

for every $\varphi \in S(\mathbb{R}^N)$. So, $\widehat{\delta_a} = e^{-2\pi i(a\cdot x)}$ in $S'(\mathbb{R}^N)$. □

Proposition 6.16 *The Fourier transform* $\mathcal{F}: S'(\mathbb{R}^N) \to S'(\mathbb{R}^N)$ *given by* $\mathcal{F}(u) = \widehat{u}$ *is a linear isomorphism.*

Proof It is easy to see that it is linear. Let us see in first place that it is injective. If $\mathcal{F}(u) = 0$, then $u(\widehat{\varphi}) = 0$ for every $\varphi \in S(\mathbb{R}^N)$. Taking into account that $\mathcal{F}: S(\mathbb{R}^N) \to S(\mathbb{R}^N)$ is surjective (see Theorem 6.20), this implies that $u(\psi) = 0$ for all $\psi \in S(\mathbb{R}^N)$ and, then, $u = 0$. This shows the injectivity.
To see that $\mathcal{F}$ is surjective, fix some $v \in S'(\mathbb{R}^N)$ and define $u : S(\mathbb{R}^N) \to \mathbb{K}$ by $u(\varphi) = v(\mathcal{F}^{-1}\varphi)$, that is, $u = v \circ \mathcal{F}^{-1}$, which (again by Theorem 6.20) belongs to $S'(\mathbb{R}^N)$. Then

$$\widehat{u}(\varphi) = u(\widehat{\varphi}) = v(\mathcal{F}^{-1}\mathcal{F}\varphi) = v(\varphi),$$

for all $\varphi \in S(\mathbb{R}^N)$. This shows that $\mathcal{F} : S'(\mathbb{R}^N) \to S'(\mathbb{R}^N)$ is surjective and completes the proof. □

Proposition 6.17 *Let* $u \in S'(\mathbb{R}^N)$*, and then:*

1. $\widehat{\partial^\alpha u} = (2\pi i)^{|\alpha|} w^\alpha \widehat{u}$.
2. $\partial^\alpha \widehat{u} = (-2\pi i)^{|\alpha|}\mathcal{F}\big(x^\alpha u\big)$.

Proof This follows directly from the definition of the Fourier transform for tempered distributions and Proposition 6.12. First, we have

$$\begin{aligned}\widehat{\partial^\alpha u}(\varphi) &= \partial^\alpha u(\widehat{\varphi}) = u\big((-1)^{|\alpha|}\partial^\alpha \widehat{\varphi}\big) = u\Big((-1)^{|\alpha|}(-2\pi i)^{|\alpha|}\mathcal{F}(x^\alpha \varphi)\Big) \\ &= u\big(\mathcal{F}((2\pi i)^{|\alpha|}x^\alpha \varphi)\big) = \widehat{u}((2\pi i)^{|\alpha|}x^\alpha \varphi) = (2\pi i)^{|\alpha|}(x^\alpha \widehat{u})(\varphi).\end{aligned}$$

The second formula follows from the following computation:

$$\widehat{x^\alpha u}(\varphi) = \big(x^\alpha u\big)(\widehat{\varphi}) = u(x^\alpha \widehat{\varphi}) = u\Big(\tfrac{1}{(2\pi i)^{|\alpha|}}\widehat{\partial^\alpha \varphi}\Big)$$
$$= \frac{1}{(2\pi i)^{|\alpha|}} u\big(\widehat{\partial^\alpha \varphi}\big) = \frac{1}{(2\pi i)^{|\alpha|}} \widehat{u}(\partial^\alpha \varphi) = \frac{1}{(-2\pi i)^{|\alpha|}} \partial^\alpha \widehat{u}(\varphi)\,.$$

□

Our aim is to use the Fourier transform on partial differential operators (recall its definition in Example 4.19) in order to solve partial differential equations. As a consequence of Proposition 6.17, we see how the Fourier transform acts on such an operator:

Corollary 6.4

1. $\widehat{\partial^\alpha \delta} = (2\pi i)^{|\alpha|} w^\alpha$.
2. *If* $P(z) = \sum_{|\alpha|\le m} a_\alpha z^\alpha$ *and* $P(\partial) = \sum_{|\alpha|\le m} a_\alpha \partial^\alpha$, *we have*

$$\mathcal{F}\big(P(\partial)u\big) = \sum_{|\alpha| lem} a_\alpha (2\pi i)^{|\alpha|} w^\alpha \widehat{u} = P(2\pi i w)\widehat{u}\,, \tag{6.18}$$

for all $u \in S'(\mathbb{R}^N)$.

We give now some consequences of the use for the inverse Fourier transform. We fix some notation. If $u \in \mathscr{D}'(\mathbb{R}^N)$, then we write $\check{u}$ for the distribution defined as $\check{u}(\varphi) = u(\check{\varphi})$, for $\varphi \in \mathscr{D}(\mathbb{R}^N)$.

Proposition 6.18 *If* $u \in S'(\mathbb{R}^N)$, *then* $\widehat{\widehat{u}} = \check{u}$ *and* $\check{\widehat{\widehat{u}}} = u$.

Proof Both formulas follow immediately from the definitions and Corollary 6.2. If $\varphi \in S(\mathbb{R}^N)$, then we have

$$\widehat{\widehat{u}}(\varphi) = u(\widehat{\widehat{\varphi}}) = u(\check{\varphi}) = \check{u}(\varphi)\,, \text{ and } \check{\widehat{\widehat{u}}}(\varphi) = \widehat{\widehat{u}}(\check{\varphi}) = u(\check{\widehat{\widehat{\varphi}}}) = u(\varphi)\,,$$

which gives the claim. □

Example 6.24 We have $\widehat{\mathbb{1}} = \widehat{\widehat{\delta}} = \check{\delta}$. But, for $\varphi \in S(\mathbb{R}^N)$, we have $\check{\delta}(\varphi) = \delta(\check{\varphi}) = \check{\varphi}(0) = \varphi(0)$, and hence, $\check{\delta} = \delta$. This implies $\widehat{\mathbb{1}} = \delta$. □

An important result in the classical theory of functions of one variable is that a function is constant if and only if its derivative is 0. Our aim now is to see that this is also the case when we deal with distributions. We need first a result.

Lemma 6.6 *Let* $\varphi \in S(\mathbb{R})$. *Then there is* $\phi \in S(\mathbb{R})$ *such that* $\phi' = \varphi$ *if and only if* $\int_{\mathbb{R}} \varphi(x)dx = 0$.

Proof Suppose first that $\phi \in S(\mathbb{R})$ is such that $\phi' = \varphi$. Then Barrow rule gives

$$\int_{\mathbb{R}} \varphi(x)dx = \lim_{A\to+\infty} \int_{-A}^{A} \varphi(x)dx = \lim_{A\to+\infty} [\phi(A) - \phi(-A)] = 0\,.$$

Assume conversely that $\int_{\mathbb{R}} \varphi(x)dx = 0$ and define

$$\phi(x) = \int_{-\infty}^{x} \varphi(y)dy = -\int_{x}^{+\infty} \varphi(y)dy\,.$$

By the Fundamental Theorem of Calculus, we have $\phi' = \varphi$, which implies that $\phi \in C^\infty(\mathbb{R})$ and $\phi^{(k)} = \varphi^{(k-1)}$ for every $k \in \mathbb{N}$. To complete the proof, we have to see that $\phi \in S(\mathbb{R})$. First of all, since $\varphi \in S(\mathbb{R})$, we have

$$\sup_{1\le k\le m} \sup_{x\in\mathbb{R}} (1+|x|^2)^{\frac{m}{2}} |\phi^{(k)}(x)| = \sup_{0\le k\le m-1} \sup_{x\in\mathbb{R}} (1+|x|^2)^{\frac{m}{2}} |\varphi^{(k)}(x)| < \infty\,,$$

for every m. It is only left to see what happens with $k = 0$. Take

$$C = \sup_{x\in\mathbb{R}} (1+|x|^2)^{\frac{m+2}{2}} |\varphi(x)| < \infty\,.$$

Then, for $x \ge 0$, we have

$$\begin{aligned}(1+|x|^2)^{\frac{m}{2}} |\phi(x)| &\le (1+|x|^2)^{\frac{m}{2}} \int_{-\infty}^{x} |\varphi(y)|dy \\ &\le (1+|x|^2)^{\frac{m}{2}} \int_{x}^{\infty} \frac{1}{(1+|y|^2)^{\frac{m+2}{2}}} C dy \le C \int_{x}^{\infty} \frac{1}{(1+|y|^2)^{2}} dy = C_1\,.\end{aligned}$$

For $x < 0$, we proceed similarly with $\phi(x) = \int_{-\infty}^{x} \varphi(y)dy$, and this completes the proof. □

Proposition 6.19 *Let $u \in S'(\mathbb{R})$. Then $u' = 0$ if and only if u is constant.*

Proof Let us suppose first that u is constant. This means that u is the distribution defined by a constant function and then u' (as a distribution) coincides with u' (as a function). Therefore, $u' = 0$. Let us conversely assume that $u \in S'(\mathbb{R})$ is such that $u' = 0$. Our aim is to see that $u = T_c$ for some c, that is, $u(\varphi) = \int_{\mathbb{R}} c\varphi(x)dx$ for every φ. We begin by fixing $\psi_0 \in \mathscr{D}(\mathbb{R}) \subseteq S(\mathbb{R})$ with $\int_{\mathbb{R}} \psi_0(x)dx = 1$. Now, given $\varphi \in S(\mathbb{R}) \subseteq L_1(\mathbb{R})$, we define

$$A_\varphi = \int_{\mathbb{R}} \varphi(x)dx$$

and consider the function $\varphi - A_\varphi \psi_0 \in S(\mathbb{R})$. Note that

$$\int_{\mathbb{R}} (\varphi - A_\varphi \psi_0)(x)dx = 0\,,$$

which, by Lemma 6.6, implies that there is $\phi \in S(\mathbb{R})$ so that $\phi' = \varphi - A_\varphi \psi_0$. We have $\varphi = \phi' + A_\varphi \psi_0$ and (recall that $u' = 0$)

$$\begin{aligned} u(\varphi) = u(\phi') + A_\varphi u(\psi_0) &= -u'(\phi) + A_\varphi u(\psi_0) \\ &= A_\varphi u(\psi_0) = u(\psi_0) \int_{\mathbb{R}} \varphi(x)dx = \int_{\mathbb{R}} u(\psi_0)\varphi(x)dx\,. \end{aligned}$$

Denoting $c = u(\psi_0)$ gives the claim and completes the proof. □

Example 6.25 We calculate the Fourier transform of Cauchy's principal value of $\frac{1}{x}$ and the Heaviside function that, as we know, are closely related. For the sake of clarity, let us denote $u = PV\frac{1}{x}$. We know from Example 6.9 that $xu = \mathbb{1}$ in the sense of (tempered) distributions. Then Example 6.24 and (6.6) give

$$\widehat{xu} = \widehat{\mathbb{1}} = \delta = H'\,.$$

On the other hand, Propositions 6.17–2 give

$$\widehat{xu} = -\frac{1}{2\pi i}\widehat{u}'\,.$$

Therefore, $\widehat{u}' = -2\pi i H'$, and $(\widehat{u} + 2\pi i H)' = 0$. By Proposition 6.19, there is a constant C with $\widehat{u} = C - 2\pi i H$, that is,

$$\widehat{u}(\varphi) = C \int_{\mathbb{R}} \varphi(x)dx - 2\pi i \int_0^\infty \varphi(x)dx\,, \tag{6.19}$$

for every $\varphi \in S(\mathbb{R})$.
A straightforward computation from the definition (recall Example 6.6) shows that $u(\check{\varphi}) = -u(\varphi)$ for every $\varphi \in S(\mathbb{R})$. Then, using Propositions 6.11–4, we have

$$\widehat{u}(\check{\varphi}) = u(\widehat{\check{\varphi}}) = u(\check{\widehat{\varphi}}) = -u(\widehat{\varphi}) = -\widehat{u}(\varphi)\,. \tag{6.20}$$

Note that, with a change of variables, we have

$$\widehat{u}(\check{\varphi}) = C \int_{\mathbb{R}} \varphi(-x)dx - 2\pi i \int_0^\infty \varphi(-x)dx = C \int_{\mathbb{R}} \varphi(t)dt - 2\pi i \int_{-\infty}^0 \varphi(t)dt\,.$$

This, combined with (6.20) and (6.19), gives $C = \pi i$ and, then,

$$\widehat{PV\tfrac{1}{x}} = \pi i(\mathbb{1} - 2H)\,.$$

We obtain from here the Fourier transform of H. First of all, using that $\widehat{\widehat{u}} = \check{u}$ (recall Proposition 6.18), we have

$$\pi i \mathcal{F}(\mathbb{1} - 2H) = \big(PV\tfrac{1}{w}\big)^{\vee} = -\big(PV\tfrac{1}{w}\big)\,.$$

On the other hand, Example 6.24 gives

$$\pi i \mathcal{F}(\mathbb{1} - 2H) = \pi i\big(\widehat{\mathbb{1}} - 2\widehat{H}\big) = \pi i\big(\delta - 2\widehat{H}\big)\,.$$

Consequently, $2\pi i \widehat{H} = \pi i \delta + PV\frac{1}{w}$, which yields

$$\widehat{H} = \delta - \frac{1}{\pi i} PV\tfrac{1}{w}\,.$$

□

As we announced above, our aim is to use the Fourier transform to solve partial differential equations where the unknown is a tempered distribution (see Examples 6.27 and 6.28), in a similar way as we did in Example 6.19 in a more 'classical' setting. As we saw there, convolution is a fundamental tool. So, we need to extend the definition to have the convolution of a distribution and a function. As a first step (as many times so far), we check how this should work for distributions defined by functions. So, given $f \in L_1(\mathbb{R}^N)$ and $\varphi, \psi \in S(\mathbb{R}^N)$, we have by Fubini's theorem

$$\begin{aligned} T_{f*\psi}(\varphi) &= \int_{\mathbb{R}^N} (f * \psi)(x)\varphi(x)dx = \int_{\mathbb{R}^N}\int_{\mathbb{R}^N} f(t)\psi(x-t)dt\,\varphi(x)dx \\ &= \int_{\mathbb{R}^N}\int_{\mathbb{R}^N} \psi(x-t)\varphi(x)dx\, f(t)dt = \int_{\mathbb{R}^N}\int_{\mathbb{R}^N} \check{\psi}(t-x)\varphi(x)dx\, f(t)dt \\ &= \int_{\mathbb{R}^N} (\check{\psi} * \varphi)(t) f(t)dt = T_f(\check{\psi} * \varphi)\,. \end{aligned}$$

With this idea, the *convolution* of a tempered distribution $u \in S'(\mathbb{R}^N)$ and a function $\psi \in S(\mathbb{R}^N)$ is defined as

$$\big(u * \psi\big)(\varphi) = u\big(\check{\psi} * \varphi\big)\,,$$

for $\varphi \in S(\mathbb{R}^N)$. Observe that, by Proposition 6.13, we have $\check{\psi} * \varphi \in S(\mathbb{R}^N)$. Moreover, it can be shown that $u * \psi \in S'(\mathbb{R}^N)$ and that it satisfies

$$\partial^\alpha (u * \varphi) = \partial^\alpha u * \varphi = u * \partial^\alpha \varphi$$
$$u * (\varphi * \psi) = (u * \varphi) * \psi$$
$$\widehat{u * \varphi} = \widehat{u}\,\widehat{\varphi}\,.$$

Example 6.26 If $a \in \mathbb{R}$ and $\phi, \varphi \in S(\mathbb{R})$, we have

$$\begin{aligned}(\delta_a * \phi)(\varphi) = \delta_a(\check{\phi} * \varphi) = \big(\check{\phi} * \varphi\big)(a) &= \int_{\mathbb{R}^N} \check{\phi}(u)\varphi(a-u)du \\ &= \int_{\mathbb{R}^N} \phi(-u)\varphi(a-u)du = \int_{\mathbb{R}^N} \phi(t-a)\varphi(t)dt = T_{T_a\phi}(\varphi)\end{aligned}$$

(recall the notation from Proposition 6.10). That is, $\delta_a * \phi = \phi(x-a)$, as distributions. □

The reader is referred to [13, Chapter 7] for a more detailed study of the convolution of tempered distributions and functions in $S(\mathbb{R}^N)$, in particular to [13, Theorem 7.19], where the proof of the aforementioned properties can be found.

Example 6.27 (Laplacian) We consider the Laplacian, defined for $u \in S'(\mathbb{R}^N)$ as

$$\Delta u = \frac{\partial^2 u}{\partial x_1^2} + \cdots + \frac{\partial^2 u}{\partial x_n^2}\,,$$

and we want to solve the equation

$$(1-\Delta)u = f\,,$$

where $f \in S'(\mathbb{R}^N)$ (or $S(\mathbb{R}^N)$). Then our operator is $P(\partial) = 1 - \Delta$, whose symbol is $P(x) = 1 - x_1^2 - x_2^2 - \cdots - x_n^2$, for $x \in \mathbb{R}^N$. We apply the Fourier transform to the equation, taking (6.18) into account and noting that $P(2\pi i w) = 1 + 4\pi^2|w|^2$, to have

$$(1 + 4\pi^2|w|^2)\widehat{u} = \widehat{f}\,.$$

This gives

$$\widehat{u} = \frac{\widehat{f}}{1 + 4\pi^2|w|^2}$$

and

$$u = \mathcal{F}^{-1}\Big(\frac{\widehat{f}}{1+4\pi^2|w|^2}\Big) \in S'(\mathbb{R}^N) \text{ or } S(\mathbb{R}^N)\,.$$

Taking, for instance, $N = 1$ and $f = \delta$, from Examples 6.23 and 6.17, we have

$$u = \mathcal{F}^{-1}\Big(\frac{1}{1+4\pi^2|w|^2}\Big) = \frac{1}{2}e^{-|x|}\,.$$

□

Example 6.28 (Wave equation) We consider the problem

$$\left.\begin{array}{l} \frac{\partial^2 u}{\partial t^2} - c^2\Delta u = 0 \\ u(0,x) = \phi(x) \\ \frac{\partial u}{\partial t}(0,x) = \psi(x) \end{array}\right\} \tag{6.21}$$

with $\phi, \psi \in S(\mathbb{R}^N)$ and $c > 0$. We take the Fourier transform with respect to x (which we denote by $\mathcal{F}_x$) and leave t fixed to get

$$\left.\begin{array}{l} \big(\frac{\partial^2}{\partial t^2} + c^2 4\pi^2|w|^2\big)\widehat{u} = 0 \\ \widehat{u}(0,w) = \widehat{\phi}(w) \\ \frac{\partial \widehat{u}}{\partial t}(0,w) = \widehat{\psi}(w) \end{array}\right\}.$$

This is a linear ordinary differential equation, whose solution is

$$\widehat{u}(t,w) = \cos(c2\pi|w|t)\widehat{\phi}(w) + \frac{\sin(c2\pi t|w|)}{2\pi c|w|}\widehat{\psi}(w)\,.$$

Therefore,

$$\begin{aligned} u(t,x) &= \mathcal{F}_w^{-1}\Big(\cos(c2\pi|w|t)\widehat{\phi}(w) + \frac{\sin(c2\pi t|w|)}{2\pi c|w|}\widehat{\psi}(w)\Big) \\ &= \mathcal{F}_w^{-1}\Big(\cos(c2\pi|w|t)\widehat{\phi}(w)\Big) + \mathcal{F}_w^{-1}\Big(\frac{\sin(c2\pi t|w|)}{2\pi c|w|}\widehat{\psi}(w)\Big)\,, \end{aligned}$$

and, by Proposition 6.11–3,

$$u(t,x) = \mathcal{F}_w^{-1}\Big(\cos(c2\pi|w|t)\Big) *_x \phi + \mathcal{F}_w^{-1}\Big(\frac{\sin(c2\pi t|w|)}{2\pi c|w|}\Big) *_x \psi\,. \tag{6.22}$$

From now on, we assume $N = 1$. On the one hand, we have (recall Example 6.23)

$$\mathcal{F}_w^{-1}\big(\cos(2\pi ct|w|)\big) = \mathcal{F}_w^{-1}\big(\cos(2\pi ctw)\big)$$
$$= \frac{1}{2}\big[\mathcal{F}_w^{-1}(e^{2\pi icwt}) + \mathcal{F}_w^{-1}(e^{-2\pi icwt})\big] = \frac{1}{2}(\delta_{-ct} + \delta_{ct}) .$$

On the other hand, Example 6.16 yields

$$\mathcal{F}_w^{-1}\Big(\frac{\sin(2\pi ct|w|)}{2\pi c|w|}\Big) = \mathcal{F}_w^{-1}\Big(\frac{\sin(2\pi ctw)}{2\pi cw}\Big) = \frac{1}{2c}\chi_{[-ct,ct]} .$$

So we have the two inverse Fourier transforms appearing in (6.22), and

$$u(t,x) = \frac{1}{2}(\delta_{-ct} + \delta_{ct}) *_x \phi + \frac{1}{2c}\chi_{[-ct,ct]} *_x \psi .$$

We know the result of the first convolution from Example 6.26. We compute now the second one, having

$$(\chi_{[-ct,ct]} * \psi)(x) = \int_{-ct}^{ct} \psi(x-s)ds = -\int_{x+ct}^{x-ct} \psi(u)du = \int_{x-ct}^{x+ct} \psi(u)du ,$$

for $\psi \in S(\mathbb{R})$. With all this, we finally have d'Alembert's formula for the solution of (6.21)

$$u(t,x) = \frac{1}{2}(\phi(x-ct) + \phi(x+ct)) + \frac{1}{2c}\int_{x-ct}^{x+ct} \psi(y)dy .$$

□

6.5 Linear Partial Differential Operators

In this chapter, we present a few results about linear partial differential operators with constant coefficients (they were introduced in Example 4.19). Let us briefly recall that a polynomial of N variables of degree m is of the form

$$P(z) = \sum_{\substack{\alpha\in\mathbb{N}_0^N \\ |\alpha|\le m}} a_\alpha z^\alpha ,$$

where we assume that there is at least one non-zero a_α with $|\alpha| = m$. The *principal part* of P is the m-homogeneous polynomial

$$P_m(z) = \sum_{|\alpha|=m} a_\alpha z^\alpha .$$

We recall that each polynomial has an associated *partial differential operator (with constant coefficients)*

$$P(\partial) = \sum_{|\alpha|\le m} a_\alpha \partial^\alpha .$$

We already saw (recall again Example 4.19) that, if Ω is some open set in $\mathbb{R}^N$, then the operator $P(\partial) : C^\infty(\Omega) \to C^\infty(\Omega)$ is continuous. Observe that this implies that

$$P(\partial) : \mathscr{D}(K) \to \mathscr{D}(K)$$

is continuous for each compact set $K \subseteq \mathbb{R}^N$.

Our main interest now is to find out when the partial differential equation defined by a given partial differential operator has solution. That is, we wonder now under what circumstances does the partial differential equation $P(\partial)u = v$ have a solution. This can be seen under several points of view. First of all, if v is a function, we may ask to find another function u so that $P(\partial)u = v$. But this is often too restrictive, and the chances to find an answer are small. If we broaden the field where we look for solutions, our chances to find one will certainly increase. Then, we are going to interpret the equations in a *distributional sense*. That is, if v is a distribution, we want to find some distribution u so that $P(\partial)u = v$ **as distributions**.

6.5.1 Fundamental Solutions. The Malgrange–Ehrenpreis Theorem

We say that a distribution $E \in \mathscr{D}'(\mathbb{R}^N)$ is a *fundamental solution* of a partial differential operator $P(\partial)$ if

$$P(\partial)E = \delta.$$

Example 6.29 The Heaviside function $H(x)$ is a fundamental solution of the operator $P(\partial) = \frac{d}{dx}$ in $\mathscr{D}'(\mathbb{R})$ (recall (6.6)). Note that, in this case, the symbol is given by $P(z) = z$. □

We give now fundamental solution for some classical partial differential operators (see Example 4.19). A complete and detailed justification can be found in [17, Chapter 1].

Example 6.30

1. The Cauchy–Riemann operator

$$\overline{\partial} = \frac{1}{2}\Big(\frac{\partial}{\partial x} + i\frac{\partial}{\partial y}\Big) \tag{6.23}$$

(the symbol of which is $P(x, y) = \frac{1}{2}(x + iy)$) has a fundamental solution given by

$$E(z) = \frac{1}{\pi z}.$$

2. The Laplacian $\Delta = \frac{\partial^2}{\partial x_1^2} + \cdots + \frac{\partial^2}{\partial x_N^2}$ has a fundamental solution given by

$$E(x) = \frac{1}{2\pi} \log |x| \text{ if } N = 2,$$

and

$$E(x) = \frac{1}{(N-2)\omega_N} \frac{1}{|x|^{N-2}} \text{ if } N \geq 3,$$

where $\omega_N = \dfrac{2\pi^{N/2}}{\Gamma(N/2)}$ is the volume of the Euclidean unit ball of $\mathbb{R}^N$.

3. The heat operator $P(\partial) = \frac{\partial}{\partial t} - \Delta_x$ has a fundamental solution given by

$$E(x, t) = H(t)\Big(\frac{1}{2\sqrt{\pi t}}\Big)e^{\frac{-|x|^2}{4t}}.$$

4. The wave operator $P(\partial) = \frac{\partial^2}{\partial t^2} - \frac{\partial^2}{\partial x^2}$ has a fundamental solution given by

$$E(x, t) = \frac{1}{2}H(t - x)H(t + x) = \begin{cases} \frac{1}{2} & \text{if } t - |x| > 0, \\ 0 & \text{if } t - |x| < 0. \end{cases}$$

□

Our aim in this section is to show that these operators always have a fundamental solution. This is an important theorem, due to Malgrange [10] and Ehrenpreis [5].

Theorem 6.31 (Malgrange–Ehrenpreis) *Every non-constant linear partial differential operator with constants coefficient has a fundamental solution.*

The original proof uses the Hahn–Banach theorem. Shortly after, Trèves gave in [16] a constructive proof, based on 'Hörmander's staircase' and partitions of unity. Rosay in [11] gave an elementary proof, based on the L_2 theory. We follow here the proof of Wagner [19] that gives an explicit formula for the fundamental solution (see (6.26) later). Before getting there, we need several lemmas.

Lemma 6.7 *Let $P(z)$ be a non-constant polynomial of N variables. Then, the set*

$$V_{\mathbb{R}}(P) = \{x \in \mathbb{R}^N : P(x) = 0\}$$

has null Lebesgue measure in $\mathbb{R}^N$.

Proof If $N = 1$, the result is trivial, since it is a finite set. We proceed by induction, and assume that the result holds for polynomials of N variables. Take now $P(x_1 \ldots, x_{N+1})$ a polynomial of $N+1$ variables, and write

$$P = \prod_{j=1}^{l} Q_j ,$$

where each Q_j is irreducible. Note that

$$V_{\mathbb{R}}(P) = \bigcup_{j=1}^{l} V_{\mathbb{R}}(Q_j) ,$$

so we may assume without loss of generality that P is irreducible.
For each fixed $a \in \mathbb{R}$, we consider the polynomial of N variables given by $P_a(x_1, \ldots, x_N) = P(x_1, \ldots, x_N, a)$. We want to see now that $P_a \neq 0$ for every a. Suppose that this is not the case, and there is some a for which $P(x', a) = 0$ for every $x' = (x_1, \ldots, x_N) \in \mathbb{R}^N$. If P has degree m, taking the Taylor expansion of the function $\lambda \in \mathbb{R} \rightsquigarrow P(x', \lambda)$ around a, we get

$$\begin{aligned} P(x', x_{N+1}) &= P(x', a) + \sum_{j=1}^{m} \frac{1}{j!} \frac{\partial^j}{\partial x_{N+1}^j} P(x', a)\,(x_{N+1} - a) \\ &= \sum_{j=1}^{m} \frac{1}{j!} \frac{\partial^j}{\partial x_{N+1}^j} P(x', a)\,(x_{N+1} - a) = (x_{N+1} - a) S(x', a) , \end{aligned}$$

contradicting the fact that P is irreducible. So, $P_a \neq 0$ for every $a \in \mathbb{R}$, and we can use Fubini's theorem and the induction hypothesis to get

$$m_{N+1}(V_{\mathbb{R}}(P)) = \int_{-\infty}^{\infty} m_N\big(\{x' \in \mathbb{R}^N : P(x', a) = 0\}\big) da = \int_{-\infty}^{\infty} m_N(V_{\mathbb{R}}(P_a)) da = 0 .$$

□

Corollary 6.5 *Let $P(z)$ be a polynomial of degree $m \geq 1$ of N variables. Then, for each $\lambda \in \mathbb{R}$ and $\eta \in \mathbb{R}^N$, the set*

$$\{w \in \mathbb{R}^N : P(2\pi i w + \lambda\eta) = 0\}$$

has null Lebesgue measure in $\mathbb{R}^N$.

Proof Just observe that the mapping $w \rightsquigarrow P(2\pi i w + \lambda\eta) = \sum_{|\alpha| \leq m} a_\alpha (2\pi i w + \lambda\eta)^\alpha$ is a polynomial of N variables, and apply Lemma 6.7. □

Lemma 6.8 *The Fourier transform satisfies, for each $u \in S'(\mathbb{R}^N)$:*

1. $\mathcal{F}^{-1}(u) = (\mathcal{F}u)\check{}$.
2. $P(\partial)\mathcal{F}^{-1}u = \mathcal{F}_w^{-1}\big(P(2\pi i w)u\big)$.

Proof For the first assertion, we write $\mathcal{F}u = \widehat{u}$. From Proposition 6.18, we know that $\widehat{\widehat{u}} = u$. Taking the Fourier transform in both sides of the equality yields the claim.

For the second statement, applying (6.18) to the distribution $\mathcal{F}u \in S'(\mathbb{R}^N)$, we have

$$\mathcal{F}(P(\partial)\mathcal{F}^{-1}u) = P(2\pi i w)u\,,$$

which immediately gives our claim. □

Lemma 6.9 *Let $P(z)$ be a non-constant polynomial of N variables. For every $\zeta \in \mathbb{C}^N$ and $T \in \mathcal{D}'(\mathbb{R}^N)$, we have*

$$P(\partial)\big(e^{\zeta x}T\big) = e^{\zeta x} P(\zeta + \partial)(T)\,.$$

Proof We proceed in two steps. In the first one, let us assume that $T = \varphi \in \mathcal{D}(\mathbb{R}^N)$. It is easy to check by induction that for every $j = 1, \ldots, N$ and $\alpha_j \in \mathbb{N}_0$, we have

$$\partial_j^{\alpha_j}(e^{\zeta x}\varphi) = e^{\zeta x}(\zeta_j + \partial_j)^{\alpha_j}\varphi\,.$$

Then

$$\partial^\alpha(e^{\zeta x}\varphi) = e^{\zeta x}(\zeta + \partial)^\alpha\varphi\,,$$

for every $\alpha \in \mathbb{N}_0^N$, and therefore,

$$P(\partial)\big(e^{\zeta x}\varphi\big) = \sum_\alpha a_\alpha \partial^\alpha(e^{\zeta x}\varphi) = \sum_\alpha a_\alpha e^{\zeta x}(\zeta + \partial)^\alpha\varphi = e^{\zeta x} P(\zeta + \partial)(\varphi)\,.$$

For our second step, recall first that $\big(\mathcal{D}(\mathbb{R}^N), \mathcal{D}'(\mathbb{R}^N)\big)$ is a dual pair (see Remark 6.1). Also, since clearly $\mathcal{D}(\mathbb{R}^N) \subseteq L_{1,\mathrm{loc}}(\Omega)$, it can be seen as a subspace of $\mathscr{D}'(\mathbb{R}^N)$. Then Proposition 3.9 gives that $\mathscr{D}(\mathbb{R}^N)$ is $\sigma\big(\mathscr{D}'(\mathbb{R}^N), \mathscr{D}(\mathbb{R}^N)\big)$-dense in $\mathscr{D}'(\mathbb{R}^N)$.

Take now an arbitrary $T \in \mathscr{D}'(\mathbb{R}^N)$, and choose a net $(\varphi_i)_{i\in I} \subseteq \mathscr{D}(\mathbb{R}^N)$ such that $T(\psi) = \lim_i \varphi_i(\psi)$ for every $\psi \in \mathscr{D}(\mathbb{R}^N)$ (we shall later see that in fact we can take a sequence, see Corollary 6.6, but for our purposes now this is enough). Before we proceed, let us note that $(-\partial)^\alpha = (-1)^{|\alpha|}\partial^\alpha$ and, then, for any distribution u, we have

$$\langle P(\partial)u, \psi\rangle = \sum_\alpha a_\alpha (\partial^\alpha u)\psi = \sum_\alpha a_\alpha (-1)^{|\alpha|} u(\partial^\alpha \psi) = u\big(P(-\partial)\psi\big).$$

This, using what we have just obtained in the first step, gives

$$\begin{aligned}\langle P(\partial)(e^{\zeta x}T), \psi\rangle &= \langle e^{\zeta x}T, P(-\partial)\psi\rangle = \langle T, e^{\zeta x}P(-\partial)\psi\rangle = \lim_i \langle \varphi_i, e^{\zeta x}P(-\partial)\psi\rangle \\ &= \lim_i \langle P(\partial)(e^{\zeta x}\varphi_i), \psi\rangle = \lim_i \langle e^{\zeta x}P(\zeta+\partial)(\varphi_i), \psi\rangle = \lim_i \langle \varphi_i, P(-\partial+\zeta)(e^{\zeta x}\psi)\rangle \\ &= \langle T, P(-\partial+\zeta)(e^{\zeta x}\psi)\rangle = \langle e^{\zeta x}P(\partial+\zeta)T, \psi\rangle .\end{aligned}$$

Since this holds for every $\psi \in \mathscr{D}(\mathbb{R}^N)$, this completes the proof. □

Lemma 6.10 *Let $\lambda_0, \lambda_1, \ldots, \lambda_m \in \mathbb{C}$ be pairwise different. Then, the system of linear equations*

$$\sum_{j=0}^m a_j\lambda_j^k = \begin{cases} 0 & \text{for } k = 0, \ldots, m-1 \\ 1 & \text{for } k = m \end{cases} \tag{6.24}$$

has a unique solution given by

$$a_j = \prod_{\substack{k=0\\k\neq j}}^m \frac{1}{\lambda_j - \lambda_k}$$

for $j = 0, 1, \ldots, m$.

Proof Let us note in first place that the matrix of the system

$$\begin{pmatrix} 1 & 1 & \cdots & 1 \\ \lambda_0 & \lambda_1 & \cdots & \lambda_m \\ \vdots & \vdots & & \vdots \\ \lambda_0^m & \lambda_1^m & \cdots & \lambda_m^m \end{pmatrix}$$

is a Vandermonde matrix that, being all $\lambda_i \neq \lambda_j$ for $i \neq j$, is nonsingular. Therefore, it has a unique solution $(a_0, a_1, \ldots, a_m) \in \mathbb{C}^{m+1}$. In order to find it, we consider the polynomial

$$p(z) = \prod_{j=0}^{m} (z - \lambda_j) ,$$

of degree m, and whose roots are exactly $\lambda_0, \lambda_1, \ldots, \lambda_m$. Fix $M > 2 \max_j |\lambda_j|$, take any $0 \leq k \leq m$, and use Cauchy's residue theorem with the function $f(z) = \frac{z^k}{p(z)}$ to get

$$\int_{|z|=M} \frac{z^k}{p(z)} dz = 2\pi i \sum_{j=0}^{m} \operatorname{Res}(f, \lambda_j) = 2\pi i \sum_{j=0}^{m} \frac{\lambda_j^k}{p'(\lambda_j)} .$$

This gives

$$\sum_{j=0}^{m} \frac{\lambda_j^k}{p'(\lambda_j)} = \frac{1}{2\pi i} \int_0^{2\pi} \frac{(Me^{it})^k}{\prod_{j=0}^{m} (Me^{it} - \lambda_j)} i M e^{it} dt$$

$$= \frac{1}{2\pi} \frac{M^{k+1}}{M^{m+1}} \int_0^{2\pi} \frac{e^{i(k+1)t}}{\prod_{j=0}^{m} (e^{it} - \lambda_j / M)} dt .$$

Since

$$\left| e^{it} - \frac{\lambda_j}{M} \right| \geq 1 - \frac{|\lambda_j|}{M} > \frac{1}{2}$$

for each $j = 0, \ldots, m$ and every $t \in [0, 2\pi]$, we have

$$\left| \int_0^{2\pi} \frac{e^{i(k+1)t}}{\prod_{j=0}^{m} (e^{it} - \lambda_j / M)} dt \right| \leq \int_0^{2\pi} \frac{1}{\left| \prod_{j=0}^{m} (e^{it} - \lambda_j / M) \right|} dt \leq 2^{m+2} \pi ,$$

for every $k = 0, \ldots, m$. Now, if $0 \leq k \leq m - 1$, this gives

$$\left| \sum_{j=0}^{m} \frac{\lambda_j^k}{p'(z)} \right| \leq 2^{m+1} \frac{M^{k+1}}{M^{m+1}} .$$

Since the right-hand side of the inequality tends to 0 as $M \to \infty$, we have

$$\sum_{j=0}^{m} \frac{\lambda_j^k}{p'(\lambda_j)} = 0 \text{ for } k = 0, \ldots, m - 1 . \tag{6.25}$$

Now, for $k = m$, we have

$$\sum_{j=0}^{m} \frac{\lambda_j^m}{p'(\lambda_j)} = \frac{1}{2\pi} \int_0^{2\pi} \frac{e^{i(m+1)t}}{\prod_{j=0}^{m}(e^{it} - \lambda_j/M)} dt \, .$$

A straightforward application of Lebesgue's dominated convergence theorem then yields

$$\sum_{j=0}^{m} \frac{\lambda_j^m}{p'(\lambda_j)} = 1 \, .$$

This and (6.25) show that $\left(\frac{1}{p'(\lambda_0)}, \frac{1}{p'(\lambda_1)}, \dots, \frac{1}{p'(\lambda_m)}\right)$ is the solution of (6.24). Noting that

$$p'(z) = \sum_{i=0}^{m} \prod_{\substack{k=0 \\ k \neq i}}^{m} (z - \lambda_k)$$

finally gives the conclusion. □

We can now give Wagner's formula for the fundamental solution of a linear partial differential operator with constant coefficients. This clearly proves Theorem 6.31.

Theorem 6.32 *Let P be a polynomial of degree m and N variables, and $\eta \in \mathbb{R}^N$ be such that $P_m(\eta) \neq 0$. Take $\lambda_0, \lambda_1, \dots, \lambda_m \in \mathbb{R}$ pairwise different and define*

$$a_j = \prod_{\substack{k=0 \\ k \neq j}}^{m} \frac{1}{\lambda_j - \lambda_k} \, .$$

Then

$$E = \frac{1}{P_m(2\eta)} \sum_{j=0}^{m} a_j e^{\lambda_j \eta x} \mathcal{F}_w^{-1}\left(\frac{\overline{P(2\pi i w + \lambda_j \eta)}}{P(2\pi i w + \lambda_j \eta)} \right) \tag{6.26}$$

is a fundamental solution of $P(\partial)$.

Proof Let us take for the moment an arbitrary $\lambda \in \mathbb{R}$ and define

$$S(w) = \frac{\overline{P(2\pi i w + \lambda \eta)}}{P(2\pi i w + \lambda \eta)} \, .$$

By Lemma 6.5, this is well defined for almost all $w \in \mathbb{R}^N$ (and then, clearly $|S(w)| = 1$). Then $S \in L_\infty(\mathbb{R}^N)$ and, by Proposition 6.14, $S \in S'(\mathbb{R}^N)$.
Take any $T \in S'(\mathbb{R}^N)$, and $\zeta \in \mathbb{C}^N$. Using Lemmas 6.8 and 6.9, we obtain

$$P(\partial)\big(e^{\zeta x}\mathcal{F}^{-1}(T)\big) = e^{\zeta x}\big(P(\partial+\zeta)\mathcal{F}^{-1}(T)\big) = e^{\zeta x}\mathcal{F}_w^{-1}\big(P(2\pi i w+\zeta)T\big).$$

We apply this to $T = S(w)$ and $\zeta = \lambda\eta \in \mathbb{R}^N$, to get

$$\begin{aligned} &P(\partial)\left(e^{\lambda\eta x}\mathcal{F}_w^{-1}\left(\tfrac{\overline{P(2\pi i w+\lambda\eta)}}{P(2\pi i w+\lambda\eta)}\right)\right)\\ &= e^{\lambda\eta x}\mathcal{F}_w^{-1}\left(P(2\pi i w+\lambda\eta)\frac{\overline{P(2\pi i w+\lambda\eta)}}{P(2\pi i w+\lambda\eta)}\right) = e^{\lambda\eta x}\mathcal{F}_w^{-1}\big(\overline{P(2\pi i w+\lambda\eta)}\big). \end{aligned}$$

If $P(z) = \sum_{|\alpha|\le m} a_\alpha z^\alpha$, we write $\overline{P}(z) = \sum_{|\alpha|\le m} \overline{a_\alpha} z^\alpha$ and

$$\begin{aligned} \overline{P(2\pi i w+\lambda\eta)} &= \sum_{|\alpha|\le m} \overline{a_\alpha(2\pi i w+\lambda\eta)^\alpha}\\ &= \sum_{|\alpha|\le m} \overline{a_\alpha}(-2\pi i w+\lambda\eta)^\alpha = \overline{P}(-2\pi i w+\lambda\eta). \end{aligned}$$

Using again Lemmas 6.8 and 6.9 (note that $\overline{P}$ is again a polynomial of N variables and degree m), we have

$$\begin{aligned} \mathcal{F}_w^{-1}\big(\overline{P(2\pi i w+\lambda\eta)}\big) &= \mathcal{F}_w^{-1}\big(\overline{P}(-2\pi i w+\lambda\eta)\big)\\ &= \overline{P}(-\partial+\lambda\eta)\big(\mathcal{F}_w^{-1}(\mathbb{1})\big) = \overline{P}(-\partial+\lambda\eta)\delta. \end{aligned}$$

Recall that $\psi\delta = \psi(0)\delta$ for every $\psi \in C^\infty(\mathbb{R}^N)$, which immediately gives $e^{\lambda\eta x}\delta = \delta$. Taking all this into account, and using once again Lemma 6.9, we get

$$\begin{aligned} &P(\partial)\left(e^{\lambda\eta x}\mathcal{F}_w^{-1}\left(\tfrac{\overline{P(2\pi i w+\lambda\eta)}}{P(2\pi i w+\lambda\eta)}\right)\right) = e^{\lambda\eta x}\overline{P}(-\partial+\lambda\eta)\delta = \overline{P}(-\partial+2\lambda\eta)(e^{\lambda\eta x}\delta)\\ &= \overline{P}(-\partial+2\lambda\eta)\delta = \sum_{|\alpha|\le m}\overline{a_\alpha}(-\partial+2\lambda\eta)^\alpha\delta\\ &= \sum_{|\alpha|=m}\overline{a_\alpha}(-\partial+2\lambda\eta)^\alpha\delta + \sum_{|\alpha|\le m-1}\overline{a_\alpha}(-\partial+2\lambda\eta)^\alpha\delta\\ &= \lambda^m\Big(\sum_{|\alpha|=m}\overline{a_\alpha}(2\eta)^\alpha\Big)\delta + \sum_{k=0}^{m-1}\lambda^k T_k, \end{aligned}$$

where each $T_k \in \mathscr{E}'(\mathbb{R}^N)$ is a linear combination of derivatives ∂^α (where $1 \le |\alpha| \le m$) of δ. Note also that $\sum_{|\alpha|=m} \overline{a_\alpha}(2\eta)^\alpha = \overline{P_m(2\eta)}$. With this and Lemma 6.10, we finally have

$$\begin{aligned}
P(\partial)E &= \frac{1}{\overline{P_m(2\eta)}} \sum_{j=0}^{m} a_j P(\partial)\Big(e^{\lambda_j \eta x} \mathcal{F}_w^{-1}\Big(\frac{\overline{P(2\pi i w + \lambda_j \eta)}}{P(2\pi i w + \lambda_j \eta)}\Big)\Big) \\
&= \frac{1}{\overline{P_m(2\eta)}} \Big(\sum_{j=0}^{m} \lambda_j^m a_j \overline{P_m(2\eta)}\delta + \sum_{k=0}^{m-1} \lambda_j^k T_k\Big) \\
&= \sum_{j=0}^{m} \lambda_j^m a_j \delta + \frac{1}{\overline{P_m(2\eta)}} \sum_{k=0}^{m-1} T_k \Big(\sum_{j=0}^{m} \lambda_j^k a_j\Big) = \delta\,,
\end{aligned}$$

which completes the proof. □

6.5.2 *Solutions of Linear PDEs*

Our aim now is to find conditions under which a linear partial differential equation with constant coefficients has a solution. To be more precise, the problem that we address now is to find conditions that ensure that, if P is a polynomial of N variables and g is some C^∞-function, then the PDE

$$P(\partial)f = g \tag{6.27}$$

has a solution in some sense. This will come as a consequence of the Malgrange–Ehrenpreis Theorem 6.31. To begin with, we need some basic facts on convolution of distributions and functions that we briefly recall here. We refer the interested reader to [7], [13], or [18, Chapter 27] for a complete account and detailed proofs.

If $\varphi \in C^\infty(\mathbb{R}^N)$ and $T \in \mathscr{D}'(\mathbb{R}^N)$ are such that at least one of them has compact support, then the function defined for $x \in \mathbb{R}^N$ by

$$(T * \varphi)(x) = \langle T_y, \varphi(x-y)\rangle$$

(meaning that, for a fixed x, the distribution T acts on the function $y \rightsquigarrow \varphi(x+y)$) is a well defined function in $C^\infty(\mathbb{R}^N)$ that satisfies

$$\operatorname{supp}(T * \varphi) \subseteq \operatorname{supp}\varphi + \operatorname{supp} T\,.$$

Proposition 6.20 *If $T \in \mathscr{E}'(\mathbb{R}^N)$ (has compact support), then the map $C_T : C^\infty(\mathbb{R}^N) \to C^\infty(\mathbb{R}^N)$ given by $C_T(\varphi) = T * \varphi$ is linear and continuous.*

Proposition 6.21 *If $T \in \mathscr{D}'(\mathbb{R}^N)$ and $\varphi, \psi \in \mathscr{D}(\mathbb{R}^N)$, then $(T * \varphi) * \psi = T * (\varphi * \psi)$.*

Proposition 6.22 *Let $\varphi \in \mathscr{D}(\mathbb{R}^N)$ be such that $\varphi \geq 0$, $\int_{\mathbb{R}^N} \varphi(x)dx = 1$, and $\operatorname{supp}\varphi = \{x \in \mathbb{R}^N : |x| \leq 1\}$. For each $\varepsilon > 0$, we consider the function given by $\varphi_\varepsilon(x) = \frac{1}{\varepsilon^N}\varphi\left(\frac{x}{\varepsilon}\right)$. For every $T \in \mathscr{D}'(\mathbb{R}^N)$, we have:*

1. $T * \varphi_\varepsilon \in C^\infty(\mathbb{R}^N)$.
2. $\operatorname{supp}(T * \varphi_\varepsilon) \subseteq \operatorname{supp} T + \{x \in \mathbb{R}^N : |x| \leq \varepsilon\}$.
3. $\lim_{\varepsilon \downarrow 0^+}(T * \varphi_\varepsilon)(\psi) = T(\psi)$ *for every* $\psi \in \mathscr{D}(\mathbb{R}^N)$.

We denote $T_\varepsilon = T * \varphi_\varepsilon$ and call them the *regularizations* of T. Just taking $\varepsilon = 1/n$, we obtain the following:

Corollary 6.6 *Let $\Omega \subseteq \mathbb{R}^N$ be open, and $T \in \mathscr{D}'(\Omega)$. Then there exists a sequence $(S_j)_j \subseteq C^\infty(\Omega)$ such that $\lim_{j\to\infty} S_j(\psi) = T(\psi)$ for every $\psi \in \mathscr{D}(\Omega)$.*

We can now consider the convolution of two distributions.

Theorem 6.33 *Let $S \in \mathscr{E}'(\mathbb{R}^N)$ and $T \in \mathscr{D}'(\mathbb{R}^N)$. There is a well defined distribution, denoted $S * T$ such that*

$$S * T(\varphi) = \big\langle S_x, \langle T_y, \varphi(x+y)\rangle\big\rangle = \big[S * (T * \check{\varphi})\big](0)$$

for every $\varphi \in \mathscr{D}(\mathbb{R}^N)$.

Proposition 6.23 *Let T be any distribution; then $\delta * T = T$. Also, if T and S are two distributions, one of them having compact support, then:*

1. $T * S = S * T$.
2. $(S * T) * R = S * (T * R)$ *for every* $R \in \mathscr{E}'(\mathbb{R}^N)$.
3. $\partial^\alpha(S * T) = S * (\partial^\alpha T) = (\partial^\alpha S) * T$ *for every* $\alpha \in \mathbb{N}_0^N$.
4. $\operatorname{supp}(S * T) \subseteq \operatorname{supp} T + \operatorname{supp} S$. *In particular, if T and S have compact support, so also has $T * S$.*

We can now give our first result on the existence of solutions of linear partial differential equations with constant coefficients (6.27).

Proposition 6.24 *Let P be a non-zero polynomial of N variables:*

1. *For each $T \in \mathscr{E}'(\mathbb{R}^N)$, there is $S \in \mathscr{D}'(\mathbb{R}^N)$ such that*
$$P(\partial)S = T\,.$$
2. *For each $g \in \mathscr{D}(\mathbb{R}^N)$, there is $f \in C^\infty(\mathbb{R}^N)$ such that*
$$P(\partial)f = g\,.$$
3. *Let $\Omega, \Omega' \subseteq \mathbb{R}^N$ be open and such that $\overline{\Omega'}$ is compact and contained in Ω. Then, for each $g \in C^\infty(\Omega)$, there is $f \in C^\infty(\Omega)$ such that $(P(\partial)f)|_{\Omega'} = g|_{\Omega'}$.*

The last statement is known as the *theorem of local solvability*.

Proof Let E be a fundamental solution of $P(\partial)$ (which, by Theorem 6.31, exists). If $T \in \mathscr{E}'(\mathbb{R}^N)$, let $S = E * T$, and use Proposition 6.23 to get

$$P(\partial)S = P(\partial)(E * T) = (P(\partial)E) * T = \delta * T = T,$$

which yields 1. Exactly the same argument shows that 2 holds.
In order to see 3, let us take $\varphi \in \mathscr{D}(\mathbb{R}^N)$ with $\varphi(x) = 1$ for every x in some neighbourhood of $\overline{\Omega'}$. Given $g \in C^\infty(\mathbb{R}^N)$, the product φg is in $\mathscr{D}(\mathbb{R}^N)$ and, by 2, we can find $\tilde{f} \in C^\infty(\mathbb{R}^N)$ with $P(\partial)\tilde{f} = \varphi g$. If we denote the restriction of $\tilde{f}$ to Ω, we have that $P(\partial)f(x) = \varphi(x)g(x) = g(x)$ for every $x \in \Omega'$. This completes the proof. □

Remark 6.6 Since $\mathscr{D}(\Omega)$ is dense in $C^\infty(\Omega)$ for each open $\Omega \subseteq \mathbb{R}^N$, Proposition 6.24 yields that the continuous linear operator $P(\partial) : C^\infty(\Omega) \to C^\infty(\Omega)$ has dense range. We are interested now in seeing if this operator is surjective. Note that, if this is the case, then the Eq. (6.27) has a solution in $C^\infty(\Omega)$ for every $g \in C^\infty(\Omega)$.

In order to study the surjectivity of the operator $P(\partial) : C^\infty(\Omega) \to C^\infty(\Omega)$, we introduce the following notation:

$$\mathcal{N}_P(\Omega) = \{f \in C^\infty(\Omega) \colon P(\partial)f = 0\}.$$

This is just the kernel of the operator $P(\partial)$, that is, the set of solutions of the homogeneous equation $P(\partial) = 0$. With this notation, $\mathcal{N}_P(\mathbb{R}^N)$ is the space of global solutions of the homogeneous equation.

It is easy to check that $P(\partial)(e^{\zeta x}) = P(\zeta)e^{\zeta x}$ for every $\zeta \in \mathbb{C}^N$. Then, $e^{z_0 x} \in \mathcal{N}_P(\mathbb{R}^N)$ for every $z_0 \in \mathbb{C}^N$ with $P(z_0) = 0$.

The following result was proved by Malgrange in 1953. We refer to [7, Theorem 7.3.6] for a proof of it.

Theorem 6.34 *Let $\Omega \subseteq \mathbb{R}^N$ be open and convex and P a non-zero polynomial of N variables. For each $f \in \mathcal{N}_P(\Omega)$, there is a sequence $(h_k)_k \subseteq \mathcal{N}_P(\mathbb{R}^N)$ such that* $\lim_{k\to\infty} h_k = f$ *in $C^\infty(\Omega)$.*

We use this result and a Mittag-Leffler method to show that, whenever the set Ω is convex, the operator $P(\partial) : C^\infty(\Omega) \to C^\infty(\Omega)$ is surjective (that is, the equation in (6.27) has a solution for every g).

Theorem 6.35 *Let $\Omega \subseteq \mathbb{R}^N$ be open and convex and P a non-zero polynomial of N variables. For each $g \in C^\infty(\Omega)$, there is $f \in C^\infty(\Omega)$ such that $P(\partial)f = g$.*

Proof Let $(K_n)_n$ be a fundamental sequence of compact, convex subsets of Ω such that

$$K_1 \subseteq \mathring{K}_2 \subseteq K_2 \subseteq \mathring{K}_3 \subseteq K_3 \subseteq \cdots \subseteq \Omega.$$

For each $j \in \mathbb{N}$, we take some open set U_j with $K_j \subseteq U_j \subseteq \mathring{K}_{j+1}$ and (recall Proposition 6.2) consider $\psi_j \in \mathscr{D}(\Omega)$ such that $\psi_j(x) = 1$ for every $x \in U_j$ and $\psi_j(x) = 0$ for $x \in \Omega \setminus K_{j+1}$. Define

$$\varphi_1 = \psi_1 \text{ and } \varphi_j = \psi_j - \psi_{j-1} \text{ for } j \geq 2 .$$

Given any $x \in \Omega$, there is some n_0 with $x \in K_{n_0} \setminus K_{n_0-1}$. Then $\psi_j(x) = 1$ for every $j \geq n_0$, and $\varphi_j = 0$ for every $j \geq n_0 + 1$. Hence,

$$\sum_{j=1}^{\infty} \varphi_j(x) = \sum_{j=1}^{n_0} \varphi_j(x) = \psi_1(x) + \sum_{j=2}^{n_0} \big(\psi_j(x) - \psi_{j-1}(x)\big) = \psi_{n_0}(x) = 1 .$$

In other words, the family is locally finite and $\sum_{j=1}^{\infty} \varphi_j = 1$.
For each $j \in \mathbb{N}$, we define $g_j : \mathbb{R}^N \to \mathbb{C}$ by $g_j(x) = g(x)\varphi_j(x)$ if $x \in \Omega$ and 0 otherwise. Clearly, $g_j \in \mathscr{D}(\mathbb{R}^N)$ and, by Proposition 6.24–2, there is $f_j \in C^\infty(\mathbb{R}^N)$ with

$$P(\partial) f_j = g\varphi_j . \tag{6.28}$$

Note that for each fixed $j \geq 2$ we have $P(\partial) f_j = 0$ on $\mathring{K}_j$ (an open, convex subset of Ω), and then by Theorem 6.34, there exists $h_j \in C^\infty(\mathbb{R}^N)$ such that

$$P(\partial) h_j = 0 \tag{6.29}$$

and (recall the definition of the seminorms in (4.8))

$$\sup_{x \in K_{j-1}} \sup_{|\alpha| < j-1} \left| \partial^\alpha (f_j - h_j) \right| < \frac{1}{2^j} .$$

Letting $h_1 = 0$, we define

$$f = \sum_{j=1}^{\infty} (f_j - h_j) .$$

Let us see that this series converges in $C^\infty(\Omega)$, which then implies that f is well defined and belongs to $C^\infty(\Omega)$. Take $n > m$ and note that $K_m \subseteq K_{k-1}$ for every $k > n$. Then, for $p \in \mathbb{N}$, we have

$$\sup_{x \in K_m} \sup_{|\alpha| \leq m} \left| \sum_{j=n+1}^{n+p} \partial^\alpha (f_j - h_j)(x) \right| < \sum_{j=n+1}^{n+p} \frac{1}{2^j} .$$

This implies our claim and then $f \in C^\infty(\Omega)$. Finally, since (6.28) and (6.29) hold for every j, we have

$$P(\partial)f = \sum_{j=1}^{\infty} \big(P(\partial)f_j - P(\partial)h_j\big) = \sum_{j=1}^{\infty} g\varphi_j = g\sum_{j=1}^{\infty} \varphi_j = g\,,$$

which completes the proof. □

For a detailed study of partial differential operators $P(\partial)$ on $\mathscr{D}'(\Omega)$ and on $C^\infty(\Omega)$, we refer the reader to [8, Chapter X].

We finish by considering a notion introduced by Schwartz. A non-zero polynomial of N variables P and its associated partial differential operator $P(\partial)$ are called *hypoelliptic* on an open set $\Omega \subseteq \mathbb{R}^N$ if for each open, non-empty $U \subseteq \Omega$ and each $u \in \mathscr{D}'(\Omega)$ such that $P(\partial)u \in C^\infty(U)$ we have $u \in C^\infty(U)$. If the polynomial P is hypoelliptic on Ω, then every distribution $u \in \mathscr{D}'(\Omega)$ that is solution of the homogeneous equation $P(\partial)u = 0$ belongs to $C^\infty(\Omega)$.

Theorem 6.36 *A non-zero polynomial of N variables P is hypoelliptic on $\mathbb{R}^N$ if and only if $P(\partial)$ has a fundamental solution whose restriction to $\mathbb{R}^N \setminus \{0\}$ belongs to $C^\infty(\mathbb{R}^N \setminus \{0\})$.*

Proof Assume that P is hypoelliptic on $\mathbb{R}^N$, and let E be a fundamental solution of $P(\partial)$. Since $P(\partial)E = \delta$ and the restriction of δ to $\mathbb{R}^N \setminus \{0\}$ belongs to $C^\infty(\mathbb{R}^N \setminus \{0\})$, the hypoellicity of P gives the conclusion.
To prove the converse, let E be a fundamental solution whose restriction to $\mathbb{R}^N \setminus \{0\}$ belongs to $C^\infty(\mathbb{R}^N \setminus \{0\})$. Fix some open set $U \subseteq \mathbb{R}^N$ and take $u \in \mathscr{D}'(\mathbb{R}^N)$ such that $f = P(\partial)u \in C^\infty(U)$. We want to show that $u \in C^\infty(U)$. To do this, take $x_0 \in U$ and $g \in \mathscr{D}(U)$ such that $g(x) = 1$ for every x in some neighbourhood U_0 of x_0 with $U_0 \subseteq U$. Then

$$\begin{aligned} P(\partial)(gu) = \sum_{|\alpha|\le m} a_\alpha \partial^\alpha(gu) &= gP(\partial)u \\ &+ \sum_{0<|\alpha|\le m} a_\alpha \sum_{\gamma<\alpha} \binom{\alpha}{\gamma} \partial^{\alpha-\gamma} g \partial^\gamma u = gP(\partial)u + v\,. \end{aligned}$$

Note that $v \in \mathscr{E}'(\mathbb{R}^N)$ and, since $\partial^{\alpha-\gamma}g$ is zero on U_0 for every $\gamma < \alpha$, we have $\operatorname{supp} v \subseteq \mathbb{R}^N \setminus U_0$. Then

$$E * \big(P(\partial)(gu)\big) = P(\partial)E * gu = \delta * gu = gu\,,$$

and

$$gu = E * \big(gP(\partial)u + v\big) = E * (gf) + E * v\,. \tag{6.30}$$

Since $gf \in \mathscr{D}(U)$, we get $E * (gf) \in C^\infty(\mathbb{R}^N)$. It remains to see that also $E * v$ is a C^∞-function on some neighbourhood of x_0. Given $\varepsilon > 0$, take $\varphi \in \mathscr{D}(\mathbb{R}^N)$ such that $\varphi_\varepsilon(x) = 1$ for $|x| < \varepsilon/2$ and $\operatorname{supp}\varphi_\varepsilon \subseteq B(0, \varepsilon)$. Then

$$E * v = (\varphi_\varepsilon E) * v + \big((1 - \varphi_\varepsilon)E\big) * v$$

and (recall Proposition 6.23)

$$\operatorname{supp}\big((\varphi_\varepsilon E) * v)\big) \subseteq \overline{B(0, \varepsilon)} + \operatorname{supp} v \subseteq \overline{B(0, \varepsilon)} + \big(\mathbb{R}^N \setminus U_0\big)\,.$$

Then there is some open neighbourhood $V \subseteq U_0$ of x_0 such that $\operatorname{supp}\big((\varphi_\varepsilon E) * v\big) \subseteq \mathbb{R}^N \setminus V$.

On the other hand, $(1 - \varphi_\varepsilon)E$ is a C^∞-function, as by assumption the restriction of E to $\mathbb{R}^N \setminus \{0\}$ belongs to $C^\infty(\mathbb{R}^N \setminus \{0\})$; hence, $(1 - \varphi_\varepsilon)E * v$ is a C^∞-function. This altogether (see (6.30)) shows that $gu \in C^\infty(V)$. Since $u|_V = gu|_V$, we have that u is C^∞ in some neighbourhood of x_0 and this completes the proof. □

Concerning the classical partial differential operators in Example 6.30, let us mention that the Cauchy–Riemann operator, the Laplacian, and the heat operator are hypoelliptic, while the wave operator is not. Indeed, given any C^2 function f on $\mathbb{R}$ that is not C^∞, the function $u(x, t) = f(x + t) - f(x - t)$ is a solution of the homogeneous wave equation, as can be easily checked.
We refer the interested reader to [8, Chapter XI] for a detailed study of hypoelliptic partial differential operators and their relation with elliptic operators.

6.6 Exercises

6.1 Show that $f(x) = |x^2 - 1| = (x^2 - 1)(2H(x - 1) - 2H(x + 1) + 1)$, where $H(x)$ is the Heaviside function. Use this formula to determine $f'(x)$ and $f''(x)$.

6.2 Show that the convergence of a sequence $(f_n)_n$ to 0 in $\mathscr{D}(\Omega)$, $C^\infty(\Omega)$ or $L_p(\Omega)$ $(1 \leq p \leq \infty)$ implies

$$\langle f_n, \varphi\rangle = \int f_n(x)\varphi(x)dx \to 0, \qquad \text{as } n \to \infty\,,$$

for all $\varphi \in \mathscr{D}(\Omega)$.

6.3 Let $\varphi \in \mathscr{D}(\mathbb{R}^N)$. Show that, if $\delta(\varphi) = 0$, then $\varphi\delta = 0$.

6.4 Consider $u \in \mathscr{D}'(\mathbb{R}^n)$ and $\varphi \in \mathscr{D}(\mathbb{R}^n)$. Decide whether or not each one of the following statements holds for arbitrary u and φ:

1. If $u(\varphi) = 0$, then $\varphi u = 0$.
2. If $\varphi u = 0$, then $u(\varphi) = 0$.

6.5 We define $T : \mathscr{D}(]0,+\infty[) \longrightarrow \mathbb{K}$ by

$$T(\varphi) = \sum_{m=1}^{\infty} \varphi^{(m)}(1/m)\,.$$

Show that $T \in \mathscr{D}'(]0,+\infty[)$, but T is not of finite order. Conclude that there is not $T_0 \in \mathscr{D}'(\mathbb{R})$ such that $T_0\psi = T\psi$ for every $\psi \in \mathscr{D}(]0,+\infty[)$.

6.6 Given $T \in \mathscr{D}'(\Omega)$ and $n \in \mathbb{N}$, we define $(x^n T)(\varphi) = T(x^n \varphi)$, for $\varphi \in \mathscr{D}(\Omega)$:

1. Show that $x^n T \in \mathscr{D}'(\Omega)$.
2. Compute $x^n \delta_0^{(k)}$ for $k \in \mathbb{N}$.

6.7 Given $\lambda \in \mathbb{C}$ with $\mathrm{Re}(\lambda) > -1$, we define the functions

$$x_+^\lambda = \begin{cases} x^\lambda & \text{for } x > 0, \\ 0 & \text{for } x \le 0 \end{cases} \quad \text{and } x_-^\lambda = \begin{cases} |x|^\lambda & \text{for } x < 0, \\ 0 & \text{for } x \ge 0\,. \end{cases}$$

Show that they define distributions on $\mathbb{R}$. Compute the derivative of $T = x_+^\lambda$ with $-1 < \lambda < 0$.

6.8 Prove that Remark 6.4 holds true.

6.9 Compute the Fourier transform of:

1. $f(x) = \begin{cases} 1 - \dfrac{|x|}{R} & \text{if } |x| \le R \\ 0 & \text{otherwise}\,. \end{cases}$
2. $f(x) = \begin{cases} 1 - x^2 & \text{if } |x| \le 1 \\ 0 & \text{otherwise}\,. \end{cases}$
3. Given $a > 0$, $f(x) = \begin{cases} e^{-ax} & \text{if } x \ge 0 \\ 0 & \text{otherwise}\,. \end{cases}$
4. $f(x) = xe^{-a|x|}$ with $a > 0$.
5. $f(x) = e^{-a|x|}\cos(bx)$, with $a > 0$.
6. $f(x) = \dfrac{1}{1+x^2}$.
7. $f(x) = \dfrac{x}{1+x^2}$.
8. $f(x) = \cos(ax)$.
9. $f(x) = \sin^3(x)$.
10. $f(x) = \dfrac{\sin(ax)}{\pi x}$.
11. $(\delta_a)^{(k)}$ for $k = 0, 1, 2, 3, \ldots$.
12. x^k for $k = 0, 1, 2, 3, \ldots$.

6.10 Let $f, g \in S(\mathbb{R})$ satisfy $f * g = 0$. Does it follow that $f = 0$ or $g = 0$?

6.11 Let $f \in S(\mathbb{R}^2)$ and define $g(x) = f(x, 0)$. Find the relation between $\widehat{f}$ and $\widehat{g}$.

6.12 Use the inversion formula to find the Fourier transform of

$$f(x) = \frac{2a}{a^2 + 4\pi^2 x^2} .$$

This can also be done directly solving the integral using residues.

6.13 Let $F(w)$ be the Fourier transform of $f(x) \in L_1(\mathbb{R})$. Compute the Fourier transform of the following functions:

1. $f(x) = f(x-1) + f(-x-1)$.
2. $f(x) = f(3x-6)$.
3. $f_3(x) = f''(x+1)$.

6.14 It is possible to improve the inversion formula for the Fourier transform for piecewise functions that are in $L_1(\mathbb{R})$ and in $L_\infty(\mathbb{R})$ (see [4, pages 228–231]). Use this result, Plancherel's formula and the Fourier transform of $\chi_{[-1,1]}$ to compute the following integrals:

1. $\displaystyle\int_{-\infty}^{+\infty} \frac{\sin x}{x} dx.$
2. $\displaystyle\int_{-\infty}^{+\infty} \frac{\sin^2 x}{x^2} dx.$

6.15 Give a direct proof, using the definitions, of the following statements:

1. $P(\partial) = \dfrac{d}{dx}$ is hypoelliptic on $\mathbb{R}$.
2. $P(\partial) = \dfrac{\partial}{\partial x}$ is not hypoelliptic on $\mathbb{R}^2$.

6.16 Show that if $T \in \mathscr{D}'(\mathbb{R}^N)$ and $\operatorname{supp} T = \{a\}$ for some $a \in \mathbb{R}^N$, then there are $k \in \mathbb{N}_0$ and $c_\alpha \in \mathbb{K}$ for $|\alpha| \leq k$ so that

$$T = \sum_{|\alpha| \leq k} c_\alpha \, \partial^\alpha \delta_a .$$

6.17 Let $T \in S'(\mathbb{R}^N)$ be a tempered distribution such that $\Delta T = 0$ (where Δ is the Laplace operator). Prove that T is a polynomial.
Hint: Use the Fourier transform and the previous exercise.

6.18 Let $a \in \mathbb{R}$. Prove that a fundamental solution of the differential operator $P(\partial) = \dfrac{d}{dx} + a$ is $E = e^{-a} H$ (where H is the Heaviside function).

6.19 Find a fundamental solution of $P(\partial) = \dfrac{\partial}{\partial x}$ on $\mathbb{R}^2$.

6.20 The Schrödinger partial differential operator is

$$P(\partial) = \frac{1}{i}\frac{\partial}{\partial t} - \Delta_x\,,$$

for $(t, x) \in \mathbb{R}^{n+1}$. A fundamental solution is given by

$$E = H(t)\exp\big(-i(n-2)\tfrac{\pi}{4}\big)(4\pi t)^{-\frac{n}{2}}\exp\big(\tfrac{|x|^2}{4it}\big)$$

(see [17, p. 43–44]). Show that E is not a C^∞-function in any neighbourhood of a point $(0, x_0)$ and the Schrödinger is not hypoelliptic.

6.21 Let $P(\partial)$ be a linear partial differential operator in $\mathbb{R}^2$ such that the equation

$$\left\{\begin{array}{l} P(\partial)u = f \text{ on } \mathbb{R}^2 \\ u(x,0) = 0 = \dfrac{\partial u}{\partial t}(x,0) \end{array}\right\} \tag{6.31}$$

has a unique solution $u \in C^\infty(\mathbb{R}^2)$ for each $f \in C^\infty(\mathbb{R}^2)$. This happens for the wave equation $P(\partial) = \dfrac{\partial^2}{\partial t^2} - \dfrac{\partial^2}{\partial x^2}$, for example. Show that $R : C^\infty(\mathbb{R}^2) \to C^\infty(\mathbb{R}^2)$ given by $Rf = u$ (the unique solution of (6.31)) defines a continuous linear operator and that $P(\partial)R$ is the identity operator on $C^\infty(\mathbb{R}^2)$.
Hint: Use the closed graph theorem and the uniqueness of the solution of (6.31).

6.22 Let $f \in S(\mathbb{R}^N)$, and $P(\partial)$ be a non-constant partial differential operator on $\mathbb{R}^N$ with constant coefficients. Show that if $P(\partial)f = 0$, then $f = 0$.
Hint: Use the Fourier transform and that the set $\{z \in \mathbb{C}^N : P(z) \neq 0\}$ is dense in $\mathbb{C}^N$.

6.23 Consider the differential operator on $\mathbb{R}$ given by $P(\partial) = \dfrac{d^2}{dx^2} + a\dfrac{d}{dx} + b$. Let f, g be solutions of $P(\partial)u = 0$ such that $f(0) = g(0)$ and $f'(0) - g'(0) = 1$. Show that

$$G(x) = \begin{cases} f(x) \text{ if } x \leq 0 \\ g(x) \text{ if } x > 0 \end{cases}$$

is a fundamental solution of $P(\partial)$.

6.24 Use polar coordinates to show that $f : \mathbb{R}^2 \to \mathbb{C}$ given by $f(x, y) = \frac{1}{x+iy}$ satisfies $\overline{\partial} f = \pi\delta$ (recall that $\overline{\partial}$ is the Cauchy–Riemann operator, see (6.23)).

Further Reading

[1–3, 6–9, 13–15, 17, 18, 20]

References

1. Brezis, H.: Functional Analysis. In: Sobolev Spaces and Partial Differential Equations. Universitext. Springer, New York (2011)
2. Cerdà, J.: Linear functional analysis. In: Graduate Studies in Mathematics, vol. 116. American Mathematical Society, Providence, RI; Real Sociedad Matemática Española, Madrid (2010). https://doi.org/10.1090/gsm/116
3. Duistermaat, J.J., Kolk, J.A.C.: Distributions. In: Cornerstones. Birkhäuser Boston, Inc., Boston, MA (2010). https://doi.org/10.1007/978-0-8176-4675-2. Theory and applications, Translated from the Dutch by J. P. van Braam Houckgeest
4. Duren, P.: Invitation to classical analysis. In: Pure and Applied Undergraduate Texts, vol. 17. American Mathematical Society, Providence, RI (2012)
5. Ehrenpreis, L.: Solution of some problems of division. I. Division by a polynomial of derivation. Am. J. Math. **76**, 883–903 (1954). https://doi.org/10.2307/2372662
6. Evans, L.C.: Partial differential equations. In: Graduate Studies in Mathematics, vol. 19, 2nd edn. American Mathematical Society, Providence, RI (2010). https://doi.org/10.1090/gsm/019
7. Hörmander, L.: The analysis of linear partial differential operators. I. In: Classics in Mathematics. Springer, Berlin (2003). https://doi.org/10.1007/978-3-642-61497-2. Distribution theory and Fourier analysis, Reprint of the second (1990) edition
8. Hörmander, L.: The analysis of linear partial differential operators. II. In: Classics in Mathematics. Springer, Berlin (2005). https://doi.org/10.1007/b138375. Differential operators with constant coefficients, Reprint of the 1983 original
9. Horváth, J.: Topological Vector Spaces and Distributions. vol. I. Addison-Wesley Publishing Co., Reading, Mass.-London-Don Mills, Ont (1966)
10. Malgrange, B.: Equations aux dérivées partielles à coefficients constants. I. Solution élémentaire. C. R. Acad. Sci. Paris **237**, 1620–1622 (1953)
11. Rosay, J.P.: A very elementary proof of the Malgrange-Ehrenpreis theorem. Am. Math. Mon. **98**(6), 518–523 (1991). https://doi.org/10.2307/2324871
12. Rudin, W.: Real and Complex Analysis, 3rd edn. McGraw-Hill Book Co., New York (1987)
13. Rudin, W.: Functional analysis, 2nd edn. In: International Series in Pure and Applied Mathematics. McGraw-Hill, Inc., New York (1991)
14. Schwartz, L.: Méthodes mathématiques pour les sciences physiques. In: Enseignement des Sciences. Hermann, Paris (1961)
15. Schwartz, L.: Théorie des distributions. Publications de l'Institut de Mathématique de l'Université de Strasbourg, IX–X. Hermann, Paris (1966). Nouvelle édition, entiérement corrigée, refondue et augmentée
16. Trèves, F.: Solution élémentaire d'équations aux dérivées partielles dépendant d'un paramètre. C. R. Acad. Sci. Paris **242**, 1250–1252 (1956)
17. Trèves, F.: Basic Linear Partial Differential Equations. Dover Publications, Inc., Mineola, NY (2006). Reprint of the 1975 original
18. Trèves, F.: Topological Vector Spaces, Distributions and Kernels. Dover Publications, Inc., Mineola, NY (2006). Unabridged republication of the 1967 original
19. Wagner, P.: A new constructive proof of the Malgrange-Ehrenpreis theorem. Am. Math. Mon. **116**(5), 457–462 (2009). https://doi.org/10.4169/193009709X470362
20. Zuily, C.: Problems in distributions and partial differential equations. In: North-Holland Mathematics Studies, vol. 143. North-Holland Publishing Co., Amsterdam; Hermann, Paris (1988). https://doi.org/10.1016/S0304-0208(08)70020-3. Translated from the French

References

1. Ahlfors, L.: Complex Analysis—An Introduction to the Theory of Analytic Functions of One Complex Variable, 3rd edn. AMS Chelsea Publishing, Providence, RI (2021). Reprint of the 1978 original
2. Apostol, T.M.: Mathematical Analysis, 2nd edn. Addison-Wesley Publishing Co., Reading, Mass./London-Don Mills, Ont. (1974)
3. Bachman, G., Narici, L.: Functional Analysis. Dover Publications, Inc., Mineola, NY (2000). Reprint of the 1966 original
4. Banach, S.: Théorie des Opérations Linéaires. Éditions Jacques Gabay, Sceaux (1993). Reprint of the 1932 original
5. Bayart, F., Matheron, E.: Dynamics of linear operators. In: Cambridge Tracts in Mathematics, vol. 179. Cambridge University Press, Cambridge (2009). https://doi.org/10.1017/CBO9780511581113
6. Bermúdez, T., Kalton, N.J.: The range of operators on von Neumann algebras. Proc. Am. Math. Soc. **130**(5), 1447–1455 (2002). https://doi.org/10.1090/S0002-9939-01-06292-X
7. Bogachev, V.I., Smolyanov, O.G.: Topological vector spaces and their applications. In: Springer Monographs in Mathematics. Springer, Cham (2017). https://doi.org/10.1007/978-3-319-57117-1
8. Bracci, F., Contreras, M.D., Díaz-Madrigal, S.: Continuous semigroups of holomorphic self-maps of the unit disc. In: Springer Monographs in Mathematics. Springer, Cham (2020). https://doi.org/10.1007/978-3-030-36782-4
9. Brezis, H.: Functional analysis. In: Sobolev Spaces and Partial Differential Equations. Universitext. Springer, New York (2011)
10. Cerdà, J.: Linear functional analysis. In: Graduate Studies in Mathematics, vol. 116. American Mathematical Society, Providence, RI; Real Sociedad Matemática Española, Madrid (2010). https://doi.org/10.1090/gsm/116
11. Conway, J.B.: Functions of one complex variable. In: Graduate Texts in Mathematics, vol. 11, 2nd edn. Springer, New York/Berlin (1978)
12. Conway, J.B.: A course in functional analysis. In: Graduate Texts in Mathematics, vol. 96. Springer, New York (1985). https://doi.org/10.1007/978-1-4757-3828-5
13. Defant, A., García, D., Maestre, M., Sevilla-Peris, P.: Dirichlet series and holomorphic functions in high dimensions. In: New Mathematical Monographs, vol. 37. Cambridge University Press, Cambridge (2019). https://doi.org/10.1017/9781108691611
14. Diestel, J.: Sequences and series in Banach spaces. In: Graduate Texts in Mathematics, vol. 92. Springer, New York (1984). https://doi.org/10.1007/978-1-4612-5200-9

J. Bonet et al., *Function Spaces and Operators between them*,
RSME Springer Series 11, https://doi.org/10.1007/978-3-031-41602-6

15. Duistermaat, J.J., Kolk, J.A.C.: Distributions. In: Cornerstones. Birkhäuser Boston, Inc., Boston, MA (2010). https://doi.org/10.1007/978-0-8176-4675-2. Theory and applications, Translated from the Dutch by J. P. van Braam Houckgeest
16. Dunford, N., Schwartz, J.T.: Linear Operators. Part I. Wiley Classics Library. John Wiley & Sons, Inc., New York (1988). Reprint of the 1958 original.
17. Duoandikoetxea, J.: Fourier analysis. In: Graduate Studies in Mathematics, vol. 29. American Mathematical Society, Providence, RI (2001). Translated and revised from the 1995 Spanish original by David Cruz-Uribe
18. Duren, P.: Invitation to classical analysis. In: Pure and Applied Undergraduate Texts, vol. 17. American Mathematical Society, Providence, RI (2012)
19. Edwards, R.E.: Functional Analysis. Dover Publications, Inc., New York (1995). Theory and applications, Corrected reprint of the 1965 original
20. Ehrenpreis, L.: Solution of some problems of division. I. Division by a polynomial of derivation. Am. J. Math. **76**, 883–903 (1954). https://doi.org/10.2307/2372662
21. Eisner, T., Farkas, B., Haase, M., Nagel, R.: Operator theoretic aspects of ergodic theory. In: Graduate Texts in Mathematics, vol. 272. Springer, Cham (2015). https://doi.org/10.1007/978-3-319-16898-2
22. Evans, L.C.: Partial differential equations. In: Graduate Studies in Mathematics, vol. 19, 2nd edn. American Mathematical Society, Providence, RI (2010). https://doi.org/10.1090/gsm/019
23. Fabian, M., Habala, P., Hájek, P., Montesinos, V., Zizler, V.: Banach space theory. In: CMS Books in Mathematics/Ouvrages de Mathématiques de la SMC. Springer, New York (2011). https://doi.org/10.1007/978-1-4419-7515-7
24. Fabian, M., Habala, P., Hájek, P., Montesinos Santalucía, V., Pelant, J., Zizler, V.: Functional analysis and infinite-dimensional geometry. In: CMS Books in Mathematics/Ouvrages de Mathématiques de la SMC, vol. 8. Springer, New York (2001). https://doi.org/10.1007/978-1-4757-3480-5
25. Floret, K., Wloka, J.: Einführung in die Theorie der lokalkonvexen Räume. In: Lecture Notes in Mathematics, No. 56. Springer, Berlin/New York (1968)
26. Gamelin, T.W.: Complex analysis. In: Undergraduate Texts in Mathematics. Springer, New York (2001). https://doi.org/10.1007/978-0-387-21607-2
27. Grosse-Erdmann, K.G., Peris Manguillot, A.: Linear Chaos. Universitext. Springer, London (2011). https://doi.org/10.1007/978-1-4471-2170-1
28. Grothendieck, A.: Topological vector spaces. In: Notes on Mathematics and its Applications. Gordon and Breach Science Publishers, New York/London/Paris (1973). Translated from the French by Orlando Chaljub
29. Hardy, G.H., Littlewood, J.E., Pólya, G.: Inequalities. Cambridge Mathematical Library. Cambridge University Press, Cambridge (1988). Reprint of the 1952 edition
30. Heuser, H.G.: Functional Analysis. A Wiley-Interscience Publication. John Wiley & Sons, Ltd., Chichester (1982). Translated from the German by John Horváth
31. Hörmander, L.: The analysis of linear partial differential operators. I. In: Classics in Mathematics. Springer, Berlin (2003). https://doi.org/10.1007/978-3-642-61497-2. Distribution theory and Fourier analysis, Reprint of the (1990) second edition
32. Hörmander, L.: The analysis of linear partial differential operators. II. In: Classics in Mathematics. Springer, Berlin (2005). https://doi.org/10.1007/b138375. Differential operators with constant coefficients, Reprint of the (1983) original
33. Horváth, J.: Topological Vector Spaces and Distributions, vol. I. Addison-Wesley Publishing Co., Reading, Mass./London-Don Mills, Ont. (1966)
34. Jameson, G.J.O.: A First Course on Complex Functions. Chapman and Hall, Ltd., London (1970)
35. Jameson, G.J.O.: Topology and Normed Spaces. Chapman and Hall, London; Halsted Press, New York (1974)
36. Jarchow, H.: Locally convex spaces. In: Mathematische Leitfäden. B. G. Teubner, Stuttgart (1981)

37. Kadets, V.: A Course in Functional Analysis and Measure Theory. Universitext. Springer, Cham (2018). https://doi.org/10.1007/978-3-319-92004-7. Translated from the 2006 Russian edition
38. Kelley, J.L.: General topology. In: Graduate Texts in Mathematics, No. 27. Springer, New York/Berlin (1975). Reprint of the 1955 edition
39. Köthe, G.: Topological Vector Spaces. I. In: Die Grundlehren der mathematischen Wissenschaften, Band 159. Springer New York Inc., New York (1969). Translated from the German by D. J. H. Garling
40. Köthe, G.: Topological vector spaces. II. In: Grundlehren der Mathematischen Wissenschaften [Fundamental Principles of Mathematical Sciences], vol. 237. Springer, New York/Berlin (1979)
41. Krengel, U.: Ergodic theorems. In: De Gruyter Studies in Mathematics, vol. 6. Walter de Gruyter & Co., Berlin (1985). https://doi.org/10.1515/9783110844641. With a supplement by Antoine Brunel
42. Lindenstrauss, J., Tzafriri, L.: Classical Banach spaces. I. In: Ergebnisse der Mathematik und ihrer Grenzgebiete, Band 92. Springer, Berlin/New York (1977). Sequence spaces
43. Lindenstrauss, J., Tzafriri, L.: Classical Banach spaces. II. In: Ergebnisse der Mathematik und ihrer Grenzgebiete [Results in Mathematics and Related Areas], vol. 97. Springer, Berlin/New York (1979). Function spaces
44. Lotz, H.P.: Uniform convergence of operators on L^∞ and similar spaces. Math. Z. **190**(2), 207–220 (1985). https://doi.org/10.1007/BF01160459
45. Malgrange, B.: Equations aux dérivées partielles à coefficients constants. I. Solution élémentaire. C. R. Acad. Sci. Paris **237**, 1620–1622 (1953)
46. Meise, R., Vogt, D.: Introduction to functional analysis. In: Oxford Graduate Texts in Mathematics, vol. 2. The Clarendon Press, Oxford University Press, New York (1997). Translated from the German by M. S. Ramanujan and revised by the authors
47. Pérez Carreras, P., Bonet, J.: Barrelled locally convex spaces. In: North-Holland Mathematics Studies, vol. 131. North-Holland Publishing Co., Amsterdam (1987). Notas de Matemática, 113
48. Queffelec, H., Queffelec, M.: Diophantine approximation and Dirichlet series. In: Texts and Readings in Mathematics, vol. 80. Hindustan Book Agency, New Delhi; Springer, Singapore (2020). https://doi.org/10.1007/978-981-15-9351-2. Second edition
49. Ransford, T.J.: A short elementary proof of the Bishop-Stone-Weierstrass theorem. Math. Proc. Camb. Philos. Soc. **96**(2), 309–311 (1984). https://doi.org/10.1017/S0305004100062204
50. Robertson, A.P., Robertson, W.: Topological vector spaces. In: Cambridge Tracts in Mathematics, vol. 53, 2nd edn. Cambridge University Press, Cambridge/New York (1980)
51. Rosay, J.P.: A very elementary proof of the Malgrange-Ehrenpreis theorem. Am. Math. Mon. **98**(6), 518–523 (1991). https://doi.org/10.2307/2324871
52. Rudin, W.: Principles of mathematical analysis, 3rd edn. In: International Series in Pure and Applied Mathematics. McGraw-Hill Book Co., New York/Auckland-Düsseldorf (1976)
53. Rudin, W.: Real and Complex Analysis, 3rd edn. McGraw-Hill Book Co., New York (1987)
54. Rudin, W.: Functional analysis, 2nd edn. In: International Series in Pure and Applied Mathematics. McGraw-Hill, Inc., New York (1991)
55. Schaefer, H.H., Wolff, M.P.: Topological vector spaces. In: Graduate Texts in Mathematics, vol. 3, 2nd edn. Springer, New York (1999). https://doi.org/10.1007/978-1-4612-1468-7
56. Schwartz, L.: Méthodes mathématiques pour les sciences physiques. In: Enseignement des Sciences. Hermann, Paris (1961)
57. Schwartz, L.: Théorie des distributions. In: Publications de l'Institut de Mathématique de l'Université de Strasbourg, IX–X. Hermann, Paris (1966). Nouvelle édition, entiérement corrigée, refondue et augmentée
58. Shapiro, J.H.: Composition operators and classical function theory. In: Universitext: Tracts in Mathematics. Springer, New York (1993). https://doi.org/10.1007/978-1-4612-0887-7

59. Stromberg, K.R.: An Introduction to Classical Real Analysis. AMS Chelsea Publishing, Providence, RI (1981). https://doi.org/10.1090/chel/376. (2015, corrected reprint of the 1981 original)
60. Taylor, A.E., Lay, D.C.: Introduction to Functional Analysis, 2nd edn. Robert E. Krieger Publishing Co., Inc., Melbourne, FL (1986)
61. Trèves, F.: Solution élémentaire d'équations aux dérivées partielles dépendant d'un paramètre. C. R. Acad. Sci. Paris **242**, 1250–1252 (1956)
62. Trèves, F.: Basic Linear Partial Differential Equations. Dover Publications, Inc., Mineola, NY (2006). Reprint of the 1975 original
63. Trèves, F.: Topological Vector Spaces, Distributions and Kernels. Dover Publications, Inc., Mineola, NY (2006). Unabridged republication of the 1967 original
64. Valdivia, M.: Topics in locally convex spaces. In: Notas de Matemática [Mathematical Notes], vol. 85. North-Holland Publishing Co., Amsterdam/New York (1982)
65. Vretblad, A.: Fourier analysis and its applications. In: Graduate Texts in Mathematics, vol. 223. Springer, New York (2003). https://doi.org/10.1007/b97452
66. Wagner, P.: A new constructive proof of the Malgrange-Ehrenpreis theorem. Am. Math. Mon. **116**(5), 457–462 (2009). https://doi.org/10.4169/193009709X470362
67. Werner, D.: Einführung in die höhere Analysis, corrected edn. Springer-Lehrbuch. [Springer Textbook]. Springer, Berlin (2009). https://doi.org/10.1007/978-3-540-79696-1
68. Wilansky, A.: Modern Methods in Topological Vector Spaces. McGraw-Hill International Book Co., New York (1978)
69. Wilansky, A.: Topology for Analysis. Robert E. Krieger Publishing Co., Inc., Melbourne, FL (1983). Reprint of the 1970 edition
70. Yosida, K.: Functional analysis. In: Classics in Mathematics. Springer, Berlin (1995). https://doi.org/10.1007/978-3-642-61859-8. Reprint of the sixth (1980) edition
71. Zuily, C.: Problems in distributions and partial differential equations. In: North-Holland Mathematics Studies, vol. 143. North-Holland Publishing Co., Amsterdam; Hermann, Paris (1988). https://doi.org/10.1016/S0304-0208(08)70020-3. Translated from the French

Index

Symbols

J. Bonet et al., *Function Spaces and Operators between them*,
RSME Springer Series 11, https://doi.org/10.1007/978-3-031-41602-6